FOUNDATION AND STRUCTURAL PROBLEMS

Solved by Microcomputer

FOUNDATION AND STRUCTURAL PROBLEMS

Solved by Microcomputer

Redmond Holloway
CEng, BE, FIStructE, MICE, MIEI

OXFORD
BSP PROFESSIONAL BOOKS
LONDON   EDINBURGH   BOSTON
MELBOURNE PARIS BERLIN VIENNA

Copyright © Redmond Holloway 1991

BSP Professional Books
A division of Blackwell Scientific
  Publications Ltd
Editorial offices:
Osney Mead, Oxford OX2 0EL
25 John Street, London WC1N 2BL
23 Ainslie Place, Edinburgh EH3 6AJ
3 Cambridge Center, Cambridge,
  MA 02142, USA
54 University Street, Carlton,
  Victoria 3053, Australia

All rights reserved. No part of this
publication may be reproduced, stored
in a retrieval system, or transmitted
in any form or by any means, electronic,
mechanical, photocopying, recording
or otherwise without the prior
permission of the publisher.

First published 1991

Set by Setrite Typesetters Ltd
Printed and bound in Great Britain by
  Hartnolls, Bodmin, Cornwall

DISTRIBUTORS

Marston Book Services Ltd
PO Box 87
Oxford OX2 0DT
(*Orders*: Tel: 0865 791155
        Fax: 0865 791927
        Telex: 837515)

USA
  Blackwell Scientific Publications, Inc.
  3 Cambridge Center
  Cambridge, MA 02142
  (*Orders*: Tel: (800) 759−6102)

Canada
  Oxford University Press
  70 Wynford Drive
  Don Mills
  Ontario M3C 1J9
  (*Orders*: Tel: (416) 441−2941)

Australia
  Blackwell Scientific Publications
  (Australia) Pty Ltd
  54 University Street
  Carlton, Victoria 3053
  (*Orders*: Tel: (03) 347−0300)

British Library
Cataloguing in Publication Data
Holloway, Redmond
  Foundation and structural problems.
  1. Structures. Design. Application of
  microcomputer systems
  I. Title
  624.17710285416

ISBN 0−632−02924−2

Whilst every care has been taken to ensure the accuracy of the contents of this book, neither the author nor the publisher can accept any liability for loss occasioned by the use of the information given or damage resulting from inaccuracies.

To my dear and patient wife, Irene.

Contents

Preface		xi
Notation index		xvii
Part 1	*General*	1
1	**Design basis**	3
	Method of analysis	3
	Limit state design	4
	Output checks and data preparation	4
2	**Structural equations and notation**	6
	Compatibility and equilibrium equations	6
	Matrix notation	16
	Tabular notation	17
3	**Matrix inversion and Gaussian elimination**	19
	Matrix inversion	19
	MATIN program	20
	Gaussian elimination	23
	PIVOT program	25
	Programs in use	26
	Program selection	26
4	**Subroutines and file merging**	28
	Subroutines	28
	Files	29
	Use of a printer	31
5	**Release-deformation flexibility analysis**	34
	Description	34
	Special features	35
	Sign convention	36
	Procedures	37

Part 2 Foundations and temporary works   47

6 Differential settlement   49
 Settlement moments   49
 Sign determination   51
 General tabular matrix   52
 Settlement effects and reactions   58
 SBEAM program   59

7 Beams and footings on an elastic foundation   64
 Subgrade reaction   64
 Footings with distributed loading   66
 Unequally distributed loading example   69
 Single concentrated load   73
 Concentrated load at different positions   75

8 Continuous foundations   80
 Twin column loads   80
 Multiple column loads   82
 Data processing   85

9 Foundations: applied moments and varying section   86
 Applied moments   86
 Combination of loads and moments   88
 Comparison with rigid foundation   90
 Foundations of varying width   92
 Foundation beams of varying depth   96
 Special case   96

10 Raft foundations   101
 Transverse bending   101
 Node types   101
 Column on raft foundation   108
 General matrix and examples   110

11 Lateral loads on piles   114
 Piles in an elastic medium   114
 Laterally loaded pile   117
 Keying in a large matrix   120
 Pile groups   121
 Pile caps   121
 Piles subject to bending   124

12 Sheet piling and strutted abutments   126
 Cantilever walls   126

	Embedded length	128
	Anchored sheet piling	131
	Varying soil conditions	139
	Diaphragm walls	139
	Alternative analytical methods	140
	Strutted abutments	140
13	**Shoring and braced excavations**	145
	Raking and flying shores	145
	Trench sheeting and cofferdams	149
	Strutted retaining walls	157
14	**Formwork and falsework**	159
	Strutted formwork	159
	Falsework and centring	162
	Strutted beams	164
	Statically indeterminate structures	169

Part 3 Lattices and grillages   171

15	**Lattice girders and trusses**	173
	N-girder	173
	Common roof truss	178
	Indeterminate trusses	180
16	**Beam frameworks**	183
	Beam framework types	183
	Manual analysis	184
	Beam frameworks with bending	188
	Analytical procedure	191
	Frame with nodal loads and bending	193
	Flow chart for XBEAM program	195
17	**Solving beam frameworks**	200
	Using the XBEAM program	200
	Symmetrically loaded frame	203
	Deflections	206
18	**Rectangular grillages**	209
	Uniform grillages	209
	Matrix size	211
	GRID program	214
	One-way grids	216
	Partitioned grillages	218
	Errors in output	227

19	**Grillage analysis examples**	228
	One-way grid example	228
	Two-way grid printout	231
	Space frames	234
20	**Torsion: cranked beams and cantilevers**	235
	Torsion in circular sections	235
	Rectangular sections	236
	End cantilevers	239
	Cantilever deflection	240
	Cranked beams	241
	Cantilever balcony	249
21	**Torsion in beam frameworks and grillages**	251
	Three-member beam framework	251
	Four-member beam framework	254
	Asymmetrical three-member frame	262
	Rectangular grillages	262
	Allowance for torsion	269

Part 4 Rigid frames 271

22	**Portal frames and box culverts**	273
	Symmetrical portal frames	273
	Box culverts	275
	Multi-bay portals	277
	Portals subject to sway	279
	Portal with cantilever arm	283
	Portals with inclined columns	284
	Pitched portals	289
	Lateral loads on pitched portals	293
	Plastic design	297
23	**Multi-storey and multi-bay frames**	300
	Symmetrical frames	300
	Lateral loads	303
	Uniform portals and rigid frames	310
	Multi-bay frames	312
	Vierendeel girders	313

Appendix A: Problems	325
Appendix B: Programs	373
Bibliography	378
Index	379

Preface

Introduction

This book attempts to bridge the gap between the traditional methods of analysis and the modern computer-based approach. It describes what might be called a 'partially computerised' design system intermediate between 'hand calculation' and 'fully computerised' methods. The first essential in all cases is to acquire a proper grasp of fundamental theory and its relation to structural behaviour. Once this is achieved, one can look to ways in which the computer can assist in obtaining results more efficiently and speedily using simple programs or routines; this is the stage described in the book. The next 'fully computerised' stage requires a specialised knowledge of machine code programming, assembly language, binary and hexadecimal notation, Boolean algebra, masking techniques, interrupts, logical file handling, graphics creation, data generation, editing functions and post processing, among other subjects, to take full advantage of the computer's remarkable powers. This is likely to remain the domain of the computer specialist rather than that of the practising engineer in the author's view. The book, therefore, is aimed primarily at the student or practising engineer equipped with a microcomputer who wishes to use it for solving common structural problems using software with which he is thoroughly familiar and which he can develop to his or her own requirements.

Scope

The book describes in Parts 1 and 2 how to deal with analytical problems commonly encountered in the design of foundations, soil retaining structures and temporary works including sheet piling, laterally loaded piles, continuous footings, falsework, coffer dams, trench shores, strutted beams, raft foundations and concrete formwork. Examples have been selected to represent the type of practical problems met with in consultants' or design offices.

General structural problems are dealt with in Parts 3 and 4 which cover analysis of lattice girders, roof trusses, beam frameworks and grillages with and without torsion, rigid portal frames, multi-storey and multi-bay frames and box culverts.

More emphasis than usual has been placed on design of foundations and temporary works, since recent failures in civil engineering construction have highlighted the necessity for checking these as thoroughly as other features. Correct analysis of foundations and earthworks is as important as that of the superstructure, of course, if the building or bridge is to perform satisfactorily, although textbooks commonly segregate the two for some reason. Basic procedures are outlined in the book to facilitate analysis and encourage computer checks on works that are often treated as of minor importance, therefore. A modified flexibility method of analysis is used for this purpose, as described in Chapter 1.

Computer programs

Computer programs listed in the book make use of standard numerical subroutines for the solution of simultaneous equations. They are written in elementary BASIC and fully explained, so both programs and the concepts on which they are based are comprehended by the reader. This contrasts with the situation with commercial software which is usually efficient but seldom simple or understandable to the user and, indeed, may be deliberately shrouded in secrecy. The dangers in a blind acceptance of imperfectly understood output (the 'black box' syndrome) are a matter of concern to many teachers and practising engineers, especially in view of some unfortunate mishaps in recent years.

It has always been held that an engineer should be familiar with any method of analysis he employs. Part of his training is not merely to know the various formulae used in analysis, but also to understand their derivation so that they can be used with confidence. Use of computers should not nullify this principle. In the book this has been borne in mind and, since all problems are tackled from first principles, the computer output should be completely understood.

Simple software of this nature, although limited in scope and incapable of utilising the modern computer's potentialities to the full, has its advantages. Programs are comparatively short and can be keyed in easily for storage. The number of variables to input can be kept to a minimum, unlike comprehensive programs covering a variety of structures, so they may be run with few interruptions. So-called 'user friendly' messages are omitted to enable the program to run continuously as far as possible; explanatory comments and headings are confined mainly to the program specification for the same reason.

Data preparation time is minimised also, which is considered a most important factor. Thus, the designer has more time to concentrate on the implications of the results or to indulge in comparative design studies. Programs are not 'interactive' but it is left to the user to decide, from a study of the output, what alterations to make for satisfactory results. The emphasis, in fact, is on the user retaining full control of the program. Once a feeling of confidence is achieved, the programs should form a useful means of checking more sophisticated output.

The importance of checking output from imperfectly understood 'bought-in' software cannot be overstressed and is referred to again in Chapter 1. Failure to satisfy himself on this score means that the engineer must rely primarily on the software supplier for the safety of the structure. This will not only pose contractual and legal difficulties but will in time lead to a downgrading of the engineer's professional standing should this become common practice.

Program usage

It will be observed that the computer is given a subsidiary, although highly important, role in the solving of problems. The author's aim has been to reduce engineering analysis to a simple, understandable process using the microcomputer to cope with the tedious or intricate mathematical steps. Some knowledge of elementary BASIC programming and familiarity with matrix notation is assumed as well as a grasp of simple structural mechanics. As no special mathematical skills are required, it is hoped that the book will be of some value to non-structural students and technicians as well as to practising engineers. For this reason, every effort has been made to avoid the use of jargon or highly technical language in the text.

Some knowledge of matrix algebra is required to follow the derivation of the matrix inversion program (MATIN), but preference has been given to the use of a tabular or 'spreadsheet' format in expressing equations in condensed form. This is in line with computer data practice in other disciplines. Avoidance of matrix algebra is seen as a help in focusing attention on the relationship between the equations and what they represent in the structure. Thus, solutions are obtained not by manipulation of matrices, transformations or other mathematical devices, but from consideration of actual displacements, sways or rotations as they occur in the structure. Line diagrams, in which significant angular rotations and linear displacements are clearly identified, are used extensively for the same reason.

Because programs are written in elementary BASIC without recourse to assembly language or machine code routines, they may be run on most microcomputers with only minor changes to accommodate non-standard BASIC statements, such as instructions to clear the screen. REM statements are included where these occur in the printed listings. One reason for this portability is the absence of screen graphics. Where a frame is described by a co-ordinate system, it is almost essential to have a line diagram displayed on the screen to check that the frame geometry has been input correctly, as errors in co-ordinates can occur very easily. Where, as in the book, the frame members are described by their relationship to a fixed 'reference element', the data can be verified simply by visual inspection. Graphics are not of great value for checking the input, therefore.

Similarly, when the output consists of bending moment values, for instance, it is assumed in all cases that the reader has the ability to construct the 'free' bending moment diagram in each span for the given load system. By imposing the output values on this, one can immediately draw the final bending moment diagram in permanent record form, as illustrated by several examples in the book. Furthermore, once the moments are known, it is not a difficult task to determine manually the shear forces and reactions, as described in Chapter 6, and to present them in diagrammatic form, if required. Extension of the programs to display such diagrams on the screen has not been considered as of much real benefit, therefore, despite their undoubted attraction to computer users.

The main computer listings will be found in Appendix B. Subsidary listings and sub-routines are shown at the end of the chapter describing their use. Although six program listings are included in the book, all problems in fact, can be resolved with no more than three of these — the others illustrate alternative means of obtaining the same results. Thus, the amount of keyboard work involved in implementing the method of analysis described is not severe. (Indeed, one program (XBEAM) would suffice for all problems, but this would be contrary to the principle adopted in the book of making the computer program appropriate to the task in hand.) The ability to obtain results with comparative ease using the simple listings in the book will, it is hoped, generate a greater feeling of confidence and understanding when the reader moves on to more advanced computer-aided design systems.

Teaching of structural theory

A concern that graduate engineers are giving 'calculations' more emphasis than 'seeing and feeling' structural behaviour initiated a recent study into the teaching of the Theory of Structures (23). One of the main inferences drawn from the study was that the concepts of structural theory appeared difficult for students to grasp for some reason. Despite the stress in modern curricula on flexibility and stiffness matrix methods, the survey revealed the outstanding popularity in practice of the Hardy Cross moment distribution method — often regarded by academics as 'archaic'. Many of the features of this attractive method of analysis will be found included in the list in Chapter 5 pertaining to the release-deformation method, used throughout the book. Indeed, the latter evolved from an original attempt to produce a 'computerised moment distribution method' which was found to be too restricted in scope.

The importance of acquiring a sound appreciation of structural behaviour is endorsed by the following quotation from the study group's report.

'Lack of appreciation of structural behaviour e.g. the inability to appraise how a structure will deform under load, is leading to an increase in design mistakes. Sophisticated and detailed analysis which ignores the most basic principles of structural behaviour has led to structural inadequacy and even collapse.'

One remedy for this sad state of affairs lies in greater attention to actual deformations and displacements in structures, which is a primary concern of this book.

Software

Readers who wish to avoid the trouble of keying in the listings or the time inevitably involved in putting error-free programs on disk can obtain software from the following supplier:

> Prom Management Ltd
> 4, Ranelagh
> Dublin 6
> Republic of Ireland

Disks supplied are suitable for IBM or compatible PCs using DOS-plus or MS DOS and incorporate both listed and merged programs as well as examples of their use.

It should be noted that, while every care has been taken in the preparation of programs listed in the book, no guarantee can be given that they provide the correct solution to a particular problem; this remains the user's responsibility.

Acknowledgement

The author would like to express his thanks to Julia Burden of Blackwell Scientific Publications for her valuable advice and help.

Redmond Holloway

Notation index

A	Angular displacement equation
	Cross-section area
a	Length, grid spacing
b	Width, grid spacing
c	Soil cohesion
	Elastic foundation factor, a^2b
d	Depth
E	Young's modulus of elasticity
	Equilibrium equation
e	Eccentricity
	Change in length
F	Force, action
f	Stress
G	Shear modulus of elasticity
g	Subgrade reaction factor, $\dfrac{6EI}{qa^2}$
H	Height of column
	Horizontal reaction
h	Height above ground level
I	Second moment of inertia
J	Second polar moment of inertia
K	Relative flexibility ratio
	e.g. $K_1 = \dfrac{L_1}{I_1}\dfrac{I_n}{L_n} = \dfrac{k_n}{k_1}$
k	Stiffness ratio, $\dfrac{I}{L}$
L	Linear displacement equation
	Span
M	Bending moment
N	Number of equations, matrix columns, etc.
n	Reference element suffix
O	Origin, nodal point
P	Stiffness ratio factor, $\dfrac{6E_nI_n}{L_n}$
p	Pressure intensity, subgrade reaction

q	Coefficient of subgrade reaction
R	Reaction, resultant
r	Stiffness parameter, $\dfrac{E_n A_n}{L_n}$
	Ratio, e.g. $\dfrac{a}{b}, \dfrac{b}{b_0}$, etc.
S	Relative flexibility parameter
	e.g. $S_1 = \dfrac{L_1}{E_1 A_1} \dfrac{E_n A_n}{L_n}$
	Strut spacing
T	Torsion, thrust
	Torsional displacement equation
V	Vertical reaction
	Number of vectors
W	Total distributed load
	Concentrated load
w	Loading intensity
X, Y	Co-ordinate axes
$\bar{x}$	Distance to centre of gravity
α	Angle
β	Angle
	Torsional factor for rectangular sections
γ	Factor for angular displacement in bending
	Rankine distribution factor
	$= \dfrac{1 - \sin \phi}{1 + \sin \phi}$
Δ	Vertical deflection, displacement
λ	Sway, lateral displacement
	Poisson's ratio
ϕ	Angle of internal friction
	Unit torsional rotation
θ	Angular rotation
Ω	Torsional constant $= 6 E_n I_n \phi_n$

PART 1
GENERAL

1 Design basis

Method of analysis

The book concentrates on one particular method of analysis and shows how this can be applied to the design of engineering works. This has been called the 'release-deformation' method, since it is based on the introduction of 'hinges' or releases at specific points and an examination of the resulting deformed shape of the structure. The discontinuities are noted on a diagram, as well as the forces or moments applied to eliminate them and restore the structure to its pre-release state. Their relationship enables a series of equations to be written down which are then put into a compact tabular matrix or 'spreadsheet' format. The coefficient matrix and column vectors are fed as DATA into an appropriate computer program which, when run, provides the required results. Statements are included for printing the data and critical stages of the analysis for checking purposes.

The analytical method adopted in the book is a variation of the conventional flexibility method, but has the advantage of stressing the affinity between the behaviour of the structure and the derived equations rather than relying on somewhat abstruse mathematical concepts. It is easier for the non-mathematician to grasp; furthermore the sign convention, often a crucial factor, is based on actual physical conditions displayed in graphical form in the deformation diagram. The coefficient matrices and vectors are expressed in general terms for each basic structural form. The data for a particular problem can be obtained directly from these expressions, so that the analysis is virtually automatic in most cases.

A useful feature is in the use of one member as a 'reference element'. This device simplifies data preparation since only relative E or I values are required and trusses or frames can be described succintly without recourse to a cumbersome and error-prone co-ordinate system, as required in most other methods. Not being a recognised method of analysis, the release-deformation method must offer some benefits to justify the trouble needed to master it. These are listed in full in Chapter 5, in addition to the points touched on here.

The book reflects the growing popularity of flexibility methods in the computer analysis of structures, in contrast to earlier interest in stiffness methods. In the study of continuum structures, such as raft foundations on an elastic medium, a finite difference approach is adopted which should form a useful introduction to the more sophisticated finite element methods on which many commercial software packages are based. Grillage analysis, including torsional effects, is covered fairly extensively in Part 3. The Winkler 'spring' theory has been adopted in the analysis of various foundation and retaining wall problems using moduli of sub-grade reaction.

While the listed programs are restricted in scope, they do illustrate the factors to be considered in purchasing or using more complex software as well as providing a means of checking results using inexpensive computer equipment. Redundant structures, such as grillages or beams on an elastic medium, are quite difficult to check by 'rule of thumb' methods, so a means of carrying out a rapid analysis that will not be too far off the mark may be very welcome in such circumstances.

Limit state design

Problems in the book are solved by the method of linear elastic structural analysis, although a brief description of the plastic method of design is included in Chapter 22, since this is often favoured for analysis of portal frames. The analytical method advocated in the book is suited to limit state design which, of course, has now largely superseded the permissible stress method. Some confusion can arise due to the fact that limit states comprise two main classifications:

(1) ultimate limit states, which are those associated with collapse or other forms of structural failure which may endanger people's safety;
(2) serviceability limit states, which correspond to states beyond which the specified service criteria are no longer met. These include excessive deformation, deflection or vibration.

Analyses for serviceability requirements are invariably based on elastic methods. Elastic analysis may be used to check ultimate strength capacity also; however, redistribution of moments obtained by elastic means may be permitted by the relevant Specification or Code of Practice, for example BS 8110 for reinforced concrete. This feature is not dealt with in the book. Likewise, design values shown for forces (or actions) and material properties are assumed to have the relevant partial load factors applied in each case. Changes in partial factors can be accommodated by simple alteration of the load vectors when different load combinations are being examined, but this aspect is not covered in the text.

Output checks and data preparation

The importance of carrying out some check on computer output is now generally recognised, as is the fallacy that computers will always produce 'correct' solutions to problems. Among possible sources of error are hardware architecture, operating systems software, accumulated small precision or 'round-off' errors, applications software 'bugs', incorrect data processing (e.g. mixed units, wrong sequences) and keyboard 'typing' mistakes. Although every effort may be made by the computer manufacturers and software suppliers to overcome these by built-in checks or editing procedures, the danger of being landed with incorrect output remains. The situation is aggravated by the false sense of security induced by the large mass of information and apparent (but unreal) degree of precision evident in most computer printouts. This should not blind one to the need for some form of independent check.

A common recommendation is that a 'hand check' be performed. This is not really practical for many forms of construction such as continuum structures, rigid frames and grillages or indeterminate frameworks. A separate computer analysis may not resolve the problem, since there is no guarantee that this is not merely a 'rerun' or does not incorporate similar errors, apart from the matter of expense and inconvenience. The answer appears to lie in a computer analysis using 'user-comprehensible' software, i.e. software embodying a fully comprehensible program listing and data whose derivation is clearly understood by the user. This may provide approximate results only in some cases, but nonetheless sufficiently accurate for checking purposes. The book aims to fulfil this function, as well as providing an efficient 'stand-alone' method of analysis for simpler conventional structures.

Problems are assumed to be solved once the data have been correctly assembled for solution by computer, as described in the text. This enables the reader to check progress without actually requiring access to a computer. Thus, although results obtained from the computer run are given in all examples, much more emphasis is placed on data preparation than on recording the solutions. The main objective is to show how to formulate the problem for solution by computer rather than to cover computer analysis as such.

Means of manipulating databases and printing in data using a spreadsheet format are described in outline. In this regard one may speculate that with the growing, almost universal use of computers in structure analysis, future examination questions may be phrased on the lines 'Nominate a computer program and list the data to find the moments in the accompanying figure, rather than the traditional 'Determine the moments in the accompanying figure'. In the same way, one can see more emphasis placed on ability to assemble data and express structural problems in schematic form and less on mathematical prowess. The tabular presentation used throughout the book is intended to ensure that data are presented correctly to the microcomputer for solution, as well as illustrating the way in which particular problems may be tackled.

2 Structural equations and notation

This book is concerned with linear structural analysis which is based on the assumption that a linear relationship exists between the applied actions and the resulting displacements. The relationship can be expressed mathematically in sets of equations that may be classified as:

(1) compatibility equations
(2) equilibrium equations
(3) load-displacement equations.

The first set contains forces and displacements, the second set contains forces (actions) and moments and the third set provides the relationship between the two. Compatibility and equilibrium equations are considered below under the general headings of (1) linear, (2) angular and (3) torsional displacements.

Compatibility and equilibrium equations

(1) Linear displacements

(a) A member of length L that undergoes a change in length, e, is subjected to a strain equal to e/L. Since, by Hooke's Law, stress equals E times strain, where E = Young's modulus,

$$\frac{F}{A} = E \frac{e}{L}$$

$$\text{or } F = \frac{EA}{L} e \qquad (2.1)$$

where F = axial load on the member
A = cross-sectional area of the member.

In the release-deformation method described later, one member in the structure is selected at the outset as the reference element. For a reference element with basic properties E_n, A_n and L_n,

$$F_n = \frac{E_n A_n}{L_n} e_n = re_n, \quad \text{where } r = \frac{E_n A_n}{L_n}.$$

STRUCTURAL EQUATIONS AND NOTATION

For any other member with properties E_1, A_1 and L_1, subject to an axial load F_1,

$$F_1 = \frac{E_1 A_1}{L_1} e_1 = \frac{re_1}{S_1}, \quad \text{where } S_1 = \frac{E_n A_n}{L_n} \frac{L_1}{E_1 A_1}$$

$$F_1 S_1 = re_1 \tag{2.2}$$

Displacements are usually considered along the XX and YY axes only and designated Δ_X and Δ_Y respectively. For instance, in Fig. 2.1 member reference 1 has end nodes 1 and 2. Under the action of an applied force, joints 1 and 2 are assumed to move to points 1' and 2'. Displacements Δ_{1X} and Δ_{1Y} are indicated at joint 1 and Δ_{2X} and Δ_{2Y} at joint 2.

Relative movement in the XX direction = $\Delta_{1X} - \Delta_{2X}$

Relative movement in the YY direction = $\Delta_{2Y} - \Delta_{1Y}$

Change in length of 12 = $e_{12} = \dfrac{F_1 L_1}{E_1 A_1} = (\Delta_{1X} - \Delta_{2X}) \cos \alpha + (\Delta_{2Y} - \Delta_{1Y}) \sin \alpha$

$$(\Delta_{1X} - \Delta_{2X}) \cos \alpha + (\Delta_{2Y} - \Delta_{1Y}) \sin \alpha - \frac{F_1 L_1}{E_1 A_1} = 0 \tag{2.3}$$

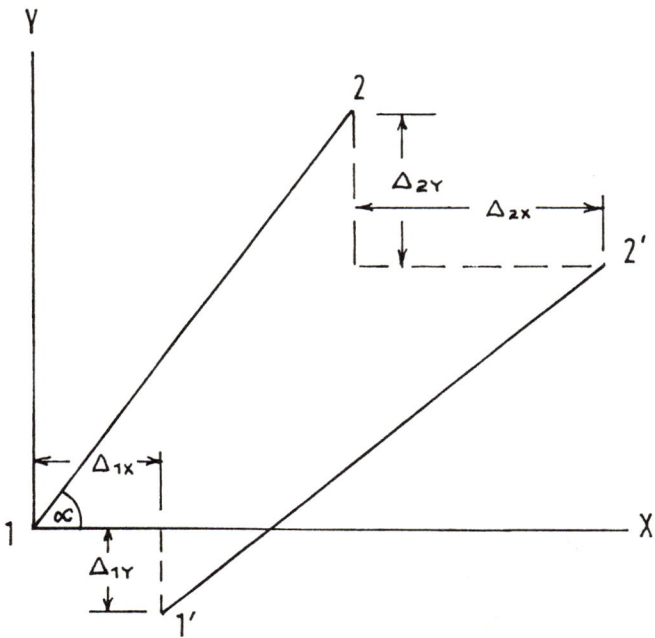

Fig. 2.1.

Multiplying across by the stiffness parameter, $r = \dfrac{E_n A_n}{L_n}$, where the suffix, n, denotes the reference element,

$$r (\Delta_{1X} - \Delta_{2X}) \cos \alpha + r (\Delta_{2Y} - \Delta_{1Y}) \sin \alpha - F_1 S_1 = 0 \quad (2.4)$$

where S_1 is the relative flexibility factor for member $1 = \dfrac{L_1}{E_1 A_1} \cdot \dfrac{E_n A_n}{L_n}$.

These are the general compatibility equations for linear displacements. Equations of this type are designated L.12, L.21, etc. or L.AB, L.BA, etc. in the examples, the terms after L (for linear equation) being the ends of the member considered in turn.

For a horizontal member, where $\alpha = 0$,

$$r (\Delta_{1X} - \Delta_{2X}) - F_1 S_1 = 0 \quad (2.5)$$

For a vertical member, where $\alpha = 90°$,

$$r (\Delta_{2Y} - \Delta_{1Y}) - F_1 S_1 = 0 \quad (2.5a)$$

The sign convention to be used with the above equations is based on the rule that where the force in the member under consideration tends to close the discontinuity it is regarded as positive. If it tends to increase the discontinuity, it is negative. Thus it is essential to know (or assume) the direction of the force in the member at the particular joint, i.e. whether it acts as a strut or a tie. Arrows are used to indicate the direction of the member forces at each node in the usual way. If the displacement is in the opposite direction to the resolved member force, it is positive. If not, it is considered negative. For instance, if node 1 is displaced to 1', as shown in Fig. 2.2, then in considering member 12 with force F_{12}, the displacement Δ_{1X} is positive and Δ_{1Y} negative. In considering member 13 with force

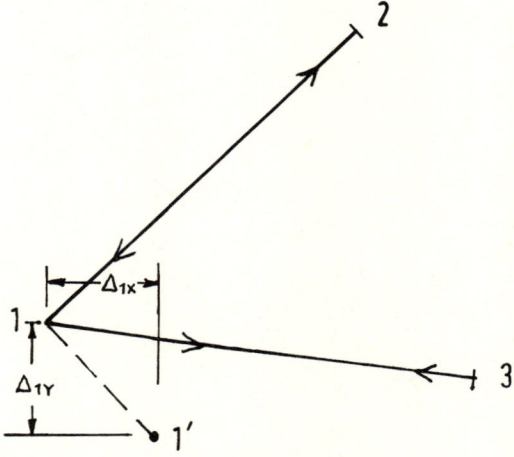

Fig. 2.2.

F_{13}, the displacement Δ_{1X} is negative and Δ_{1Y} is also negative. (See Chapter 13 for further examples.)

(b) Equilibrium equations are designated E.1X, E.2X, etc. or E.1Y, E.2Y, etc. depending along which axis the forces are resolved. Referring to Fig. 2.3, we can obtain the equilibrium equations by simple resolution of the actions as follows:

E.1X: $\quad\quad\quad F_{12} \cos \theta_{12} + F_{13} \cos \theta_{13} = R \cdot \cos \phi \quad\quad\quad$ (2.6)
E.1Y: $\quad\quad\quad F_{12} \sin \theta_{12} - F_{13} \sin \theta_{13} = R \cdot \sin \phi \quad\quad\quad$ (2.7)

where R is the resultant of the external forces acting at the nodal intersection of members 12 and 13.

Positive values are taken as towards the right on the XX axis and upwards along the YY axis. (Although the method of analysis can be used quite satisfactorily for the solution of three-dimensional space frames, plane structures only are considered in the book.)

(2) Angular displacements

(A) DUE TO IMPOSED LOADS

Compatibility equations for angular displacements due to bending can be obtained for a beam in which moments are applied at the supports, as shown in Fig. 2.4. Under the influence of these moments, the beam will

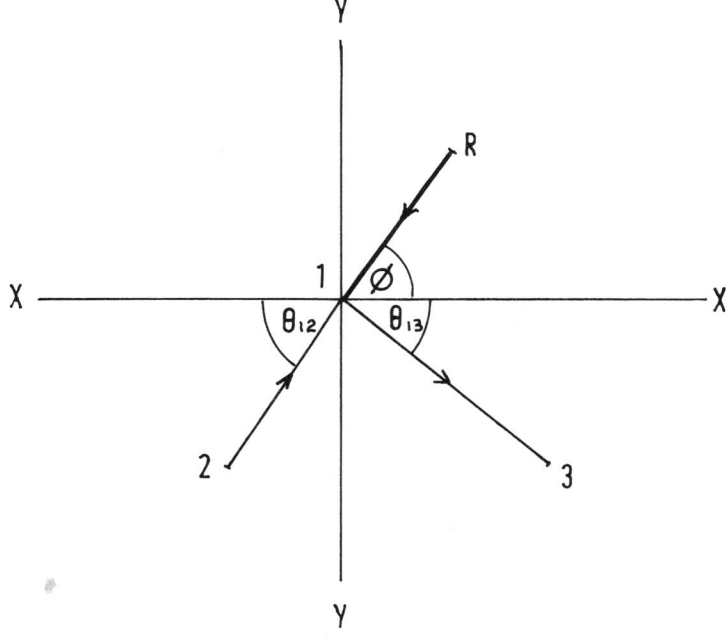

Fig. 2.3.

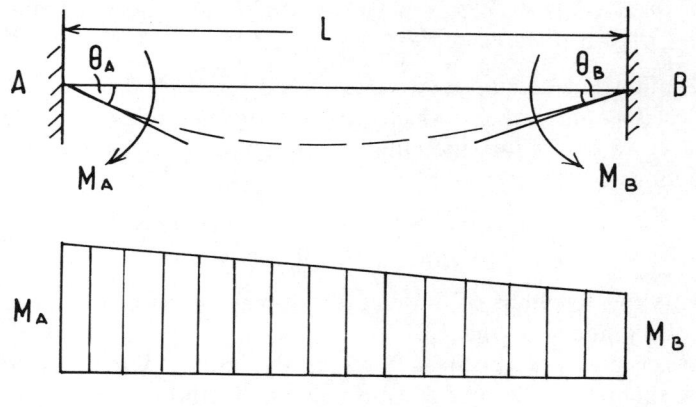

Fig. 2.4.

assume the deflected shape shown by the dashed line with tangential angles at the supports equal to θ_A and θ_B.

The area of the bending moment (BM) diagram in Fig. 2.4 is

$$\frac{M_A L}{2} + \frac{M_B L}{2}$$

By Mohr's moment-area theorem (Wang and Salmon, 1984), taking moments about B,

$$L \theta_A = \frac{A\bar{x}}{EI} = \frac{1}{EI}\left(\frac{M_A L}{2}\frac{2L}{3} + \frac{M_B L}{2} \cdot \frac{L}{3}\right)$$
$$= \frac{L^2}{EI}\left(\frac{M_A}{3} + \frac{M_B}{6}\right)$$

The equation for angular change, therefore, is

$$\theta_A = \frac{M_A L}{3EI} + \frac{M_B L}{6EI}$$

Multiplying across by $\dfrac{6EI}{L}$ gives the angular compatibility equation

A.A: $\qquad 2M_A + M_B = \dfrac{6EI}{L}\theta_A = P\theta_A, \quad \text{where } P = \dfrac{6EI}{L} \qquad$ (2.7a)

Similarly, we can obtain an equation for the angular change at B,

A.B: $\qquad 2M_B + M_A = P\theta_B$

The equations are designated A.A, A.B, etc., the second letter indicating the support or node under consideration. The angular changes, θ_A and θ_B, can be expressed in terms of load, W, and span, L, from slope/deflection considerations (see below). That is,

$$2M_A + M_B = \gamma WL \qquad (2.8)$$
$$M_A + 2M_B = \gamma WL \qquad (2.9)$$

where γ is a factor dependent on the type of loading.

One way of determining the factor, γ, is by the conjugate beam method. For instance, for a uniformly distributed load where the central BM diagram ordinate = WL/8, the area of the BM diagram is

$$\frac{2}{3} L \cdot \frac{WL}{8} = \frac{WL^2}{12}$$

From the conjugate beam theory (Wang and Salmon, 1984), treating the BM diagram as a load,

$$\text{reaction at A} = \frac{WL^2}{24} = EI \, \theta_A$$

$$\theta_A = \theta_B = \frac{WL^2}{24EI} \qquad (2.9a)$$

$$\text{or } \frac{6EI}{L} \theta = P\theta = \frac{WL}{4} = \gamma WL \qquad (2.9b)$$

so that, for this loading case, $\gamma = 1/4$. Values of γ for conventional types of loading obtained in a similar fashion are shown in Table 2.1. In each case W = the total load on the span

$$\frac{6EI}{L} \theta = P\theta = \gamma WL \qquad (2.9c)$$

Equations (2.8) and (2.9) refer to a single uniform span. In continuous beams where spans may vary in section, a relative flexibility factor needs to be introduced to take account of P = 6EI/L varying from span to span. For this case the equations become:

A.A: $\qquad 2M_A + M_B = \gamma KWL \qquad (2.10)$
A.B: $\qquad M_A + 2M_B = \gamma KWL \qquad (2.11)$

where γ = factor from Table 2.1, K = relative flexibility factor, W = imposed load and L = span.

In the release-deformation method, the moments M_A, etc. are applied to eliminate the angular changes induced by the imposed loads at release, i.e. in the opposite direction to the moments shown in Fig. 2.4 (see Fig. 2.5).

(B) DUE TO DIFFERENTIAL SETTLEMENT

Angular changes at beam supports may be brought about by differential settlements. Thus, in Fig. 2.6 the settlements Δ_A and Δ_B produce angular changes at A and B equal to $(\Delta_B - \Delta_A)/L$. To correct these and restore the beam to its initial state, moments M_A and M_B are applied on release, as shown in the figure. Their direction is determined by the requirement

Table 2.1 Values of γ for various load patterns

Loading	At A	At B
(uniform load)	$\dfrac{1}{4}$	$\dfrac{1}{4}$
(point load at centre)	$\dfrac{3}{8}$	$\dfrac{3}{8}$
(point load at a, b)	$\dfrac{ab}{L^3}(L+b)$	$\dfrac{ab}{L^3}(L+a)$
(triangular, symmetric)	$\dfrac{5}{16}$	$\dfrac{5}{16}$
(triangular, asymmetric)	$\dfrac{4}{15}$	$\dfrac{7}{30}$

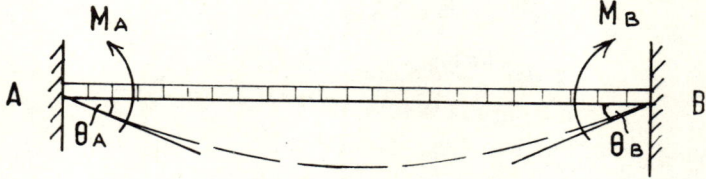

Fig. 2.5.

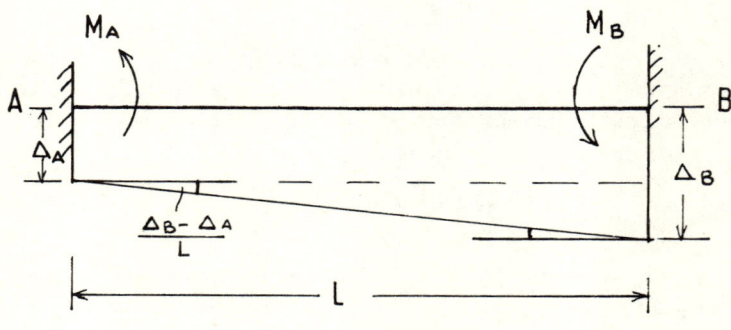

Fig. 2.6.

that the moments should counteract the angular change at each support, as marked in the diagram. The equations, obtained in the same way as in the previous section, are:

A.A: $\quad \dfrac{M_A L}{3EI} + \dfrac{M_B L}{6EI} = \dfrac{\Delta_B - \Delta_A}{L}$

$2M_A + M_B = \dfrac{P}{L}(\Delta_B - \Delta_A)$, where $P = \dfrac{6EI}{L}$ \hfill (2.12)

Similarly,

A.B: $\qquad M_A + 2M_B = \dfrac{P}{L}(\Delta_B - \Delta_A)$ \hfill (2.13)

These equations enable the moments due to differential settlement to be obtained for an unloaded beam. Note that the direction of M_B is opposite to that in Fig. 2.5 used in the previous section. Usually settlement and bending moments due to imposed loads have to be combined, so the sign on the right-hand side of equation 2.13 should be reversed to take account of this. The procedure is described more fully in Chapter 6.

(C) DUE TO SWAY

The 'sway' phenomenon in rigid frames causes angular changes at the nodes, as illustrated in Fig. 2.7. The fixed-ended member, AB, (it may be a beam or column) suffers a relative displacement, λ, at end B. The angular change at A and B equals θ.

For small angles, $\tan \theta = \dfrac{\lambda}{H} = \theta$

To restore the angular displacements on release of the end supports, equal moments must be applied at A and B, as shown in Fig. 2.7. Using the same notation as before,

A.A: $\qquad \dfrac{M_A H}{3EI} + \dfrac{M_B H}{6EI} = \theta = \dfrac{\lambda}{H}$

Normally sway effects are considered in conjunction with moments due to imposed loads, in which case the direction of the applied release

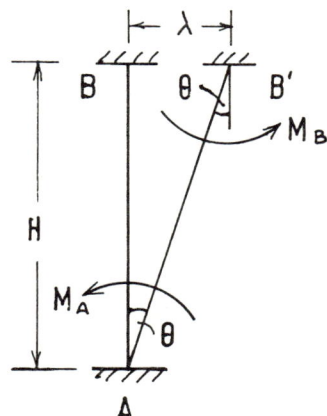

Fig. 2.7.

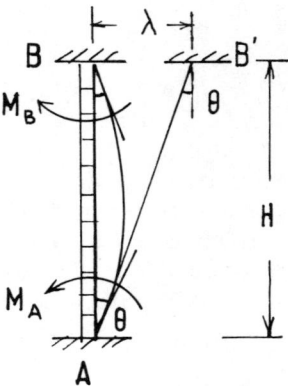

Fig. 2.8.

moments is as shown in Fig. 2.8. With this convention, assuming the moments applied at A and B to correct sway equal to M_1, then $M_A = M_1$ and $M_B = -M_1$ so that we obtain

A.A.: $\qquad \dfrac{M_1 H}{3EI} - \dfrac{M_1 H}{6EI} = \dfrac{\lambda}{H}$

$$2M_1 H - M_1 H = \dfrac{6EI}{H}\lambda = P\lambda$$

$$M_1 = \dfrac{P}{H}\lambda \qquad (2.14)$$

Thus, a sway vector made up of inverses of the column heights or girder bay lengths may be used for determining the values of the sway moments. This fact is used in the solution of rigid frames and portals subject to vertical and horizontal loading.

(D) DUE TO NODAL DEFLECTIONS

Nodal deflections are another source of angular change in grillages or frameworks subject to loading at an angle (normally acting vertically) to the plane of the structure. As with support settlements, the changes can be represented by the terms Δ/L, where Δ is the vertical displacement of the node. For example, assuming fixed supports at A, B, C and D in Fig. 2.9, the angular changes at release due to nodal deflection, Δ_1, are $\dfrac{\Delta_1}{a}$ at A and B and $\dfrac{\Delta_1}{b}$ at C and D. The moments applied on release of the joints are shown in Fig. 2.9. As before,

A.A.: $\qquad \dfrac{M_A a}{3EI} + \dfrac{M_1 a}{6EI} = \dfrac{\Delta_1}{a}$

STRUCTURAL EQUATIONS AND NOTATION 15

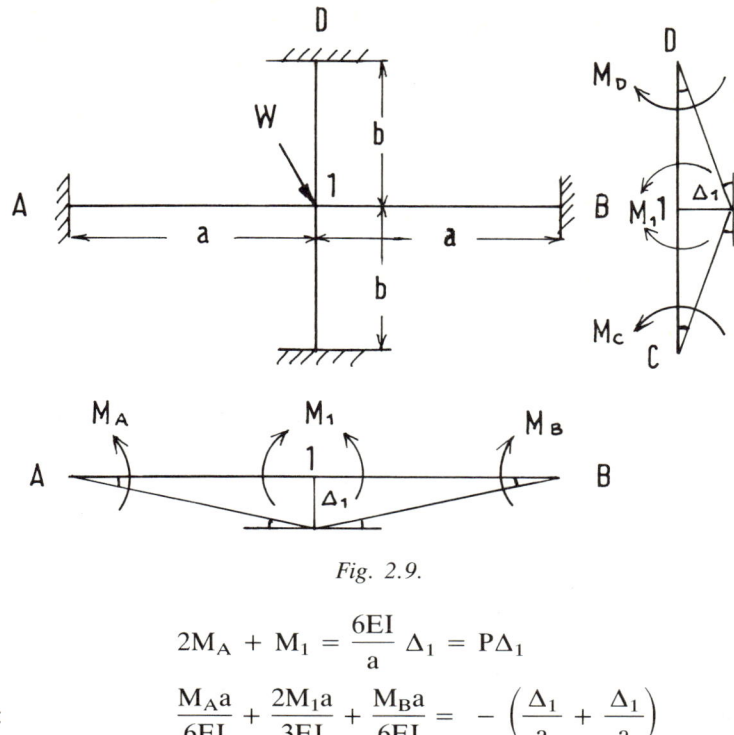

Fig. 2.9.

$$2M_A + M_1 = \frac{6EI}{a} \Delta_1 = P\Delta_1$$

A.1: $\dfrac{M_A a}{6EI} + \dfrac{2M_1 a}{3EI} + \dfrac{M_B a}{6EI} = -\left(\dfrac{\Delta_1}{a} + \dfrac{\Delta_1}{a}\right)$

(Negative signs are introduced since the applied moments increase the angular discontinuities at that joint.)

$$M_A + 4M_1 + M_B = -\frac{2P}{a}\Delta_1$$

By writing down the equations for each joint in turn, a coefficient matrix and set of nodal deflection vectors can be produced that enables the grillage moments to be determined, as described more fully in Chapter 15.

(E) EQUILIBRIUM EQUATIONS

Equations for equilibrium conditions at the joints or supports under the applied loads where bending occurs, as illustrated in Fig. 2.10, may be expressed in the form

E.1: $\dfrac{M_A - M_{1A}}{a} + \dfrac{M_B - M_{1B}}{b} = W$

$\therefore M_A - M_{1A} + rM_B - rM_{1B} = Wa, \quad \text{where } r = \dfrac{a}{b}$

Note how the arrows assist in determining the correct sign; if the moment induces an upwards reaction at 1 the sign is positive and vice versa.

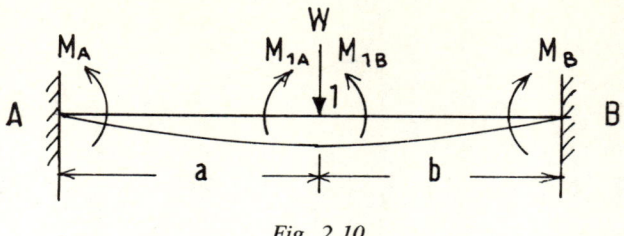

Fig. 2.10.

(3) Torsional displacements

Torsion in beams and grillages is frequently ignored, but the torsional stiffnesses of the members can result in a redistribution of moments and shears in the structure. Consider, for example, the simple frame in Fig. 2.11 with fixed supports at A and B and simply supported at C and carrying a uniformly distributed load perpendicular to its plane. Bending in member 1C will cause twisting to occur in A1B. This is fixed at each end, so the rotation at 1 relative to either end equals $T_1 L_1 \phi_1$,

where T_1 = torsion in A1B induced by the bending in 1C,
L_1 = distance to the fixed support,
ϕ_1 = rotation per unit length per unit torsion of A1B.

Torsional movement tends to reduce the bending moment in 1C at support C but introduces a moment in it at node 1. Thus, we have an extra unknown in the frame requiring an extra equation for its solution. This is obtained by considering the angular changes due to torsion in the members, as described in Chapter 20.

Matrix notation

Matrix notation enables a group of algebraic or numerical quantities to be described by a single symbol. This not only simplifies the appearance of

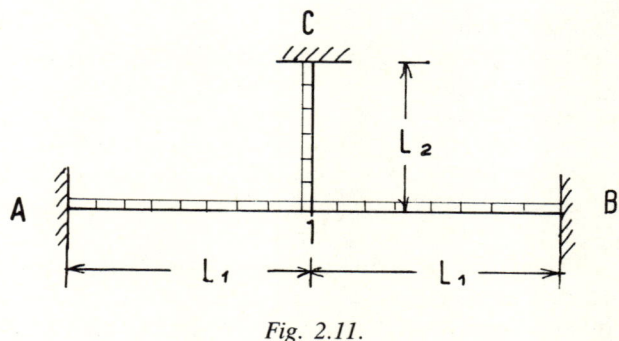

Fig. 2.11.

equations but makes them easier to handle and present to the computer for solution.

Consider, for example, the following set of equations:

Equation 1 $K_1M_A + K_2M_B + K_3M_C = W_1$
Equation 2 $K_4M_A + K_5M_B + K_6M_C = W_2$
Equation 3 $K_7M_A + K_8M_B + K_9M_C = W_3$

These may be written in the following matrix form:

$$\begin{bmatrix} K_1 & K_2 & K_3 \\ K_4 & K_5 & K_6 \\ K_7 & K_8 & K_9 \end{bmatrix} \begin{bmatrix} M_A \\ M_B \\ M_C \end{bmatrix} = \begin{bmatrix} W_1 \\ W_2 \\ W_3 \end{bmatrix}$$

or in condensed matrix form,

$$\mathbf{K \cdot M = W}$$

The expression in the first set of brackets is a *matrix*; the individual terms are *elements* or *coefficients*. Where only one column of elements is involved it is called a column matrix, column vector or simply *vector*. Thus both **M** and **W** are vectors.

To solve for the unknowns M_A, M_B and M_C, we have to solve the matrix equation

$$\mathbf{M = K^{-1}W}$$

For a given load system, W, this entails finding the inverse of the coefficient matrix, K, denoted by K^{-1}. Various numerical techniques have been devised for inverting matrices, two of which are described in the book. It should be noted that for a properly formulated problem in which the number of equations equals the number of unknowns, this results in a square matrix, the size of which may be denoted by (N × N).

Tabular notation

Engineers and computer programmers are more accustomed to presenting data in tabular or spreadsheet format. In this case, the previous set of simultaneous equations in *tabular matrix* form would appear as shown in Table 2.2, where columns two to four indicate the coefficients read row-by-row; the unknowns to which they refer are shown in the headings. The arrangement of the coefficient matrix and vector terms is unchanged, it will be seen. However, we can now enter the results for each unknown in another row below the table when these are obtained. This tabular format is simple and handy to use and has been adopted generally throughout the book to summarise sets of simultaneous equations in generalised form.

Table 2.2 Tabular matrix

Equation	M_A	M_B	M_C	$\overline{W}$
1	K_1	K_2	K_3	W_1
2	K_4	K_5	K_6	W_2
3	K_7	K_8	K_9	W_3

To solve a problem by computer, the terms in the table are evaluated and fed in as a series of data statements with the appropriate applications program. It is important that the terms are entered in the correct sequence, which is usually described in a REM statement near the commencement of the program listing. In the MATIN program and MATSUB subroutine used extensively in the book, the coefficient matrix is entered first, followed by one or more column vectors and then such other sets of terms as required to solve the problem. This can be shown very clearly by setting out the data statements in sequence in the form of a *data table*, as shown in Table 2.3.

Table 2.3 shows, against the headings in the first column, the sequence row-by-row in which the various terms are to be inserted in data statements somewhere in the computer program before it is run. (For convenience the data statements are usually placed at the end of the program.) This format is used to illustrate most of the examples given in the book. As before, results can be entered for each unknown in an extra row below the table. It is advisable to insert the name of the program opposite the answers with the values to be input when run, as shown below. This forms a useful check since, for a program employing a different subroutine such as PIVOT, which is listed in Appendix B, the sequence of data statements will be different, although the actual data used are the same.

In distinguishing between Tables 2.2 and 2.3, it will be seen that the *tabular matrix* is constant for any particular form of construction, whereas the *data table* arrangement depends on the program or subroutine selected for use. This distinction is important in ensuring that data terms are fed in correctly.

Table 2.3 Data table

Equation	M_A	M_B	M_C
1	K_1	K_2	K_3
2	K_4	K_5	K_6
3	K_7	K_8	K_9
$\overline{W}$	W_1	W_2	W_3
Results MATIN N = 3			

3 Matrix inversion and Gaussian elimination

Matrix inversion

A number of techniques are available for inverting matrices, as mentioned in the previous chapter; for example, factorisation, successive transformation, partitioning, etc. Inversion using a step-by-step partitioning method is especially suitable for microcomputers. Not only is the inversion done rapidly in comparison to most other methods but, by putting the inverse matrix elements into the same location in memory as the original elements during the inversion process, we can conserve storage space — a definite advantage where programs are run on the cheaper range of microcomputers.

If, in addition, we include a routine for back-substitution, then the program can be used to solve a set of simultaneous equations, provided certain conditions are met, as described later. A program of this type under the synoptic title of MATIN is reproduced in Appendix B. It is used to solve various problems later in the book.

A knowledge of matrix algebra is required to understand the background to the program. A matrix may be partitioned and its submatrices defined as follows:

$$\mathbf{A} = \begin{bmatrix} a_{11} & a_{12} & a_{13} & a_{14} \\ a_{21} & a_{22} & a_{23} & a_{24} \\ a_{31} & a_{32} & a_{33} & a_{34} \end{bmatrix} = \begin{bmatrix} \mathbf{A}_{11} & \mathbf{A}_{12} \\ \mathbf{A}_{21} & \mathbf{A}_{22} \end{bmatrix}$$

where $\mathbf{A}_{11}$ denotes the submatrix $\begin{bmatrix} a_{11} \\ a_{21} \end{bmatrix}$ and $\mathbf{A}_{12}$ denotes the submatrix $\begin{bmatrix} a_{12} & a_{13} & a_{14} \\ a_{22} & a_{23} & a_{24} \end{bmatrix}$ etc.

The reciprocal of a matrix $\mathbf{A}$ is denoted by $\mathbf{A}^{-1}$, called the inverse of $\mathbf{A}$. In matrix algebra, $\mathbf{A}\mathbf{A}^{-1} = \mathbf{I}$, where $\mathbf{I}$ is a unit matrix, that is one which has unit elements on the diagonal and zeroes elsewhere. If we have a square matrix, $\mathbf{A}$, of size (N × N) with inverse denoted by $\mathbf{B}$, partitioned in such a way that the submatrices on the diagonal are square, then the following equation holds good:

$$\begin{matrix} \mathbf{A} & \mathbf{B} & = & \mathbf{I} \\ \begin{bmatrix} \mathbf{A}_1 & \mathbf{A}_2 \\ \mathbf{A}_3 & \mathbf{A}_4 \end{bmatrix} & \begin{bmatrix} \mathbf{B}_1 & \mathbf{B}_2 \\ \mathbf{B}_3 & \mathbf{B}_4 \end{bmatrix} & = & \begin{bmatrix} \mathbf{I} & 0 \\ 0 & \mathbf{I} \end{bmatrix} \end{matrix}$$

Expanding the product $\mathbf{AB} = \mathbf{I}$ leads to the following:

$$\mathbf{A_1 B_1} + \mathbf{A_2 B_3} = \mathbf{I} \qquad (3.1)$$
$$\mathbf{A_1 B_2} + \mathbf{A_2 B_4} = 0 \qquad (3.2)$$
$$\mathbf{A_3 B_1} + \mathbf{A_4 B_3} = 0 \qquad (3.3)$$
$$\mathbf{A_3 B_2} + \mathbf{A_4 B_4} = \mathbf{I} \qquad (3.4)$$

These equations may be solved to express the unknown submatrices $\mathbf{B_1}$, $\mathbf{B_2}$, etc. in terms of the known submatrices $\mathbf{A_1}$, $\mathbf{A_2}$, etc. Thus, from equation (3.2),

$$\mathbf{B_2} = -\mathbf{A_1}^{-1} \mathbf{A_2} \mathbf{B_4}$$

Substituting these values for $\mathbf{B_2}$ in equation (3.4) gives

$$\mathbf{B_4} = (\mathbf{A_4} - \mathbf{A_3} \mathbf{A_1}^{-1} \mathbf{A_2})^{-1}$$

Similarly,

$$\mathbf{B_3} = -(\mathbf{A_4} - \mathbf{A_3} \mathbf{A_1}^{-1} \mathbf{A_2})^{-1} \mathbf{A_3} \mathbf{A_1}^{-1}$$
$$\mathbf{B_2} = -\mathbf{A_1}^{-1} \mathbf{A_2} (\mathbf{A_4} - \mathbf{A_3} \mathbf{A_1}^{-1} \mathbf{A_2})^{-1}$$
$$\mathbf{B_1} = \mathbf{A_1}^{-1} + \mathbf{A_1}^{-1} \mathbf{A_2} (\mathbf{A_4} - \mathbf{A_3} \mathbf{A_1}^{-1} \mathbf{A_2}) \mathbf{A_3} \mathbf{A_1}^{-1}$$

Putting

$$\alpha = (\mathbf{A_4} - \mathbf{A_3} \mathbf{A_1}^{-1} \mathbf{A_2})$$
$$\beta = \mathbf{A_3} \mathbf{A_1}^{-1}$$
$$\gamma = \mathbf{A_1}^{-1} \mathbf{A_2}$$

we can write the equations more concisely as

$$\mathbf{B_4} = \alpha^{-1}; \quad \mathbf{B_3} = -\alpha^{-1} \beta; \quad \mathbf{B_2} = -\gamma \alpha^{-1}; \quad \mathbf{B_1} = \mathbf{A_1}^{-1} + \gamma \alpha^{-1} \beta$$

The parameters on the right-hand side of the equations show that the process requires the inversion of $\mathbf{A_1}$ and α; the latter is of the same order as $\mathbf{A_4}$. If the partitioning is so arranged that $\mathbf{A_4}$ is a single element then, to solve the unknowns, we require the inversion of $\mathbf{A_1}$, which is of the order $(N-1)(N-1)$ and α, the latter being a single element or scalar quantity so that α^{-1} is merely its reciprocal. The inversion of $\mathbf{A_1}$ can be made to depend on the inversion of a submatrix of order $(N-2)(N-2)$. This reduction may be repeated until the whole inversion depends on the inversion of the first element of $\mathbf{A_1}$, i.e. its reciprocal.

MATIN program

In writing a computer program, it is usual to start with a flow chart depicting the main stages of the analysis. The flow chart in Fig. 3.1 summarises the steps needed to comply with the mathematical expressions in the previous section for inverting the matrix by partitioning. The program itself, entitled MATIN, is listed in Appendix B. Before the program is run, the matrix coefficients and constants have to be entered

MATRIX INVERSION AND GAUSSIAN ELIMINATION

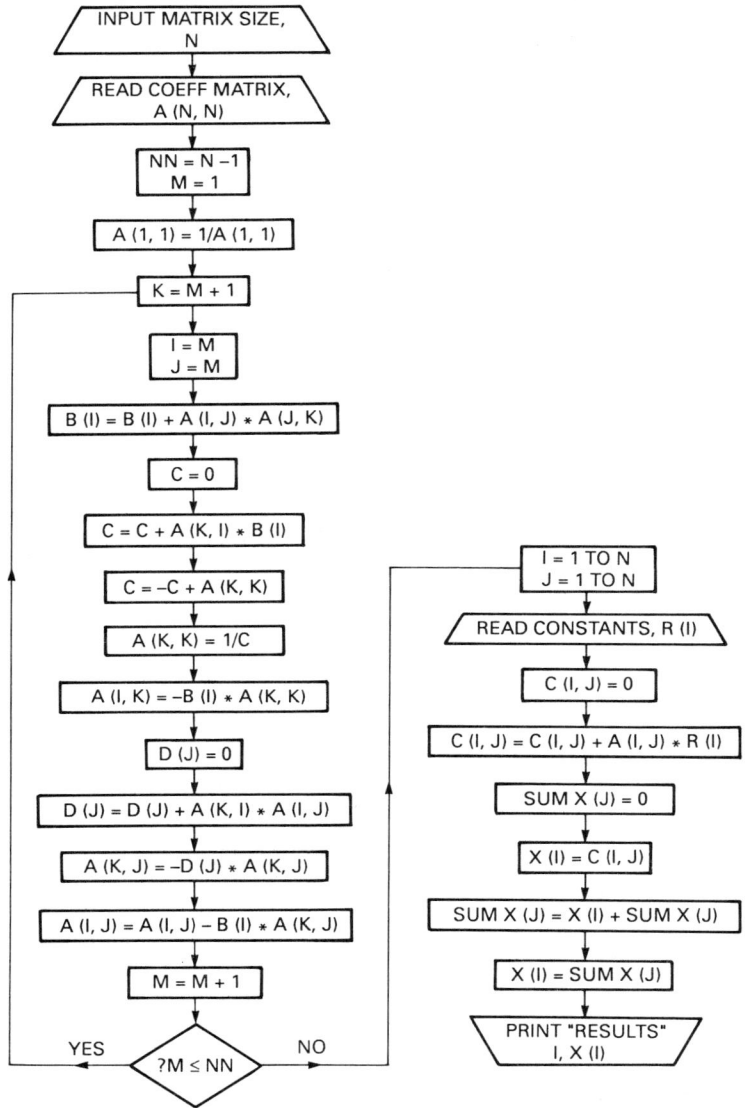

Fig. 3.1.

in DATA statements, as instructed in line 625 of the program. The matrix, **A**, to be inverted is of the order (N × N). To describe its size, the user must first input the value of N. The program then reads and prints out the coefficients. In carrying out the computation, element A_{11} is first of all inverted by replacement with its reciprocal. The variable M is used as a counter to record the size of matrix being inverted at any stage.

Thus, in inverting the (N × N) matrix, initially M = 1. This is followed after execution of the full loop by a change to M = 2 and so forth up to the final inversion where M = N − 1 = NN. At this stage the **A** matrix has been fully inverted.

The **B**, **C** and **D** matrices are used for temporary storage of the α, β and γ parameters. The A (K, J) values computed in the loop commencing with statement 795 are the elements of the B_3 matrix. The loop commencing with statement 810 computes the value of the B_2 matrix which is stored in A (I, J). These become the A_1^{-1} elements for use in the next cycle of the main loop. The program for inverting matrix **A** is finished in statement 825.

As we are using the inversion process for the purpose of solving simultaneous equations, the program continues with instructions for reading and printing out the constant terms and then multiplying the inverse by the constants so as to evaluate and print out the unknowns under the heading of "RESULTS:−".

Statements for printing out the inverse have been omitted from the program, as this is not usually of concern to the user. If the inverse is required for some reason, then add the following lines:

```
826 PRINT "MATRIX INVERSE:−"
827 FOR I = 1 to N: FOR J = 1 to N
828 PRINT A (J, I);
829 NEXT J: PRINT: NEXT I: PRINT
```

The program will work for most square matrices. It will not work if the determinant |A| = 0, in which case the matrix is said to be 'singular'. This can be the result of errors in the matrix terms or inconsistencies or redundancies in the equations on which it is based. If the difficulty arises from zero division, it may be overcome on occasion by rearranging the order of the equations.

The DIM statement in line 640 specifies the largest size of array that can be stored in the variable of that name. The size in the listing has been chosen to deal with the range of problems described in the book without reserving an unnecessarily large amount of memory storage space. For more complex problems, the dimension size can be increased merely by altering the number in line 640.

With some microcomputers it may be advisable to modify the PRINT statements to secure an unambiguous printout. For example, in place of PRINT I, X(I) one might key in PRINT I;SPC(3);X(I). This will ensure that a space appears in the printout between the index I and the results. This comment applies to further listings in the book also, of course.

The program listed in Appendix B can be used as it stands for solving problems involving a single set of simultaneous equations, as described later. Probably its greatest value, however, lies in its use as a subroutine, as described in the next chapter.

MATRIX INVERSION AND GAUSSIAN ELIMINATION

Gaussian elimination

[Optional section (see comments under 'Program selection' at the end of the chapter).]

The process of forming an upper triangular matrix is used in a method for solving simultaneous equations known as 'Gaussian elimination'. In an upper triangular matrix all coefficients below and to the left of the leading diagonal are zero. The first step in this method is to rearrange the matrix, if necessary, so as to place the largest element in the first column on the leading diagonal. The remaining coefficients in the first column are then eliminated and the equations rearranged to put the largest term in the second column on the leading diagonal. The remaining coefficients in that column are then eliminated and the process repeated until the matrix is in upper triangular form. The coefficient on the leading diagonal used during the elimination process is called the 'pivot', hence the alternative title 'pivotal condensation' for the process.

The method is illustrated in the following example which shows also how to evaluate the determinant of the matrix. Given the equations

$$2x_1 + 3x_2 + 5x_3 = 5 \text{ or } 9$$
$$3x_1 + 4x_2 + 7x_3 = 6 \text{ or } 14$$
$$x_1 + 3x_2 + 2x_3 = 5 \text{ or } -1$$

these may be expressed in matrix form thus:

$$\begin{bmatrix} 2 & 3 & 5 \\ 3 & 4 & 7 \\ 1 & 3 & 2 \end{bmatrix} \begin{bmatrix} X_1 \\ X_2 \\ X_3 \end{bmatrix} = \begin{bmatrix} 5 \\ 6 \\ 5 \end{bmatrix} \text{ or } \begin{bmatrix} 9 \\ 14 \\ -1 \end{bmatrix}$$

The first step is to rearrange the rows so as to place the largest element in the first column on the leading diagonal.

$$\begin{bmatrix} \underline{3} & 4 & 7 \\ 2 & 3 & 5 \\ 1 & 3 & 2 \end{bmatrix} = \begin{bmatrix} 6 \\ 5 \\ 5 \end{bmatrix} \text{ or } \begin{bmatrix} 14 \\ 9 \\ -1 \end{bmatrix}$$

Next pivot on the underlined element; multiply the first row by 2/3 and subtract from the second row.

$$\begin{bmatrix} \underline{3} & 4 & 7 \\ 0 & \dfrac{1}{3} & \dfrac{1}{3} \\ 1 & 3 & 2 \end{bmatrix} = \begin{bmatrix} 6 \\ 1 \\ 5 \end{bmatrix} \text{ or } \begin{bmatrix} 14 \\ -\dfrac{1}{3} \\ -1 \end{bmatrix}$$

Pivot on the underlined element; multiply first row by 1/3 and subtract from the third row.

$$\begin{bmatrix} 3 & 4 & 7 \\ 0 & \frac{1}{3} & \frac{1}{3} \\ 0 & 1\frac{2}{3} & -\frac{1}{3} \end{bmatrix} = \begin{bmatrix} 6 \\ 1 \\ 3 \end{bmatrix} \text{ or } \begin{bmatrix} 14 \\ -\frac{1}{3} \\ -5\frac{2}{3} \end{bmatrix}$$

Rearrange the rows to place the largest element in the second row on the leading diagonal.

$$\begin{bmatrix} 3 & 4 & 7 \\ 0 & 1\frac{2}{3} & -\frac{1}{3} \\ 0 & \frac{1}{3} & \frac{1}{3} \end{bmatrix} = \begin{bmatrix} 6 \\ 3 \\ 1 \end{bmatrix} \text{ or } \begin{bmatrix} 14 \\ -5\frac{2}{3} \\ -\frac{1}{3} \end{bmatrix}$$

Pivot on the underlined element; multiply the second row by 1/5 and subtract from the third row.

$$\begin{bmatrix} 3 & 4 & 7 \\ 0 & 1\frac{2}{3} & -\frac{1}{3} \\ 0 & 0 & \frac{2}{5} \end{bmatrix} = \begin{bmatrix} 6 \\ 3 \\ \frac{2}{5} \end{bmatrix} \text{ or } \begin{bmatrix} 14 \\ -5\frac{2}{3} \\ \frac{4}{5} \end{bmatrix}$$

The matrix is now in upper triangular form; hence the determinant equals $3 \times 1\frac{2}{3} \times \frac{2}{5} = 2$.

In order to determine the unknowns X_1, X_2 and X_3 one can again adopt a process of back-substitution. In this process the unknowns are determined in reverse order, that is, starting with X_3 and finishing with X_1. We can evaluate X_3 from the equations represented by the bottom row, then X_2 from the second row and X_1 from the top row. In this example, if we denote the upper triangular matrix by U and the right-hand vectors by B,

$$X_3 = \frac{B_3}{U_{33}} = \frac{2}{5} \div \frac{2}{5} = 1 \text{ or } \frac{4}{5} \div \frac{2}{5} = 2$$

$$X_2 = \frac{B_2 - U_{23}X_3}{U_{22}} = \frac{3 + \frac{1}{3} \times 1}{\frac{5}{3}} = 2 \text{ or } \frac{-\frac{17}{3} + \frac{1}{3} \times 2}{\frac{5}{3}} = -3$$

$$X_1 = \frac{B_1 - U_{12}X_2 - U_{13}X_3}{U_{11}} = \frac{6 - 4 \times 2 - 7 \times 1}{3} = -3$$

$$\text{or } \frac{14 - 4 \times -3 - 7 \times 2}{3} = 4$$

Results, therefore, are:

Case (a): $X_1 = -3$; $X_2 = 2$; $X_3 = 1$
Case (b): $X_1 = 4$; $X_2 = -3$; $X_3 = 2$ determinant = 2

PIVOT program

A program written in elementary BASIC to carry out the above operations is reproduced in Appendix B. The program, entitled PIVOT, is dimensioned so as to handle coefficient matrices up to size (20 × 20) and up to five sets of vectors at a time. The program includes an evaluation of the determinant, D. Line 785 is based on the rule that, if any pair of rows or columns are interchanged, the value of the determinant changes sign. If the determinant is zero, then no solution exists (except for homogeneous equations, with which we are not concerned). A value of D close to zero means that the equations are poorly conditioned, so that numerical difficulties are to be expected in obtaining a solution. The yardstick adopted in the program, in fact, is the ratio of the pivotal coefficient to the mean coefficient value; a ratio of less than 10^{-5} will result in the error message "Ill conditioned equations" being displayed on the screen or printer. The mean coefficient of the matrix — the sum of the absolute coefficient values divided by the square of the matrix size — is evaluated and printed out for the user's information.

Referring to the listed program, it will be seen that the number of equations, N, and vectors, V, have to be input when the program is run. In lines 675–690 the extended matrix is read from the DATA statements previously entered and then printed out, i.e. coefficient and vector terms combined row-by-row. The magnitude of the mean coefficient is calculated and printed out in lines 695–725. Conversion to upper triangular matrix form is performed in lines 730–840. The terms NP1 and NM1 are used for the values N + 1 and N − 1 respectively. The leading row is denoted by I1 and the pivotal element by IP. Rows are interchanged to place the largest coefficient in the pivotal position. A check is made on its relative magnitude in line 845. The determinant is evaluated and printed out in lines 850–865, followed by back-substitution in lines 870–930. In order to print out the results in their correct order, the terms are placed in another matrix B(I, K), before being displayed in lines 935–950.

It should be mentioned that, as with the MATIN program described earlier, PIVOT may equally well be used as a subroutine, with some minor amendments, to enable it to be called up in a main program.

Programs in use

One important point to note is the sequence in which the data must be fed into the computer. In the PIVOT program, the *extended matrix*, including the vector terms, should be keyed in row-by-row. In the MATIN program, on the other hand, the coefficient matrix is to be typed in first, followed by the vector terms. Hence the importance attached to the data table, as described in Chapter 2.

The difference can be illustrated by showing the results from both programs, assuming they have already been saved on disk or tape, in solving the set of equations given previously. Before running the PIVOT program, add the following DATA statements:

```
1000  DATA   2, 3, 5, 5, 9
1005  DATA   3, 4, 7, 6, 14
1010  DATA   1, 3, 2, 5, −1
```

Input $N = 3$ and $V = 2$. The results should agreed with those obtained manually in the previous section. Because of round-off errors, slight deviation from the true results can occur.

In using the MATIN program, it is preferable to enter the coefficients and vector terms on different lines. An example of this is shown below.

```
1000  DATA   2, 3, 5, 3, 4, 7, 1, 3, 2
1005  DATA   5, 6, 5
```

Input $N = 3$ to obtain one set of results. For the second set, overtype line 1005 before running, i.e. key in

```
1005  DATA   9, 14, −1
```

The results should agree with those obtained previously. By including the statements listed on p. 22, the matrix inverse will be given with the results as well; in this case it is:

$$\begin{array}{rrr} -6.5 & 4.5 & 0.5 \\ 0.5 & -0.5 & 0.5 \\ 2.5 & -1.5 & -0.5 \end{array}$$

A modification of MATIN to enable two sets of results to be displayed and combined in one run is described later in the book.

Program selection

In this chapter, two fundamental programs for solving sets of simultaneous equations have been described. The choice is to some extent a matter of personal preference. In the book preference has been given to MATIN and its derivatives for the reasons mentioned before, namely speed of operation and conservation of storage space. In addition, as the constants are entered in a group, they can be altered readily to deal with different magnitudes of loading or various partial load factors. By saving MATIN and the programs incorporating the MATSUB subroutine (listed later), all the problems in the next part of the book can be tackled successfully. The PIVOT and PIVOTA listings have been marked 'optional' on this account, and may be disregarded by the busy practitioner or anyone anxious to keep keyboard work to a minimum.

4 Subroutines and file merging

Subroutines

In the last chapter two programs suitable for solving simultaneous equations, namely MATIN and PIVOT, were described (the listings are contained in Appendix B). Useful as the listed MATIN and PIVOT programs are in themselves, their value increases further when written in the form of subroutines or subsidiary programs that can be called up in a main program. This is done by inserting a GOSUB statement in the main program and a RETURN statement at the end of the subroutine. This procedure can save a lot of wasteful duplication in program writing and means, too, that one can construct and check a long program in separate segments, if required.

If the two listed programs are to be used as subroutines it is *essential* to make the following alterations to enable them to be called up.

PIVOT   Alter line 955 END to 955 RETURN
MATIN   Alter line 950 END to 950 RETURN

These changes are sufficient but other amendments may be desirable to improve the output. Data input instructions may be omitted, for instance, since these will appear in the main program. A different title and style of heading may be used to indicate its new status as a subroutine. Some PRINT statements may be omitted to avoid duplication.

The changes considered advisable in MATIN are shown in the listing at the end of the chapter. The subroutine has been given the synoptic title of MATSUB. Instructions for deleting and inserting statements are given with the listing. Included is a 'flag', $F = -1$, in line 830 that allows reading of the constant terms to be skipped over when the subroutine is called up. This device is used in the XBEAM and GRID programs listed in Appendix B. The line may be omitted where this process does not form part of the main program.

After making the alterations, SAVE the subroutine under the title MATSUB so that it may be merged with a main program, as described later. This will enable a whole range of problems on grillages and beam frameworks to be tackled, as described in Part 3. It is left to the reader to make changes on similar lines to PIVOT if it is to be used as a subroutine.

Writing a subroutine

We may illustrate the use of subroutines by showing how to extend the PIVOT program so it will not merely print out the results for more than one load vector, but will combine them and print out the combination. This may be done by forming additional arrays in which the original results are stored before being summed. The following subroutine is an example of how this may be done for two sets of results.

```
1000 REM SUB-ROUTINE TO COMBINE RESULTS
1005 PRINT: PRINT "COMBINED RESULTS:-"
1010 DIM R(20,1),S(20,1),T(20,1)
1015 FOR I=1 TO N
1020 K=NP1
1025 R(I,1)=B(I,K)
1030 K=M: IF M=NP1 THEN 1040
1035 S(I,1)=B(I,K)
1040 T(I,1)=R(I,1)+S(I,1)
1045 PRINT I,T(I,1)
1050 NEXT I
1055 RETURN
```

If this is incorporated in the main program by means of a GOSUB statement, it is advisable to amend the title to show that its scope has been changed. One might also use the opportunity to streamline the output. For example, printouts of the determinants and mean coefficients might be omitted from PIVOT, as these are seldom of practical concern. The PRINT instructions for ill-conditioned equations should appear in the last line of the program, as before. These alterations to accommodate the subroutine are summarised in the instructions at the head of the PIVOTA listing at the end of the chapter. The revised program, under its title of PIVOTA, may be saved for use later in solving problems involving differential settlement. This is quite optional, however, as an alternative method employing MATIN will be described also.

Files

Where programs in the microcomputer's memory are to be saved for future use, permanent storage is usually provided on tape or disk. Cassette units for tape storage are cheap but slow and prone to error; disk drives are far more reliable and have the tremendous advantage of permitting random access. Once a disk has been formatted, the titles of all programs or files stored on it can be displayed in catalogue form and any one of them accessed almost immediately. This flexibility is a feature of all disk operating systems (DOS) and allows us to consider storage not merely of programs but information of any type that one may want to recover later for reuse. Unfortunately, the formatting procedures required to initialise the disk to support DOS files vary, so that a disk formatted on one make of computer is often unusable on another. In this situation, short programs that can be keyed in by the user and so run on any computer make have one advantage.

A file may be described as an organised collection of information that can be accessed in some way. A disk file refers to information residing on a disk. Before a program is saved it must be given a file name; this is usually up to eight characters in length, depending on the specification ('filespec') for that computer. You may then use the SAVE command when using BASIC to put the program on disk, assuming this has previously been formatted. BASIC saves the file in a compressed binary format. With the IBM PC you can use an 'A' option to save the program in ASCII or text format,

i.e. SAVE filespec, A

ASCII files take up more space, since they make use of a standard set of characters only. However, programs saved in ASCII form may be read as data files and can be merged with other files, as described in the next section.

Files are commonly divided into two main categories:

(1) program files
(2) data files.

A program file is one containing the user program which has been SAVEd under a file name on disk or tape. Examples are the programs listed in this book, once saved. These are what the computer actually uses when it executes a program. Data files, essentially, are collections of information (including text) that may be read and manipulated by the program in the computer's memory. They are useful when a program requires or produces large amounts of data, as in word processing or stock control, for example.

Files may be transferred between disk drives or from one disk to another via DOS using the COPY command. This may be used to display the disk file on screen, print it out or transfer it to another computer, if compatible. File handling procedures vary from computer to computer and file management in itself is a specialised subject. However, one task of special importance to us, since it can save a considerable amount of time and trouble, is the merging of files that may be either in memory or on disk.

Merging files

Two situations are commonly encountered in practice, as follows.

(A) COMBINING A PROGRAM AND A DATA FILE

It is assumed that a program such as MATIN for analysing a structure is already on disk or tape. Before this is run, data for the structure under examination must be entered. If this structure, which may be a Warren girder of specific size, dimensions and relative member stiffness, is one likely to be used time and again, the data for it may be written into a data file and stored under a file name, say WARREN1. To obtain the forces or displacements, we have to merge the two files, add the data for loading and run the combined program to obtain the results.

(B) COMBINING A MAIN PROGRAM AND A SUBROUTINE

In this case it is assumed that a subroutine, such as MATSUB, has already been saved and that one has keyed in the main program and saved it independently on disk. Obviously a lot of time can be saved by merging the two programs, adding whatever additional lines are required and running the combined program to get the results.

The procedure in both the above cases when using BASIC is much the same. First, it is generally advisable to have the line numbers in the data file or the subroutine higher than those in the source program. It is essential that there is no clash of numbers in any event, if all the statements are to be retained. To arrange for this, you may have to use the RENUM facility to renumber the lines in the data file or subroutine.

With the IBM PC, you should then use the SAVE (filespec), A command to save the data file or subroutine in ASCII format on disk under a new name. Next load the main program into memory and use the MERGE command followed by the name of the ASCII file. This searches for the named file and, when found, merges the program lines in memory. You may then add whatever data or additional instructions you wish before it is RUN. You may also SAVE the combined program under a new title, assuming the old files are to be retained on disk. On some microcomputers the same result is achieved by the use of different commands, such as SPOOL, LIST and EXEC. One word of warning — be sure to have copies of the files on disk before attempting to merge them in case the programs are lost in the merging process.

In later chapters two main programs are listed under the titles of XBEAM and GRID. These are to be merged with MATSUB before use in the manner described above.

Use of a printer

Almost all microcomputers can be interfaced with a printer to provide a permanent record of the results and such additional information as the user may require. In our case, this will usually consist of:

(1) headings, giving program title, job reference, identification of the structure or member being analysed and the type of loading;
(2) values to input when running the program;
(3) data, for verification;
(4) results at particular stages of the analysis, with headings;
(5) final results in the prescribed order.

A positive effort should be made to keep paperwork generated by the computer to a minimum. Because it is so very easy to output masses of information, there is a temptation to record a lot of quite superfluous data. This may look impressive to the outsider, but it is a waste of both time and materials. The best guideline is to have just sufficient information on the printout to enable a third party to check the results unaided.

It may be necessary with some microcomputers to modify the listed PRINT statements to obtain a neat, orderly display. As well as this, individual users are likely to have their own ideas on how much information should be recorded. An example of a printout obtained with the PIVOTA program in analysing a two-span beam subject to settlement is shown below, for comparison. (The basis of the analysis is described in Chapter 6.)

```
HIGH ST PROJECT: BLOCK A: BEAM C4-C6
PIVOTA PROGRAM: TYPE 2 LOADING: D5 = 40; D6 = 125
   NUMBER OF EQUATIONS? 3
   NUMBER OF VECTORS? 2

TABULAR DATA:-
2   1   0    36      19.21
1   4   1    72      19.21
0   1   2    36     -37.95

RESULTS:-
1        12
2        12
3        12

1         4.84166
2         9.5266
3       -23.7383

COMBINED RESULTS:-
1        16.8416
2        21.5266
3       -11.7383
```

The printout shows the support moments due to superimposed loading and differential settlement separately followed by the combined moments. Additional headings may be written into the program to identify these more clearly, of course; this is a matter of personal choice.

Once the support moments are known it is not a difficult matter to determine the remaining moments from the 'free' BM diagrams and to evaluate the shear forces and support reactions. The computer program may be extended to give these on the lines of many commercial software packages, but the values are so easy to determine manually that this has not been considered worthwhile. (See worked example, Chapter 6.)

MATSUB

Amend MATIN program by deleting lines 625, 925 and 940 and keying in the following:

```
600 REM ----------
605 REM - MATSUB -
610 REM ----------
615 REM SUB-ROUTINE TO SOLVE EQUATIONS USING
616 REM INVERSE OF MATRIX, SIZE (NXN)
830 IF F=-1 THEN 865: REM FLAG INDICATOR
950 RETURN
```

PIVOTA

Note – optional program

Amend PIVOT program by deleting lines 725, 740, 785, 850–865 and key in the following:

```
600 REM **********
605 REM * PIVOTA *
610 REM **********
625 REM AND INCLUDES ROUTINE FOR COMBINING RESULTS FOR 2 VECTORS.
845 IF ABS(A(N,N)/MC) < PR THEN 1060
955 GOSUB 1000: REM TO PRINT COMBINED RESULTS
960 END
1000 REM SUBROUTINE TO COMBINE RESULTS
1005 PRINT: PRINT "COMBINED RESULTS:-"
1010 DIM   R(20,1),S(20,1),T(20,1)
1015 FOR I=1 TO N
1020 K=NP1
1025 R(I,1)=B(I,K)
1030 K=M: IF M=NP1 THEN 1040
1035 S(I,1)=B(I,K)
1040 T(I,1)=R(I,1)+S(I,1)
1045 PRINT I,T(I,1)
1050 NEXT I
1055 RETURN
1060 PRINT "ILL CONDITIONED EQUATIONS": END
```

5 Release-deformation flexibility analysis

Description

The release-deformation method of linear structural analysis is essentially a variant of the well-known flexibility method in that joints in a frame are initially released and the relative displacements and discontinuities are examined and expressed in matrix format to obtain a solution. However, in the release-deformation method a solution is obtained by applying moments or forces to restore the structure to its original state, taking into account the relative flexibility of the members; thus, it may be described as a displacement method using a modified stiffness matrix.

Its main characteristic lies in the use of deformation diagrams to describe the physical behaviour of the structure under load and to determine correct sign usage. This is used in preference to the more rigorous mathematical approach adopted in the matrix flexibility method, although the basic concepts are much the same.

To investigate the structure's behaviour, a series of releases or 'hinges' is introduced at certain points, usually at joints or over supports. The deformed shape of the structure is then examined and the discontinuities noted. The forces and moments needed to eliminate these discontinuities are then estimated, taking into account the physical properties of the members. The procedure may be described in stages as follows.

(1) Introduce releases into the structure at certain specific points. Although theoretically these may be anywhere, it is usually convenient to locate them at joints or supports. If the joints are capable of displacement, temporary restraints are required.
(2) Introduction of these releases will cause rotations to occur in beams and linear displacements in trusses, for instance. The structure in its deformed state is sketched out so as to identify these discontinuities in terms of angular rotations, linear displacements or nodal movements.
(3) Forces or moments are now applied simultaneously to the released structure of such magnitude and orientation as to eliminate the discontinuities and restore it to its pre-release state. In so doing, the conditions of compatibility of deformations and equilibrium of forces must be satisfied. These are expressed in equations, enabling us to determine the unknown forces, moments and displacements.

(4) The set of simultaneous equations can be put into a compact, tabular format using flexibility parameters and matrix notation. Slope/deflection equations enable angular changes to be expressed in terms of load, span and relative flexibility. The tabulated figures provide the DATA to be fed into an appropriate computer program.

It should be observed that, in the case of trusses, the releases will be applied to members at one end only and in a selected sequence. These releases cause the direct forces in the members to be reduced to zero and the resultant shortening or lengthening produces linear discontinuities at the joints that can be evaluated.

Special features

Since it is used exclusively in the following chapters, special features of the release-deformation method are summarised below.

(1) Use of a deformation diagram to identify the discontinuities on release and the moments or forces necessary to restore the structure to its original status. This does *not* represent the actual deformed shape but the shape assumed after introduction of the releases and whatever temporary restraints are required for stability. Equations are derived from a study of this diagram rather than from theoretical concepts. Successful use depends on an understanding of how structures perform, therefore, rather than on mathematical prowess.
(2) Use of the deformation diagram in conjunction with a simple convention for determination of signs. If the assumed deformation pattern and input data are correct, the answer will have a positive sign. A negative sign indicates that the deformation is not, in fact, as assumed or that an error has occurred in the input. Negative results need investigation but there can be a perfectly valid reason for their occurrence, for example a deformation diagram applicable to vertical loading will differ from that for support settlements in a continuous beam, so negative results can be expected when these are analysed in conjunction with each other.
(3) Use of a 'reference element' in describing the structure so as to cater for differing E and I values and orientations. Typical 'reference elements' are the first span of a continuous beam, portal frame beam member or first finite element of a continuum structure. This procedure simplifies data preparation considerably since relative E and I values only are required; furthermore, instead of using a rather error-prone and clumsy co-ordinate system, trusses or lattice girder members can be described entirely by the angle they subtend relative to the selected reference element.

(4) Use of flexibility parameters to simplify calculation of the coefficient matrix terms.
(5) Vectors dealing with sway, nodal displacements or settlements are easily calculated, often consisting merely of inverses of the column heights or beam spans.
(6) Use of computer programs written in elementary BASIC for obtaining the required moments, etc., using data supplied by the release-deformation analysis. Although six programs are listed in the book, only three are required to solve all the problems covered in the text. (One comprehensive program could suffice, but would entail entering a lot of superfluous data as well as being inappropriate in many cases.)
(7) Computer programs can be understood by anyone with a knowledge of elementary BASIC programming; they are short enough to be keyed in without difficulty. Because they do not make use of machine code routines, the programs will run on any make of microcomputer with a BASIC facility.
(8) Efforts have been made to keep data preparation time to a minimum. Programs contain very few INPUT statements, so they will run without interruption most of the time. This is considered more important than the use of so-called 'computer friendly' messages. Provision has been included for recording all the data and vital stages of the analysis as well as the answers in the printouts for checking purposes.
(9) Coefficient matrices and vectors are expressed in general terms for each basic structural system. These terms may be used directly to supply the data for a particular problem, so that the analysis is a routine procedure in many cases.

Sign convention

The convention adopted in the analysis is based on the following general rules:

(1) In equilibrium equations, forces opposing the applied loads or restraints at a joint are considered positive; in resolving forces, those acting to the right when projected on the XX axis and acting upwards on the YY axis are considered positive.
(2) In linear compatibility equations, forces closing the linear discontinuity at that point are regarded as positive.
(3) In angular compatibility equations, moments closing the angular discontinuity at the point of application are regarded as positive.

Consequently, in beams carrying downward-acting loads, release moments are applied at the supports in the direction shown by the arrows in Fig.

2.3 in order to counteract the angular changes there. These are considered positive; in this event, positive moments will produce tension at the *upper* surface of the beam. Release moments applied to counteract angular changes due to settlement or nodal displacements are considered positive if they act in the same direction as the 'loads' moments; otherwise the vector terms are given a negative sign.

In foundations, where soil reactions act upwards, release moments are applied in the opposite direction to the above, as shown in Fig. 7.5. In this case, positive moments produce tension at the *lower* surface, which is the normal situation for foundations.

Cantilevers are dealt with by applying a release moment at the support (assuming a temporary prop at the end) in the *same direction* as the cantilever moment and making the vector term negative. This ensures correct sign when cantilever and span moments are combined.

Bending moment diagrams are constructed with positive values above the baseline. This means that in all cases, except for foundations, the diagram will appear on the *tension* side of the member.

Procedures

Before examining in detail the way in which foundation and grillage problems can be tackled, it may be useful to discuss the general procedures used in solving various structural forms, commencing with one of the simplest, the single-span beam.

(1) Single-span beams

The equations for single-span beams subject to vertical loads were derived in Chapter 2. Thus, for a beam with fixed supports carrying a uniformly distributed load, using equations (2.8) and (2.9) and the appropriate value of γ from Table 2.1, we obtain

A.A: $\qquad 2M_A + M_B = \dfrac{1}{4}WL$

A.B: $\qquad M_A + 2M_B = \dfrac{1}{4}WL$

Hence, $\qquad M_A = M_B = \dfrac{WL}{12}$

Similarly, for a symmetrical triangular load we obtain

$$M_A = M_B = \dfrac{WL}{9.6}$$

Bending moments at the supports for other types of loading may be obtained in the same fashion. Obviously manual calculations are quite adequate for solving such problems.

(2) Continuous beams

The procedure for continuous beams may be illustrated by reference to Fig. 5.1, which represents a continuous four span beam supporting distributed live loads. Releases are introduced at supports A, B, C, D and E causing the beams to sag and assume the shape shown by the dashed lines. Tangential lines to the deformed shapes give the changes in slope, θ, at the supports. Thus, θ_{AB} is the change in slope of AB, θ_{BA} is the change in slope of BA and so forth.

To eliminate these angular changes, moments are applied to the ends of the spans. These, denoted by M_A, M_B, etc., are shown on the figure and applied in a direction that will close the discontinuities and restore the beam to its natural alignment. (Since $M_{BA} = M_{BC}$ the notation M_B is adequate.)

The first span is taken as the reference element for the basic properties E, I and L. Since E is constant for all spans, we can describe the beam's behaviour in bending by the flexibility ratio, $k = L/I$. By selecting the first span with suffix, r, as the reference member, the relative flexibility ratio of any span, n, may be written as

$$K_n = \frac{k_n}{k_r} = \frac{L_n}{I_n} \frac{I_r}{L_r} \tag{5.1}$$

Thus, for span 1, $k_1 = \frac{L_1}{I_1}$ and $K_1 = 1$

span 2, $k_2 = \frac{L_2}{I_2}$ and $K_2 = \frac{L_2}{I_2} \frac{I_1}{L_1}$; if $I_2 = I_1$, $K_2 = \frac{L_2}{I_2}$

span 3, $k_3 = \frac{L_3}{I_3}$ and $K_3 = \frac{L_3}{I_3} \frac{I_1}{L_1}$; if $I_3 = I_1$, $K_3 = \frac{L_3}{L_1}$, etc.

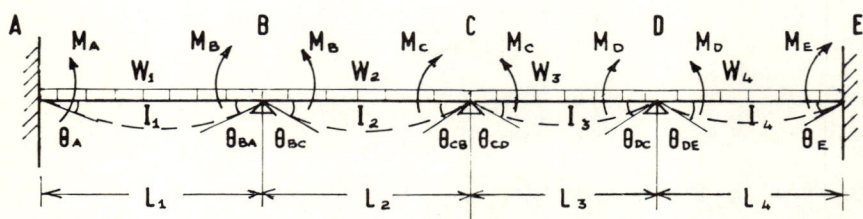

Fig. 5.1.

Compatibility equations, similar to equations (2.8) and (2.9), may be written for conditions at each of the supports, as follows:

A.A: $$M_A \frac{L_1}{3EI_1} + M_B \frac{L_1}{6EI_1} = \theta_{AB}$$

Multiplying across by $\frac{6EI_1}{L_1} = P$,

$$2M_A + M_B = \frac{6EI_1}{L_1} \theta_{AB} = P\theta_{AB}$$

But $P\theta_{AB} = \gamma W_1 L_1$ (see equations (2.7a) and (2.9a)) and for a uniformly distributed load, from Table 2.1, $\gamma = \frac{1}{4}$, so we have

A.A: $$2M_A + M_B = \frac{1}{4} W_1 L_1$$

For the angular change at support B, denoted by A.B, we have

A.B: $$M_A \frac{L_1}{6EI_1} + M_B \left(\frac{L_1}{3EI_1} + \frac{L_2}{3EI_2}\right) + M_c \frac{L_2}{6EI_2} = \theta_{BA} + \theta_{BC}$$

$$M_A + 2(1 + K_2) M_B + K_2 M_c = \gamma W_1 L_1 + \gamma K_2 W_2 L_2$$
$$= \frac{1}{4} W_1 L_1 + \frac{1}{4} K_2 W_2 L_2$$

Similarly,

A.C: $$K_2 M_B + 2(K_2 + K_3) M_c + K_3 M_D = \frac{1}{4}(K_2 W_2 L_2 + K_3 W_3 L_3)$$

A.D: $$K_3 M_c + 2(K_3 + K_4) M_D + K_4 M_E = \frac{1}{4}(K_3 W_3 L_3 + K_4 W_4 L_4)$$

A.E: $$K_4 M_D + 2 K_4 M_E = \frac{1}{4} K_4 W_4 L_4$$

These equations are expressed in tabular matrix form in Table 5.1. To solve a particular problem, the terms in Table 5.1 are evaluated and fed in as DATA in correct sequence when the selected computer program is in memory. Essentially the same procedure is used in solving for any number of spans under any system of loading, using the appropriate γ factor from Table 2.1.

From the above it will be seen that the $\overline{W}$ vector term comprises the sum of the γ WKL values for the spans either side of the support under consideration, where

γ = factor determinable from Table 2.1

K = relative flexibility factor = $\dfrac{k_r}{k_n}$

Table 5.1

Equation	Moment					$\overline{W}$
	A	B	C	D	E	
A.A	2	1	0	0	0	$\frac{1}{4}W_1L_1$
A.B	1	$2(1+K_2)$	K_2	0	0	$\frac{1}{4}(W_1L_1 + K_2W_2L_2)$
A.C	0	K_2	$2(K_2+K_3)$	K_3	0	$\frac{1}{4}(K_2W_2L_2 + K_3W_3L_3)$
A.D	0	0	K_3	$2(K_3+K_4)$	K_4	$\frac{1}{4}(K_3W_3L_3 + K_4W_4L_4)$
A.E	0	0	0	K_4	$2K_4$	$\frac{1}{4}K_4W_4L_4$

W = total load on span (or concentrated load value)
L = span.

If the end supports A and E are pinned, then $M_A = M_E = 0$, i.e. the angular change on release is zero. Equations A.A and A.E may thus be deleted, since no releases are needed there. It follows that the first and last rows and columns of Table 5.1 are omitted in dealing with simply supported ends.

Where support settlements occur in fixed-ended or continuous beams, an additional vector based on the angular changes on release due to settlement is tabulated. The method is described in detail in Chapter 6. Analysis of cranked beams, interconnected cantilevers and beams with torsion is dealt with in Chapter 20.

To see how the procedure for a continuous beam is applied in practice, consider Fig. 5.2 in which the four-span beam carries a variety of load systems. Relative moments of inertia for each span are marked on the figure. In determining the relative flexibility ratios we can let $I_1 = 1$. Therefore,

$$k_1 = \frac{I_1}{L_1} = 0.2; \quad k_2 = \frac{1.333}{10}; \quad k_3 = \frac{1.2}{6}; \quad k_4 = \frac{0.8}{8}$$

and $K_1 = 1; \quad K_2 = \frac{0.2}{0.1333} = 1.5; \quad K_3 = 1; \quad K_4 = 2$

The vector quantities to be entered in the data table may be obtained in a step-by-step calculation as shown opposite (compare with Table 5.1).

The results of these calculations are incorporated in Table 5.2. Entering these figures in DATA statements with MATIN produces the results shown in the bottom line of the table.

In Fig. 5.2 (b), the support moment diagram, consisting of a series of straight lines connecting the support moment values, is superimposed on the 'free' bending moment diagrams. This gives a new base-line with positive values above the line and negative values below, as shown hatched. The combined moments diagram is drawn to a straight line base in Fig.

RELEASE-DEFORMATION FLEXIBILITY ANALYSIS

Support	Span	γ WKL			$\overline{W}$	
A	AB	$\dfrac{4 \times 2 \times 5 \times 5}{15 \times 2}$	=	6.667	6.667	(A)
B	BA	$\dfrac{7 \times 2 \times 5 \times 5}{30 \times 2}$	=	5.833 ⎫	76.146	(B)
B	BC	$\dfrac{5 \times 3 \times 10 \times 3 \times 10}{16 \times 2 \times 2}$	=	70.313 ⎭		
C	CB	$\dfrac{5 \times 3 \times 10 \times 3 \times 10}{16 \times 2 \times 2}$	=	70.313 ⎫	110.813	(C)
C	CD	$\dfrac{2 \times 6 \times 6}{4} + \dfrac{3 \times 10 \times 6}{8}$	=	40.5 ⎭		
D	DC	$\dfrac{2 \times 6 \times 6}{4} + \dfrac{3 \times 10 \times 6}{8}$	=	40.5 ⎫	200.5	(D)
D	DE	$\dfrac{5 \times 8 \times 2 \times 8}{4}$	=	160.0 ⎭		
E	ED	$\dfrac{5 \times 8 \times 2 \times 8}{4}$	=	160.0	160.0	(E)

5.2 (c) to give the final BM diagram. This procedure is followed throughout the book, although in most cases the 'free' and support moment diagrams have been omitted since they are incorporated in the final diagram. Shear force diagrams can be drawn simply by using values obtained by ordinary statical considerations, as described for Fig. 6.6, Chapter 6.

Support moments can be obtained for different load patterns (e.g. alternative spans loaded) merely by altering the $\overline{W}$ terms before running the program.

Table 5.2 Data table

Support	Moment				
	A	B	C	D	E
A	2	1	0	0	0
B	1	5	1.5	0	0
C	0	1.5	5	1	0
D	0	0	1	6	2
E	0	0	0	2	4
$\overline{W}$	6.667	76.146	110.813	200.5	160
Results MATIN N = 5	−2.34	11.34	14.52	21.20	29.40

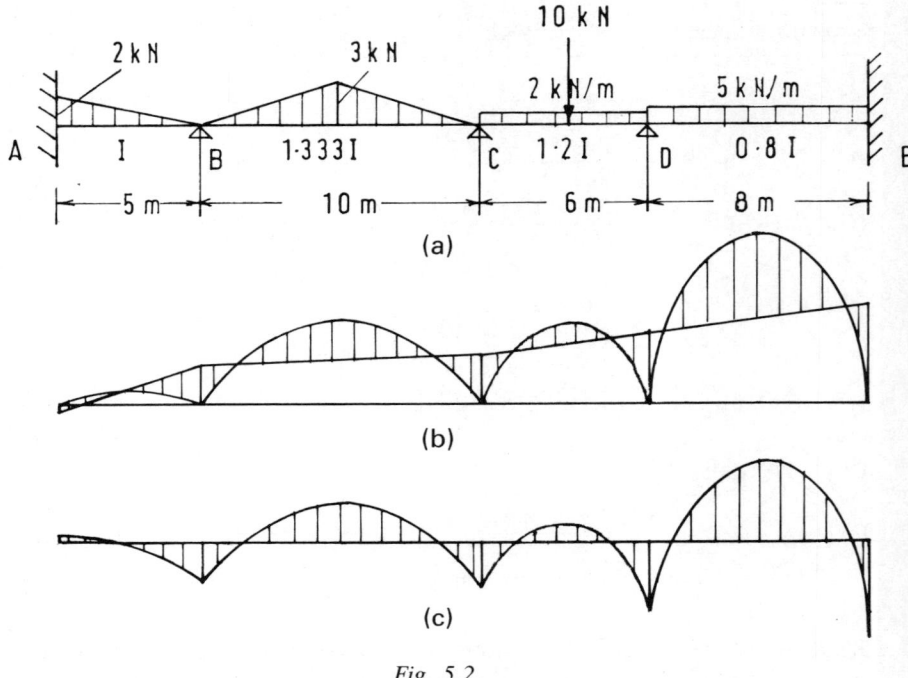

Fig. 5.2.

(3) Continuum structures and finite elements

The previous equations assume that beams are of uniform section throughout the span and that the applied loads conform to some definite pattern. Where either the beam or loading is irregular, the problem can be tackled by dividing the member into a number of small sections (finite elements) and considering each in turn. Equilibrium and compatibility equations may be written down for each node in sequence, following which the tabular data can be resolved by means of a computer program in the usual way.

Consider the general case of a beam of variable depth carrying an irregular distributed loading of intensity, w, as shown in Fig. 5.3. The simply supported beam is assumed to be divided into a number of short strips of length, a, with moments of inertia I_1, I_2, etc. Nodes are numbered 0, 1, 2, 3, etc. as shown; the loads applied to each node are designated W_0, W_1, W_2, etc. It is assumed that no bending occurs between nodes, so that angular changes on release are due entirely to nodal displacements Δ_1, Δ_2, Δ_3, etc. The bottom diagram shows the deformed shape of the beam AB when releases are introduced successively at each node, assuming these are held in their deflected position by temporary props. Moments M_1, M_2, M_3, etc. are applied so as to eliminate the angular

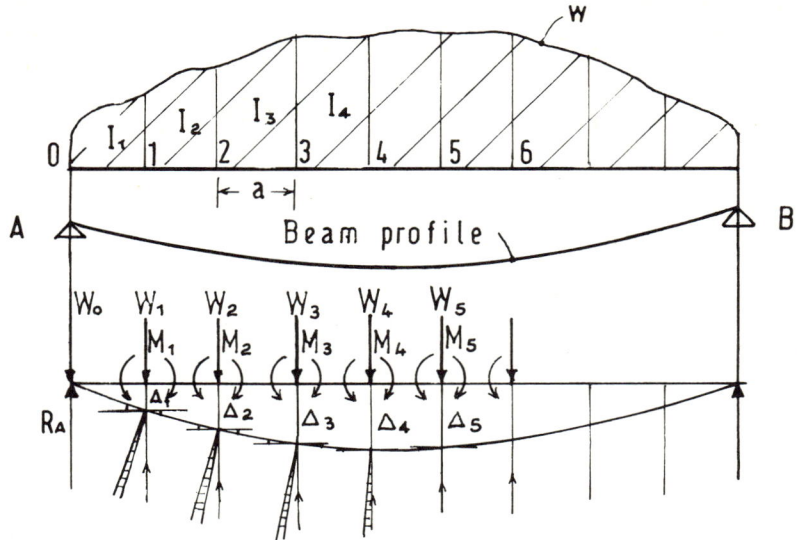

Fig. 5.3.

changes as shown in the diagram. Using the same notation as before one can obtain the following equations:

E.O: $\dfrac{M_1}{a} = R_A - W_0$

$M_1 = a(R_A - W_0)$

E.1: $\dfrac{M_1}{a} + \dfrac{M_1 - M_2}{a} = w_1 a$

$2M_1 - M_2 = W_1 a$, where $W_1 = w_1 a$

E.2: $\dfrac{M_2 - M_1}{a} + \dfrac{M_2 - M_3}{a} = w_2 a$

$-M_1 + 2M_2 - M_3 = W_2 a$

E.3: $-M_2 + 2M_3 - M_4 = W_3 a$

A.1: $\dfrac{M_1}{3EI_1}a + \dfrac{M_1}{3EI_2}a + \dfrac{M_2}{6EI_2}a = \dfrac{\Delta_1}{a} - \dfrac{\Delta_2 - \Delta_1}{a} = \dfrac{1}{a}(2\Delta_1 - \Delta_2)$

Multiplying across by $\dfrac{6EI_1}{a} = P$,

$2(1 + K_2)M_1 + K_2 M_2 + \dfrac{P}{a}(-2\Delta_1 + \Delta_2) = 0$

A.2: $\dfrac{M_1}{6EI_2}a + \dfrac{M_2}{3EI_2}a + \dfrac{M_2}{3EI_3}a + \dfrac{M_3}{6EI_3}a = \dfrac{\Delta_2 - \Delta_1}{a} - \dfrac{\Delta_3 - \Delta_2}{a}$

$$= \frac{1}{a}(-\Delta_1 + 2\Delta_2 - \Delta_3)$$

$$K_2M_1 + 2M_2(K_2 + K_3) + K_3M_3 + \frac{P}{a}(\Delta_1 - 2\Delta_2 + \Delta_3) = 0$$

A.3: $\quad K_3M_2 + 2M_3(K_3 + K_4) + K_4M_4 + \dfrac{P}{a}(\Delta_2 - 2\Delta_3 + \Delta_4) = 0$

Similar equations may be written for the other nodes. In tabular form the equations appear as shown in Table 5.3. Variations of this table will be found in later chapters dealing with continuum structures such as foundations on an elastic medium, soil-retaining structures, sheet piling, laterally loaded piles, raft foundations, etc.

(4) Trusses

In analysing triangulated structures or trusses, use is made of the equations for linear displacement given in Chapter 2. Member forces can be obtained for determinate structures from equilibrium equations alone. Joint displacements can be ascertained from equation (2.5) once the member forces and properties are known (see Chapter 13).

The solution of indeterminate structures may require use of equilibrium, linear and angular displacement equations (see trussed beam example, Chapter 14). The deformation diagram assumes considerable significance in the analysis of indeterminate structures and rigid frames because of its usefulness in identifying the discontinuities that occur on release and their appropriate sign.

(5) Rigid frames

Symmetrical rigid frames of any practical size or shape, loaded symmetrically, can be solved directly using the MATIN program. Where moments due to sway have to be considered in addition to those due to the applied loads, these 'secondary moments' must be calculated also. To do this, moments due to unit sway are calculated and multiplied by factors obtained from consideration of the unbalanced shears at the column heads. The computer program XBEAM, listed later, may be used to perform these operations. Since relative sway moments only are required in the initial stages of the analysis, sway vector terms are expressed simply as inverses of the column heights. The deformation diagram provides the key to determination of the relative displacements and their sign.

Multi-storey frames subject to sway can be tackled in the same way by using sway correction factors for each storey in turn. The computer

Table 5.3 Tabular matrix for variable-depth beam

Equation	1		2		3		4		$\overline{W}$
	M	$\dfrac{P}{a}\Delta$	M	$\dfrac{P}{a}\Delta$	M	$\dfrac{P}{a}\Delta$	M	$\dfrac{P}{a}\Delta$	
E.0	1								$a(R_A - W_0)$
A.1	$2(1+K_2)$	-2	K_2	1					0
E.1	2		-1						$W_1 a$
A.2	K_2	1	$2(K_2+K_3)$	-2	K_3	1			0
E.2	-1		2		-1				$W_2 a$
A.3			K_3	1	$2(K_3+K_4)$	-2	K_4	1	0
E.3			-1		2		-1		$W_3 a$
A.4									
E.4					Repeat enclosed terms				
etc.									

program uses the fundamental subroutine as many times as there are storeys in the frame to obtain the unit moments, sums the results in sequence and uses the subroutine again to determine the factors and finally prints out the algebraic sum of the factorised values as the final moments. Multi-bay frames are analysed similarly by inputting the appropriate equilibrium equations for conditions at the column heads and again using the XBEAM program to output the results.

(6) *Grillages and beam frameworks*

Intersecting beam frameworks and grillages loaded at right angles to their plane can be dealt with in much the same way as described in the previous section. Two programs are listed for solving these types of structures. GRID is written for structures that have the loads applied at the nodal points only. Provision is included in XBEAM for bending in members where loads are applied away from the nodes. Where torsional effects have to be taken into account, additional factors are introduced into the coefficient matrix, as described in Chapters 20 and 21.

PART 2
FOUNDATIONS AND TEMPORARY WORKS

6 Differential settlement

Settlement moments

Differential settlement of the supports can induce serious bending moments in continuous or fixed-ended beams. These usually arise where foundation loads are transmitted to soils with different settlement characteristics or where reactions at the supports vary considerably. The magnitude of these moments may be ascertained very simply as described before by introducing releases over the supports. The deformation diagram can be drawn to show the inclination of the unloaded beam after release so as to identify the angular discontinuities.

The angular change produced by a settlement, Δ, at one support of a fixed beam of span length, L, is equal to $\frac{\Delta}{L}$. If both supports settle, as in Fig. 6.1, the angular change is equal to $(\Delta_B - \Delta_A)/L$. Now we can apply moments to the released beam to counteract the angular change at each support, as shown in the figure. This gives rise to the equations derived in Chapter 2, as follows:

A.A: $\qquad 2M_A + M_B = \frac{P}{L}(\Delta_B - \Delta_A), \quad$ where $P = \frac{6EI}{L}$

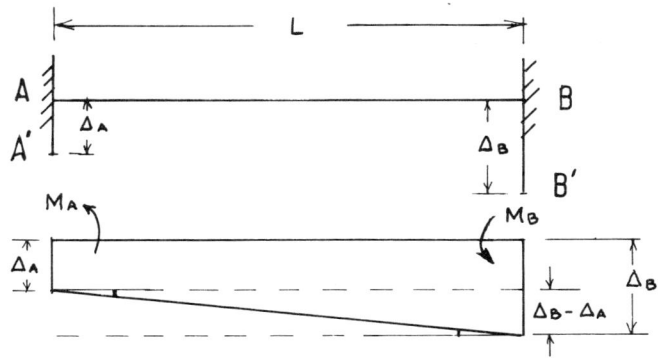

Fig. 6.1.

A.B: $\quad M_A + 2M_B = -\dfrac{P}{L}(\Delta_B - \Delta_A)$

In tabular matrix form these are expressed as shown in Table 6.1.

Table 6.1

Equation A	M_A	M_B	$\overline{\Delta}$
A	2	1	$\dfrac{P}{L}(\Delta_B - \Delta_A)$
B	1	2	$-\dfrac{P}{L}(\Delta_B - \Delta_A)$

If, in addition, the beam carries a vertical load on the span, the $\overline{W}$ 'loads' vector is required to obtain the moments due to the imposed loading, as described in Chapter 2. For a uniformly distributed load the full tabular matrix for a single span is, therefore, as shown in Table 6.2. The two sets of constants in the right-hand columns give two sets of answers which are combined to give the final solution. Thus the PIVOTA program is particularly suitable for solving problems of this sort, although we will see how one can modify MATIN to function similarly.

Table 6.2

Equation A	M_A	M_B	$\overline{W}$	$\overline{\Delta}$
A	2	1	$\dfrac{1}{4}WL$	$\dfrac{P}{L}(\Delta_B - \Delta_A)$
B	1	2	$\dfrac{1}{4}WL$	$-\dfrac{P}{L}(\Delta_B - \Delta_A)$

If the beam is continuous over a number of spans, the total angular change at a support equals the difference in end settlements divided by the span *in each direction* preceded, of course, by the appropriate sign. Thus for the two-span beam in Fig. 6.2 the total angular changes at each support are as follows:

At A: $\quad \dfrac{\Delta_B - \Delta_A}{L_1}$

At B: $\quad \dfrac{\Delta_B - \Delta_A}{L_1} + \dfrac{\Delta_B - \Delta_C}{L_2}$

At C: $\quad \dfrac{\Delta_B - \Delta_C}{L_2}$

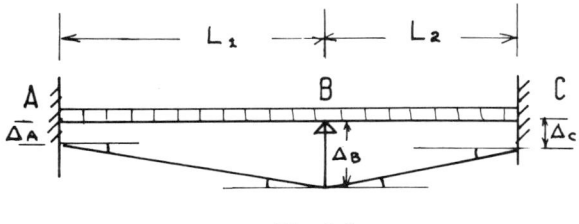

Fig. 6.2.

These terms may be presented in tabular matrix form, as shown in Table 6.3. The sign convention is explained in the next section.

Table 6.3

Equation A	M_A	M_B	M_C	$\overline{W}$	$\overline{\Delta}$
A	2	1	0	$\frac{1}{4}W_1L_1$	$P\left(\dfrac{\Delta_B - \Delta_A}{L_1}\right)$
B	1	$2(1 + K_2)$	K_2	$\frac{1}{4}(W_1L_1 + K_2W_2L_2)$	$-P\left(\dfrac{\Delta_B - \Delta_A}{L_1} + \dfrac{\Delta_B - \Delta_C}{L_2}\right)$
C	0	K_2	$2K_2$	$\frac{1}{4}K_2W_2L_2$	$P\left(\dfrac{\Delta_B - \Delta_C}{L_2}\right)$

Sign determination

The procedure for determining the correct sign to use is illustrated in Fig. 6.3. In the top diagram, the moments applied at the supports to overcome the angular changes there at release due to the applied vertical loading are shown. The settlement pattern in Fig. 6.3(b) illustrates the position where these increase beyond the middle support. As before, moments are applied to counteract the angular changes; these are marked + where their direction coincides with the equivalent moments in Fig. 6.3(a) and − where they act in the opposing direction.

Another settlement pattern is depicted in Fig. 6.3(c) with the release moments indicated similarly. These diagrams illustrate how the deformation diagrams can be used to determine the appropriate sign for the compatibility equations. The procedure can be summarised in the simple rule:

Subtract the lesser settlement from the larger one in each span. If the moment applied at the supports to eliminate settlement movement acts in the opposite direction to the 'loads' moment, insert a negative sign in front of the settlement difference divided by the span; otherwise use a positive sign. Finally, multiply by $P = 6EI_1/L_1$.

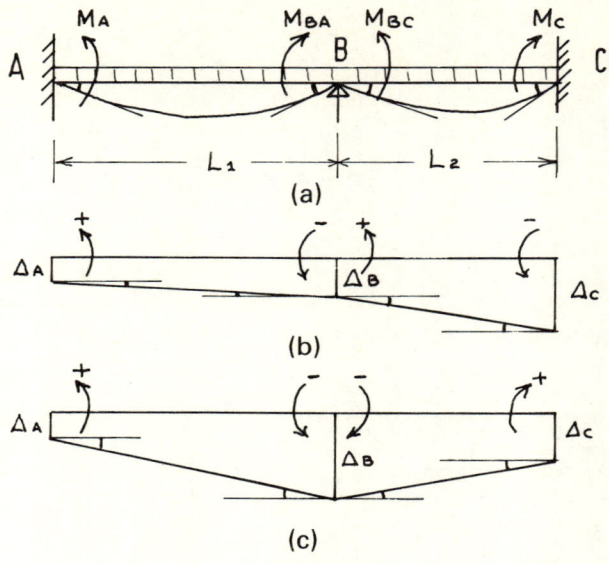

Fig. 6.3.

General tabular matrix

The four-span beam in Fig. 6.4, carrying uniformly distributed loads and subject to any differential settlement pattern, can be solved by using the angular displacement equations, as before:

A.A: $2M_A + M_B = \dfrac{1}{4}W_1L_1 \pm P\left(\dfrac{\Delta_A - \Delta_B}{L_1}\right)$

A.B: $M_A + 2(1 + K_2)M_B + K_2M_C = \dfrac{1}{4}(W_1L_1 + K_2W_2L_2) \pm P\left(\dfrac{\Delta_B - \Delta_A}{L_1} \pm \dfrac{\Delta_B - \Delta_C}{L_2}\right)$

A.C: $K_2M_B + 2(K_2 + K_3)M_C + K_3M_D = \dfrac{1}{4}(K_2W_2L_2 + K_3W_3L_3) \pm P\left(\dfrac{\Delta_C - \Delta_B}{L_2} \pm \dfrac{\Delta_C - \Delta_D}{L_3}\right)$

A.D: $K_3M_C + 2(K_3 + K_4)M_D + K_4M_E = \dfrac{1}{4}(K_3W_3L_3 + K_4W_4L_4) \pm P\left(\dfrac{\Delta_D - \Delta_C}{L_3} \pm \dfrac{\Delta_D - \Delta_E}{L_4}\right)$

A.E: $K_4M_D + 2K_4M_E = \dfrac{1}{4}K_4W_4L_4 \pm P\left(\dfrac{\Delta_E - \Delta_D}{L_4}\right)$

These equations are summarised in Table 6.4.

Table 6.4 General matrix table for beams with settlement

Equation A	Moment					$\overline{W}$	$\overline{\Delta}$
	A	B	C	D	E		
A	2	1	0	0	0	$\frac{1}{4}W_1L_1$	$\pm P\dfrac{\Delta_A - \Delta_B}{L_1}$
B	1	$2(1+K_2)$	K_2	0	0	$\frac{1}{4}(W_1L_1 + K_2W_2L_2)$	$\pm P\left(\dfrac{\Delta_B - \Delta_A}{L_1} \pm \dfrac{\Delta_B - \Delta_C}{L_2}\right)$
C	0	K_2	$2(K_2+K_3)$	K_3	0	$\frac{1}{4}(K_2W_2L_2 + K_3W_3L_3)$	$\pm P\left(\dfrac{\Delta_C - \Delta_B}{L_2} \pm \dfrac{\Delta_C - \Delta_D}{L_3}\right)$
D	0	0	K_3	$2(K_3+K_4)$	K_4	$\frac{1}{4}(K_3W_3L_3 + K_4W_4L_4)$	$\pm P\left(\dfrac{\Delta_D - \Delta_C}{L_3} \pm \dfrac{\Delta_D - \Delta_E}{L_4}\right)$
E	0	0	0	K_4	$2K_4$	$\frac{1}{4}K_4W_4L_4$	$\pm P\dfrac{\Delta_E - \Delta_D}{L_4}$

Note: Signs of $\overline{\Delta}$ terms determined as described in text.

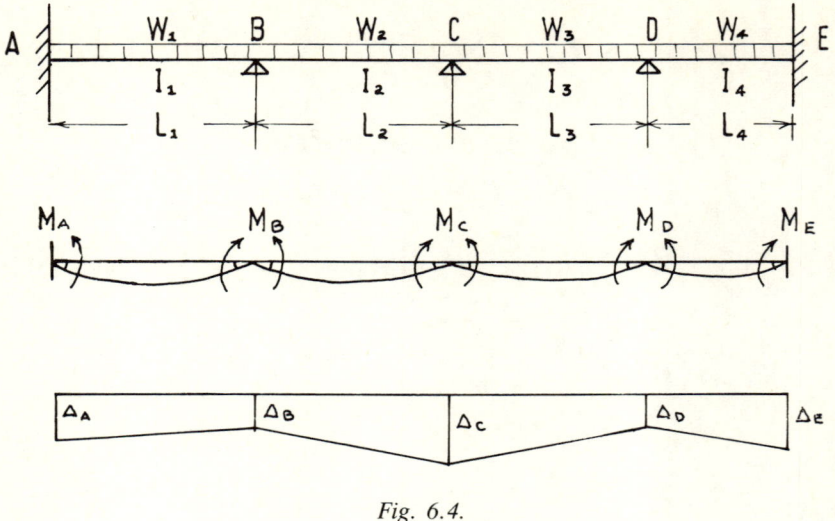

Fig. 6.4.

This table may be used to investigate continuous beams of any number of spans and with any loading or settlement pattern using the factors and sign convention described previously. We may illustrate the procedure by means of a few examples.

Example 6.1

A beam of uniform moment of inertia, $I = 52.26 \times 10^3$ mm^4 and with modulus $E = 210$ kN/mm^2 carries a distributed load of 1 kN/m. Support B settles by 50 mm and support C by 110 mm.

Thus, $\quad P = \dfrac{6EI}{L} = \dfrac{6 \times 210 \times 52.26 \times 10^3}{12 \times 10^3} = 54.87 \times 10^2$

From the deformation diagrams in Fig. 6.5 it can be seen that the 'loads' moments M_A and M_{BC} are in the same direction as the 'settlement' moments, while moments M_{BA} and M_C are reversed and therefore attract a negative sign. The right-hand sides of the equations are, therefore:

At A: $P\Delta/L = 5487 \times \dfrac{50}{12 \times 1000} = 22.86$

At B: $P\Delta/L = 5487 \times \left(-\dfrac{50}{12 \times 1000} + \dfrac{110 - 50}{12 \times 1000}\right) = -22.86 + 27.43 = 4.57$

At C: $P\Delta/L = 5487 \times \left(-\dfrac{110 - 50}{12 \times 1000}\right) = -27.43$

DIFFERENTIAL SETTLEMENT

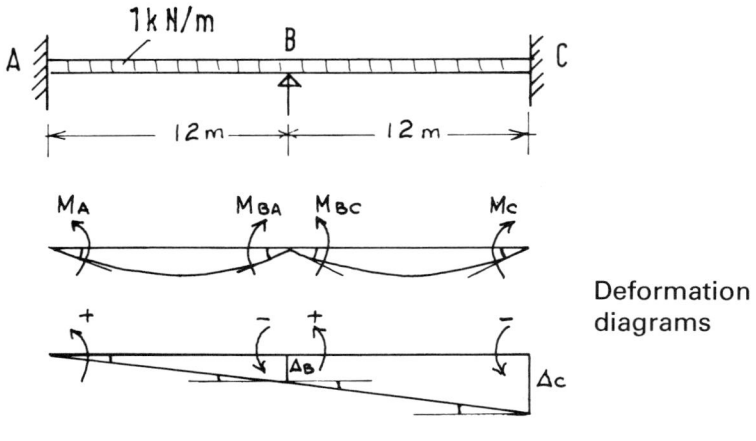

Deformation diagrams

Fig. 6.5.

The 'loads' vector term $= \dfrac{1}{4} KWL = \dfrac{1 \times 12 \times 12}{4} = 36$ at supports A and C, and
$= 72$ at support B.

Table 6.5

Equation A	M_A	M_B	M_C	$\overline{W}$	$\overline{\Delta}$
A	2	1	0	36	22.86
B	1	4	1	72	4.57
C	0	1	2	36	−27.43

Since $K_2 = 1$, the tabular matrix is as shown in Table 6.5. The figures in the table may now be entered in a series of data statements in the listed PIVOT program. Remember that in accordance with the program instructions, the figures are entered row-by-row, including the vector terms in the last two columns. By running the program with input values $N = 3$ and $V = 2$, the following results should be obtained:

	M_A	M_B	M_C
Moments due to loads	12	12	12
Moments due to settlement	10.29	2.29	−14.86
Combined moments	22.29	14.29	− 2.86

In this case the combined moments are obtained manually. In the next example a program is used to output the combined results for a different settlement pattern.

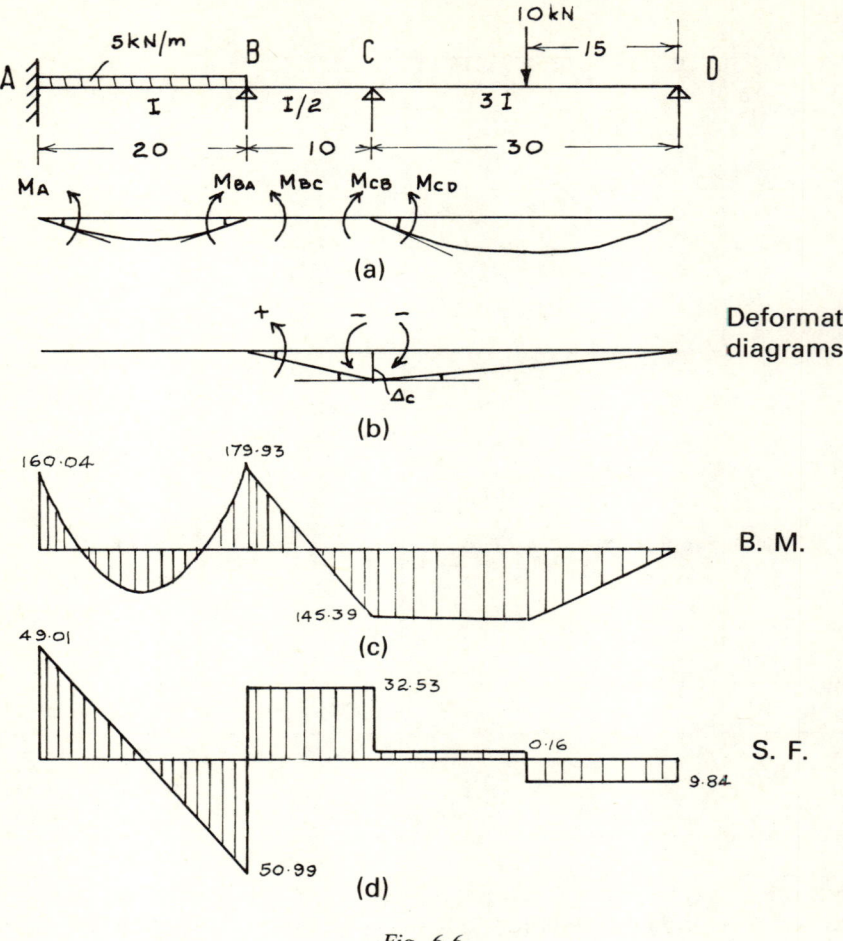

Fig. 6.6.

The above results may equally well be obtained by using MATIN, inputting the coefficients first of all followed by each vector in turn, so that the program is RUN twice.

Example 6.2

This three-span example, shown in Fig. 6.6, has one end support pinned. The moment of inertia of the beam in span AB is 2.48×10^6 mm^4 = I. Assuming Young's modulus $E = 210$ kN/mm^2 and that support C settles by 15 mm, the support moments may be determined using the following data:

$$K_2 = \frac{10}{20}\frac{2I}{I} = 1; \quad K_3 = \frac{30}{20}\frac{I}{3I} = 0.5$$

$$P = \frac{6 \times 210 \times 2.48 \times 10^6}{20 \times 10^3} = 156.24 \times 10^3$$

$\overline{W}$ values for supports A and B $= \frac{1}{4} \times 1 \times 100 \times 20 = 500$

$\overline{W}$ value for support C $= \frac{3}{8} \times \frac{1}{2} \times 10 \times 30 = \frac{900}{16} = 56.25$

From the deformation diagrams (a) and (b) one can see that negative signs are to be used for the PΔ/L terms at support C.

At support A, PΔ/L $= 0$

At support B, PΔ/L $= 156.24 \times 10^3 \times \dfrac{15}{10 \times 10^3} = 234.36$

At support C, PΔ/L $= 156.24 \times 10^3 \times -\left(\dfrac{15}{30 \times 10^3} + \dfrac{15}{10 \times 10^3}\right)$

$ = -312.48$

This provides the required information for the data statements which appear in the printout below, obtained using a PIVOTA program incorporating a subroutine for summing the results.

```
NUMBER OF EQUATIONS?                            3

NUMBER OF VECTORS?                              2

TABULAR DATA:-
2            1            0       500              0
1            4            1       500         234.36
0            1            3        56.25     -312.48

RESULTS:-
1            213.486
2             73.026
3             -5.592

1            -53.451
2            106.901
3           -139.794

COMBINED RESULTS:-
1            160.036
2            179.927
3           -145.386
```

The printout gives a record of the data for verification as well as the separate and combined moments for M_A, M_B and M_C. Knowing the 'free' bending moments, the final bending moment diagram can be drawn as shown in Fig. 6.6(c). In constructing the BM diagram, since positive moments indicate tension on the top surface, the diagram is drawn on the tension side of the beam.

Settlement effects and reactions

Use of the correct sign is most important where settlement and bending effects are combined. A change of sign in one settlement term can produce a completely different result. The deformation diagrams should be constructed with care, therefore, to ensure that they represent the given conditions correctly. In the same way, the output should be checked to see that it agrees with the assumed displacement pattern.

Where linear elastic methods of analysis are used, settlements at the supports can often be seen to have a profound effect on the bending moment values. Codes of Practice may permit a certain redistribution of moments, which reduces the peak values used for the sizing of members. Full advantage should be taken of these concessions for economy in design.

In the previous example, as in most cases described in later chapters, the computer analysis provides the values of the support moments only. These are not the only effects to be considered in the design of members, of course, but once the moments are known it is usually a simple matter to determine shears, axial forces, support reactions, etc. For shears, the general equation $V_A = (M_A - M_B)/L$, in conjunction with the externally applied forces, enables the shear force (SF) diagram to be drawn, as in Fig. 6.6(d). The arrows indicating the direction of the release moments in the figure will help to determine the correct sign to apply, bearing in mind that forces acting upwards are considered positive.

To show how differential settlement can affect the support reactions, these are calculated below for two cases in Example 6.2, namely case (a) without settlement and case (b) with settlement.

(A) WITHOUT SETTLEMENT

$$R_A = \frac{100}{2} + \frac{213.49 - 73.03}{20} = 50 + 7.02 = 57.02$$

$$R_B = \frac{100}{2} - \frac{213.49 - 73.03}{20} + \frac{73.03 + 5.59}{10} = 50 - 7.02 + 7.86$$
$$= 50.84$$

$$R_C = \frac{10}{2} - \frac{73.03 + 5.59}{10} - \frac{5.59}{30} = 5 - 7.86 - 0.19 = -3.05$$

$$R_D = \frac{10}{2} + \frac{5.59}{30} = 5.19$$

(B) WITH SETTLEMENT

$$R_A = \frac{100}{2} + \frac{160.04 - 179.93}{20} = 50 + (-0.99) = 49.01$$

$$R_B = \frac{100}{2} - \frac{160.04 - 179.93}{20} + \frac{179.93 + 145.39}{10}$$
$$= 50 + 0.99 + 32.53 = 83.52$$

$$R_C = \frac{10}{2} - \frac{179.93 + 145.39}{10} - \frac{145.39}{30} = 5 - 32.53 - 4.84 = -32.37$$

$$R_D = \frac{10}{2} + \frac{145.39}{30} = 5 + 4.84 = 9.84$$

These reactions, naturally, are the sum of the shear forces at each support, as may be seen from the SF diagram, Fig. 6.6 (d). Once the moments and shear forces are known, the stresses in the beam can be determined as described in standard textbooks on the subject. If the section is unsuitable, a heavier section may have to be selected. Assuming Young's Modulus, E, is constant, then if the change is in one span only, this will alter the K value for that span. The $\overline{\Delta}$ vector terms will not be affected unless the first ('reference') span beam size is changed. These alterations are easily made so that analysis of different sections can be carried out quite rapidly.

SBEAM program

The MATIN program and MATSUB subroutine are employed generally throughout the book for problem solving and can be used to obtain results in this case also. Those who wish to avoid the trouble of keying in and saving the listed PIVOT and PIVOTA programs will find this useful.

One way would be to run the MATIN program twice, first with the loads vector and then with the settlement vector and then sum the results manually. However, it is clearly preferable to have a program that will print out the data, beam moments, settlement moments and combined values with appropriate headings in one run. This objective forms the basis of the program listed at the end of the chapter; this is to be merged with the MATSUB subroutine before use. It has been given the synoptic title of SBEAM and calls up MATSUB twice before adding the results to show the combined moments.

We may illustrate its use by solving the problem in Fig. 6.7, in which support B is assumed to settle by 35 mm and support C by 15 mm. The beam has a moment of inertia, $I = 2 \times 10^5$ mm^4 and $E = 200$ kN/mm^2. Thus

$$K_2 = \frac{12}{9} = \frac{4}{3}; \quad K_3 = \frac{8}{9}; \quad P = \frac{6 \times 200 \times 2 \times 10^5}{9 \times 10^3} = 26.667 \times 10^3$$

60 FOUNDATION AND STRUCTURAL PROBLEMS

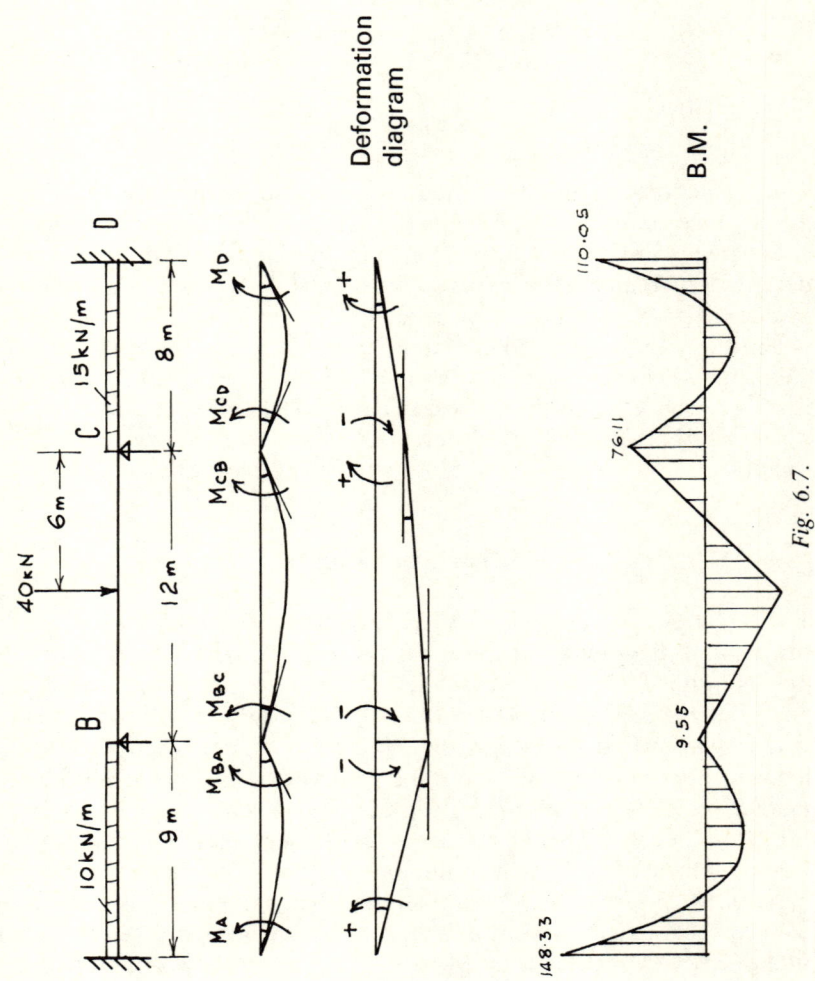

Fig. 6.7.

DIFFERENTIAL SETTLEMENT

$\overline{W}$ vector (γKWL) terms:

					$\overline{W}$
Span AB:	$\dfrac{1}{4} \times 90 \times 9$	$= 202.5$			202.5 (A)
Span BC:	$\dfrac{3}{8} \times \dfrac{4}{3} \times 40 \times 12$	$= 240$			442.5 (B)
Span CD:	$\dfrac{1}{4} \times \dfrac{8}{9} \times 15 \times 8 \times 8$	$= 213.33$			453.33 (C)
					213.33 (D)

$\overline{\Delta}$ vector ($P\Delta/L$) terms:

Support A: $\quad 26.667 \times 10^3 \times \dfrac{35}{9 \times 10^3} \qquad\qquad\qquad = \quad 103.71$

Support B: $\quad -26.667 \times 10^3 \times \left(\dfrac{35}{9 \times 10^3} + \dfrac{35-15}{12 \times 10^3}\right) = -148.16$

Support C: $\quad 26.667 \times 10^3 \times \left(\dfrac{35-15}{12 \times 10^3} - \dfrac{15}{8 \times 10^3}\right) = \quad -5.56$

Support D: $\quad 26.667 \times 10^3 \times \dfrac{15}{8 \times 10^3} \qquad\qquad\qquad = \quad 50.00$

These figures can be used to construct a data table, as shown in Table 6.6.

Table 6.6 Data table for use with SBEAM

Support	Moment			
	A	B	C	D
A	2	1	0	0
B	1	4.667	1.333	0
C	0	1.333	4.444	0.889
D	0	0	0.889	1.778
$\overline{W}$	202.5	442.5	453.33	213.33
$\overline{\Delta}$	103.7	−148.16	−5.56	50.00

The printout from the SBEAM program, which includes the tabulated data, is reproduced below to two places of decimals. Note that the data in this case consist of the coefficient terms followed in turn by the loads and settlement vectors.

```
MATRIX SIZE?           4
DATA:-
2        1         0         0
1        4.667     1.333     0
0        1.333     4.444     0.889
0        0         0.889     1.778
```

202.5
442.5
453.33
213.33

BEAM MOMENTS:−
1 70.93
2 60.63
3 66.47
4 86.75

SETTLEMENT VECTOR:−
103.71
−148.16
−5.56
50.00

SETTLEMENT MOMENTS:−
1 77.35
2 −50.99
3 9.35
4 23.44

COMBINED MOMENTS:−
1 148.28
2 9.64
3 75.83
4 110.19

The results are printed in the same sequence as the coefficient headings in the data table, namely M_A, M_B, M_C, M_D.

SBEAM in general use

Apart from combining loads and settlement moments, the SBEAM program has other uses, one of which is worth mentioning. Modern limit state Codes of Practice incorporate partial load factors that can vary with the type of loading. This often requires calculations to be carried out separately for dead and imposed loads, for instance, which attract different partial load factors. The SBEAM program can be used with different $\overline{W}$ vectors just as effectively so as to obtain moments due to dead, superimposed and total loading in one run. Altering the PRINT statements in lines 50, 70 and 80 will enable the three sets of values to be printed out with appropriate headings. Wind loading can be dealt with similarly, of course.

Further examples on differential settlement will be found in Appendix A. The most important objective in tackling these problems is to ensure that the data are properly formulated for whatever computer program you elect to use, i.e. that the items of data are calculated correctly and put in proper sequence in data statements to be incorporated in the program.

SBEAM

```
5 REM  *********
10 REM * SBEAM *
15 REM  *********
20 REM PROGRAM FOR SOLVING CONTINUOUS BEAMS WITH SETTLEMENT AT SUPPORTS
25 REM USING MATSUB SUBROUTINE.
30 REM DATA INSTRUCTIONS: ENTER COEFF MATRIX, BEAM VECTOR
35 REM AND SETTLEMENT VECTOR IN SEQUENCE
40 DIM  W(20),S(20)
45 GOSUB 630
50 PRINT "BEAM MOMENTS:-"
55 FOR I= 1 TO N
60 W(I)=X(I): PRINT I,W(I)
65 NEXT I: PRINT
70 PRINT "SETTLEMENT VECTOR:-"
75 GOSUB 840
80 PRINT "SETTLEMENT MOMENTS:-"
85 FOR I=1 TO N
90 S(I)=X(I): PRINT I,S(I)
95 NEXT I: PRINT
100 PRINT "COMBINED MOMENTS:-"
105 FOR I=1 TO N
110 R(I)=W(I)+S(I): PRINT I,R(I)
115 NEXT I
120 END
```

Note: To be merged with MATSUB with line 830 deleted before saving.

7 Beams and footings on an elastic foundation

Subgrade reaction

Footings are generally arranged so that the resultant applied load passes through the centroid of the footing. Bending moments in the foundation are often computed on the assumption that the soil pressure is uniformly distributed over the base. In reality, as Terzaghi has demonstrated, the contact pressure of footings on sand decreases from the centre outwards while for rigid footings on soft clay it may well increase towards the perimeter. He has suggested, therefore, that instead of 'contact pressure' the term 'subgrade reaction' be used where uniform distribution is considered under a rigid footing.

A rigid footing remains plane when it settles, so the subgrade reaction will have a planar distribution and can be calculated by ordinary statics. Figure 7.1 shows a gravity retaining wall of base width, b, with a resultant force, R, acting on the base at distance a from the toe. Subgrade reactions per unit length, p_1 and p_2, are given by the equations:

$$\frac{p_1 + p_2}{2} = R \qquad (7.1)$$

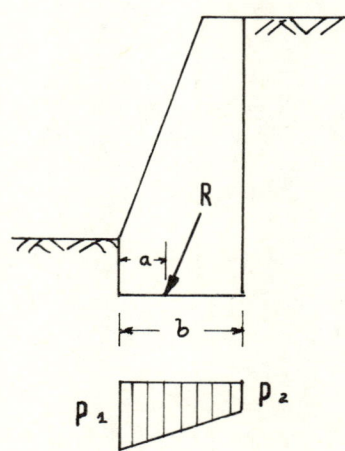

Fig. 7.1.

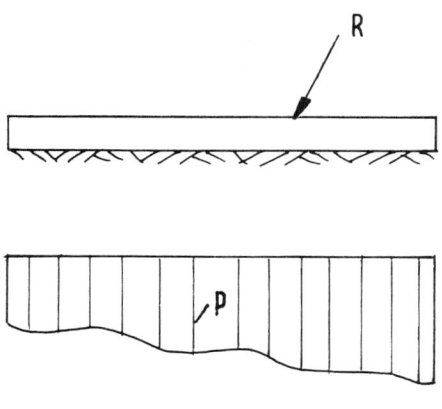

Fig. 7.2.

$$\frac{b^2}{6}(p_1 + 2p_2) = Ra \tag{7.2}$$

Where non-rigid foundations are concerned, however, the flexibility of the foundation affects the distribution of the soil reactions on the base, as shown in Fig. 7.2. Using the theory of elastic beams on a continuous elastic support, we can compute the subgrade reactions, vertical settlements and moments in the foundation. Flexibility and finite difference methods may be used to analyse footings in accordance with this theory, as described in the next section.

The subgrade reaction is assumed proportional to settlement of the foundation, giving rise to the equation

$$p = q\Delta \tag{7.3}$$

where p = subgrade reaction
 Δ = corresponding settlement
 q = coefficient of subgrade reaction.

The value of q may be ascertained from loaded plate tests on different soil types. Tests were carried out originally on a 1 ft × 1 ft (305 mm × 305 mm) square plate; the results gave the moduli of subgrade reaction for the various soil types. Typical values based on work by Terzaghi and Peck (1967) are shown in Table 7.1, where all values are given in MN/m^3, i.e. $10^3 \times kN/m^3$. To determine the coefficient of subgrade reaction, q, for footings of different sizes, the tabulated values need to be adjusted using the following multipliers:

for sands: multiply by $\left(\dfrac{b + 0.305}{2b}\right)^2$ for footings of width b in metres

for clays: multiply by $\dfrac{0.305}{b}$ for footings of width b in metres and in addition by $\frac{L + 0.152}{1.5L}$ for foundations of length L in metres.

In practice q will be found to lie in the range 5000 to 25 000 kN/m³ for normal soils.

Table 7.1 Moduli of subgrade reaction (MN/m³)

Soil material	Condition					
	Loose	Medium	Dense	Firm	Very stiff	Hard
Dry sand	15	45	175			
Submerged sand	10	30	100			
Clay				25	50	100

Footings with distributed loading

Figure 7.3 shows an elastic beam foundation of length L and width b supporting an unevenly distributed load. The beam suffers angular changes along its length dependent on the applied load and elastic properties of the soil and beam. To determine the angular changes, consider the beam divided into a number of small strips of length a between nodal points 0, 1, 2, 3, etc. Assume the external forces are gathered to the nodal points. The loads applied at each node are denoted by W_0, W_1, W_2, etc. Forces arising from the subgrade reaction likewise are denoted P_0, P_1, P_2, etc.

The subgrade reaction pressures are assumed to vary linearly between the nodal points so that the force on any strip equals the average subgrade reaction pressure multiplied by the area of the strip. With very small strip lengths it may be assumed, without too much error, that their centres of gravity occur at the centre-line of the strip. The force equations for the various nodes can be written thus:

At 0, $\quad P_0 = \frac{1}{2}\left(p_0 + \frac{p_0 + p_1}{2}\right)\frac{ab}{2} \quad = \frac{ab}{8}(3p_0 + p_1)$

At 1, $\quad P_1 = \frac{1}{2}\left(\frac{p_0 + p_1}{2} + \frac{p_1 + p_2}{2}\right)ab = \frac{ab}{4}(p_0 + 2p_1 + p_2)$

At 2, $\quad P_2 = \frac{1}{2}\left(\frac{p_1 + p_2}{2} + \frac{p_2 + p_3}{2}\right)ab = \frac{ab}{4}(p_1 + 2p_2 + p_3)$ etc.

If we now introduce releases at the nodal points, these will cause angular changes to occur at 0, 1, 2, etc. so that the beam assumes the shape shown in the deformation diagram, Fig. 7.3, with deflections Δ_0, Δ_1, Δ_2, etc. at the nodes.

BEAMS AND FOOTINGS ON AN ELASTIC FOUNDATION 67

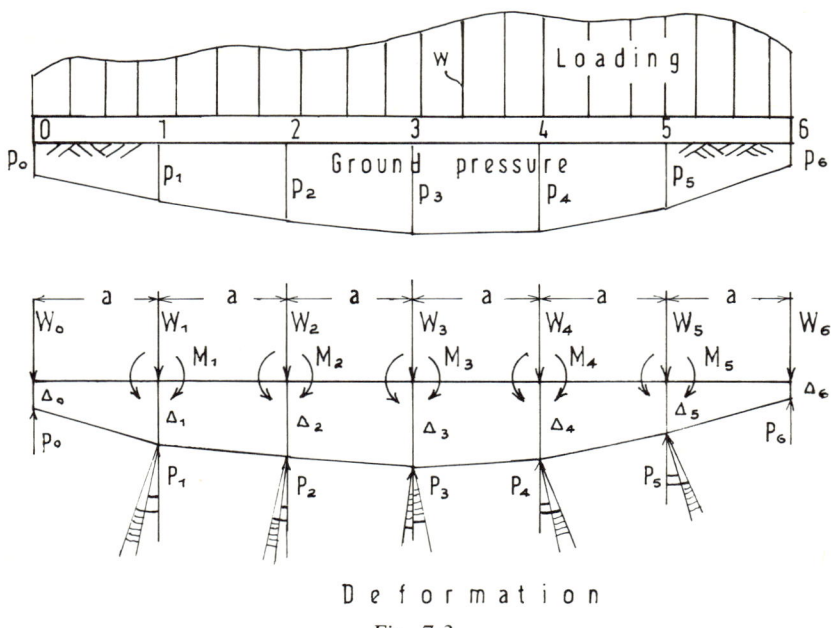

Fig. 7.3.

The angular changes at the released nodes due to these deflections may be seen from the deformation diagram. The ends are assumed to be pinned, so no angular change takes place there on release of joints 0 and 6.

Angular change

at joint 1 $= \dfrac{\Delta_1 - \Delta_0}{a} - \dfrac{\Delta_2 - \Delta_1}{a} = -\dfrac{1}{a}(\Delta_0 - 2\Delta_1 + \Delta_2)$

at joint 2 $= \dfrac{\Delta_2 - \Delta_1}{a} - \dfrac{\Delta_3 - \Delta_2}{a} = -\dfrac{1}{a}(\Delta_1 - 2\Delta_2 + \Delta_3)$

at joint 3 $= \dfrac{\Delta_3 - \Delta_2}{a} - \dfrac{\Delta_4 - \Delta_3}{a} = -\dfrac{1}{a}(\Delta_2 - 2\Delta_3 + \Delta_4)$

at joint n $= -\dfrac{1}{a}(\Delta_{n-1} - 2\Delta_n + \Delta_{n+1})$

It can be shown that the final equation holds good for any type of angular change such as those at nodes 2, 3 and 4. Where the deformation diagram assumes a 'concave' shape, however, the angular change will be preceded by a positive sign instead of the negative one shown above. The direction of the release moments will also be reversed in that case, so that type of settlement pattern will, under the convention adopted in Fig. 7.3, result in a negative moment sign appearing in the output (see case 3, Fig. 7.6).

In accordance with the release-deformation method, moments are now applied at the nodes to eliminate these angular discontinuities and restore the beam to its natural alignment. The direction of the moments is indicated by arrows in the deformation diagram labelled M_1, M_2, etc. As the end node, 0, is considered 'free', the release moment applied to the first strip at node 1 should act so as to restore the strip to its original horizontal alignment. This determines the direction of the moments applied at node 1 and subsequent nodes, indicated by arrows in Fig. 7.3. It will be noted that their direction is the opposite to that adopted in Chapter 6 for continuous beams. It follows that for foundations subject to upward reactions from the soil, positive bending moments indicate tension at the lower surface, which is the norm for foundations in practice. Bending in the strips is ignored, so the sign convention described in Chapter 5 in considering differential settlements does not apply, since there is no question of combining bending and settlement effects. Moments are applied in this case, therefore, to eliminate the net angular change at each joint. (If the assumed direction is incorrect, this will be shown up by a negative sign in the results.)

From equation (7.3) we can write

$$p_0 = q\,\Delta_0$$
$$p_1 = q\,\Delta_1$$
$$p_n = q\,\Delta_n \quad \text{etc.}$$

The compatibility equations for angular displacement at the various nodes are as follows:

A.1:
$$\frac{2M_1}{3EI}a + \frac{M_2}{6EI}a = \frac{\Delta_1 - \Delta_0}{a} - \frac{\Delta_2 - \Delta_1}{a}$$

$$4M_1 + M_2 = -\frac{6EI}{a}\left(\frac{\Delta_0 - 2\Delta_1 + \Delta_2}{a}\right)$$

$$= -\frac{6EI}{qa^2}(p_0 - 2p_1 + p_2)$$

Let $g = \dfrac{6EI}{qa^2}$

$$4M_1 + M_2 + gp_0 - 2gp_1 + gp_2 = 0$$

A.2:
$$\frac{M_1}{6EI}a + \frac{2M_2}{3EI}a + \frac{M_3}{6EI}a = \frac{\Delta_2 - \Delta_1}{a} - \frac{\Delta_3 - \Delta_2}{a}$$

$$M_1 + 4M_2 + M_3 = -\frac{6EI}{a}\left(\frac{\Delta_1 - 2\Delta_2 + \Delta_3}{a}\right)$$

$$M_1 + 4M_2 + M_3 + gp_1 - 2gp_2 + gp_3 = 0$$

A.3: $\quad M_2 + 4M_3 + M_4 + gp_2 - 2gp_3 + gp_4 = 0 \quad$ etc.

From the loading and deformation diagrams, the equilibrium equations can be constructed as follows:

E.0: $\dfrac{M_1}{a} + W_0 = P_0$

Substituting for P_0 from the previous force equation,

$$-\dfrac{M_1}{a} + \dfrac{ab}{8}(3p_0 + p_1) = W_0$$

$$-4M_1 + 3/2 a^2 b p_0 + 1/2 a^2 b p_1 = 4aW_0$$

Let $c = a^2 b$

$$-4M_1 + 3/2 c p_0 + 1/2 c p_1 = 4aW_0$$

E.1: $\dfrac{M_1}{a} + \dfrac{M_1 - M_2}{a} + P_1 = W_1$

$$\dfrac{M_1}{a} + \dfrac{M_1 - M_2}{a} + \dfrac{ab}{4}(p_0 + 2p_1 + p_2) = W_1$$

$$4M_1 + 4M_1 - 4M_2 + a^2 b p_0 + 2a^2 b p_1 + a^2 b p_2 = 4aW_1$$

$$8M_1 - 4M_2 + c p_0 + 2c p_1 + c p_2 = 4aW_1$$

E.2: $\dfrac{M_2 - M_1}{a} + \dfrac{M_2 - M_3}{a} + P_2 = W_2$

$$\dfrac{2M_2}{a} - \dfrac{M_1}{a} - \dfrac{M_3}{a} + \dfrac{ab}{4}(p_1 + 2p_2 + p_3) = W_2$$

$$8M_2 - 4M_1 - 4M_3 + a^2 b p_1 + 2a^2 b p_2 + a^2 b p_3 = 4aW_2$$

$$-4M_1 + 8M_2 - 4M_3 + c p_1 + 2c p_2 + c p_3 = 4aW_2$$

E.3: $-4M_2 + 8M_3 - 4M_4 + c p_2 + 2c p_3 + c p_4 = 4aW_3$

Similar equations may be written for the other nodes. Table 7.2 expresses the above equations in tabular matrix format. The equations may be solved, if properly conditioned, by a computer program such as MATIN or PIVOT.

As mentioned earlier, positive bending moments in this case indicate *tension at the lower surface*. Positive results are plotted above the baseline in the diagrams, so that the bending moment diagram appears on the *compression side of the baseline* treated as a neutral axis. Ground pressures have been measured downwards, however, since the pressure diagrams are also used to represent settlements. These conventions should be borne in mind in reading the diagrams in Chapters 7 to 10 on foundations.

Unequally distributed loading example

To see how we can use the MATIN program effectively, assume that the

Table 7.2 Matrix for footing with distributed loading

Equation	0 p	1 p	1 M	2 p	2 M	3 p	3 M	4 p	4 M	5 p	5 M	6 p	$\overline{W}$
E.0	$\tfrac{3}{2}c$	$\tfrac{c}{2}$	-4										$4aW_0$
A.1	g	$-2g$	4	g	1								0
E.1	c	$2c$	8	c	-4								$4aW_1$
A.2		g	1	$-2g$	4	g	1						0
E.2		c	-4	$2c$	8	c	-4						$4aW_2$
A.3				g	1	$-2g$	4	g	1				0
E.3				c	-4	$2c$	8	c	-4				$4aW_3$
A.4						g	1	$-2g$	4	g	1		0
E.4						c	-4	$2c$	8	c	-4		$4aW_4$
A.5								g	1	$-2g$	4	g	0
E.5								c	-4	$2c$	8	c	$4aW_5$
E.6										$\tfrac{c}{2}$	-4	$\tfrac{3}{2}c$	$4aW_6$

BEAMS AND FOOTINGS ON AN ELASTIC FOUNDATION 71

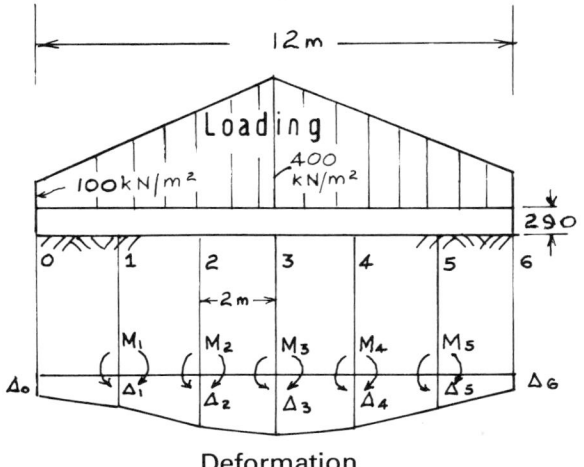

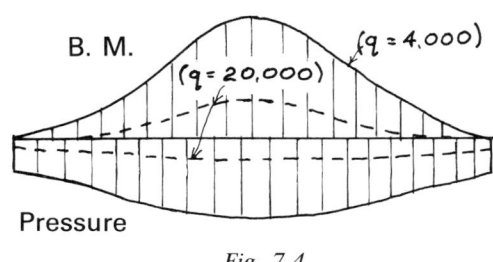

Fig. 7.4.

applied load varies in intensity from 100 kN/m² at the ends to 400 kN/m² at the centre of a footing 12 m long, 1 m wide and 290 mm thick (Fig. 7.4). If the soil stratum is a saturated sand with a modulus of subgrade reaction equal to 10 000 kN/m³, then $q = (1.305/2)^2 \times 10\,000 = 4260$, say 4000 kN/m³. For an initial trial, divide the foundation into six sections, so that $a = 2$ m and $c = 4$. The moment of inertia of the foundation, $I = 0.002$ m⁴. If we assume $E = 28 \times 10^6$ kN/m², then

$$g = \frac{6 \times 28 \times 10^6 \times 0.002}{4 \times 10^3 \times 2 \times 2} = 21$$

The $\overline{W}$ vector terms (4aW) are respectively:

$$4aW_0 = 8 \times \frac{250}{2} \times 1 = 1000; \quad 4aW_1 = 8 \times \frac{150 + 250}{2} \times 2 = 3200$$

$$4aW_2 = 8 \times 300 \times 2 = 4800; \quad 4aW_3 = 8 \times \frac{700}{2} \times 2 = 5600$$

The data table incorporating these values is shown in Table 7.3. Results obtained with MATIN are shown in the lower line opposite $q = 4000$.

Table 7.3 Data table for example in Fig. 7.4

Equation	0	1		2		3		4		5		6	
	p	p	M	p	M	p	M	p	M	p	M	p	
E.0	6	2	−4										
A.1	21	−42	4	21	1								
E.1	4	8	8	4	−4								
A.2		21	1	−42	4	21	1						
E.2		4	−4	8	8	4	−4						
A.3				21	1	−42	4	21	1				
E.3				4	−4	8	8	4	−4				
A.4						21	1	−42	4	21	1		
E.4						4	−4	8	8	4	−4		
A.5								21	1	−42	4	21	
E.5								4	−4	8	8	4	
E.6										2	−4	6	
W	1000	0	3200	0	4800	0	5600	0	4800	0	3200	1000	
MATIN N = 12	126.5	221.8	50.7	299.2	170.6	330.3	241.1	299.2	170.6	221.8	50.7	126.5	q = 4000
MATIN N = 12	98.2	211.8	3.2	312.5	40.9	356.5	71.8	312.5	40.9	211.8	3.2	98.2	q = 20000

These have been used to construct the pressure distribution and bending moment diagrams in Fig. 7.4.

To see the effect of different soil properties, we may re-run the program assuming a hard clay with a modulus of subgrade reaction of 75 000 kN/m^3. The value of q adjusted first of all for width = $(1.305/2)^2 \times 75\,000$ = 31 932 and adjusted for length of foundation = $(12.15/18) \times 31\,932$ = 21 560. Adopting a value of q = 20 000, the same tabulated values may be used as before, but with g = 42 instead of 21. The results obtained using MATIN are shown in the bottom line of Table 7.3 opposite q = 20 000.

Comparative results have been plotted in Fig. 7.4. From this it will be seen that the firmer soil produces a greater variation in the subgrade reaction pressures, with lower values at the ends of the footing and a higher value at the centre; it also results in a marked reduction in the bending moments.

Apart from bending moments, it is usually advisable to carry out a check on shear forces and settlements in foundation design. The pressure distribution diagram can be used to obtain the shear forces at critical sections. Settlements can be determined by transposing equation (7.3), namely:

$$\Delta = \frac{p}{q} \text{ in m}$$

$$= 1000 \frac{p}{q} \text{ in mm} \quad (7.4)$$

Thus, the maximum predicted settlement in this case on saturated sand (q = 4000) is 83 mm and on hard clay (q = 20 000) is 18 mm. Clearly this size of foundation is unlikely to be acceptable on the weaker soil.

A rough check on the results is always advisable. In this case it will be seen that the average pressure under the footing is about 250 kN/m^2. Multiplying this by the base area of the footing gives a load of 3000 kN. This corresponds reasonably well with that derived from the $\overline{W}$ column, which is $\Sigma 4aW/4a = 23\,600/8 = 2950$ kN.

Single concentrated load

Normally loads are transmitted to footings by columns that are so narrow in comparison to the size of the footing that they can be regarded as applying concentrated loads. In most cases the load is applied at the centre of the footing, as shown in Fig. 7.5. Settlements, which are taken to be proportional to the subgrade reactions, will result in the footing assuming the shape shown in the figure. As before, we may divide the footing beam into a number of small strips and introduce releases at the

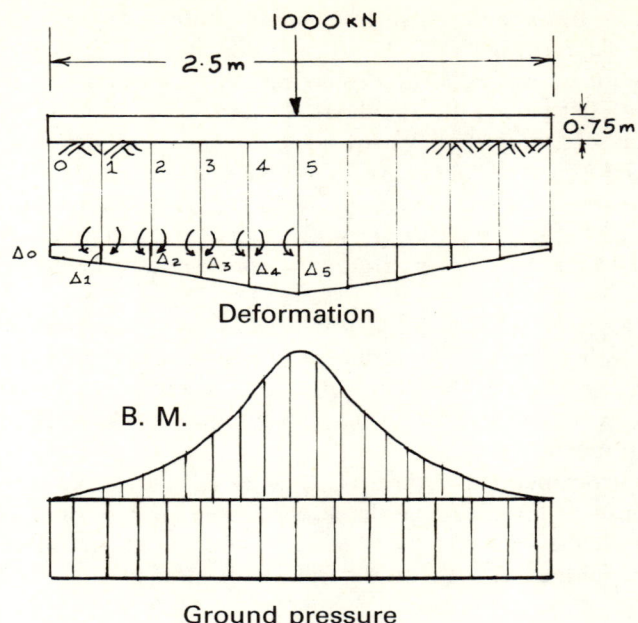

Fig. 7.5.

nodal points. The shape assumed under these conditions is shown by the deformation diagram, on which are indicated also the moments applied to eliminate the angular discontinuities due to settlement. The equations representing the conditions will be similar to those derived in the previous section, except for the loads vector terms in the equilibrium equations; these must be altered to agree with the new load system.

For a symmetrical case such as that shown in Fig. 7.5, it will be sufficient to consider a matrix for one half of the beam only. Some modification will be needed, however, to the equations for the centre and adjoining node, since conditions at the centre differ from those at the end of the beam. If, for example, the footing in Fig. 7.5 is divided into ten strips as shown, then equation E.6 will not appear in the table and the equations for node 5 should be modified as follows:

A.5: $\quad\quad\quad\quad M_4 + 2M_5 + gp_4 - gp_5 = 0$
E.5: $\quad\quad\quad\quad -4M_4 + 4M_5 + cp_4 + cp_5 = 2aW_5$

Omission of equation E.6 reduces the matrix size from (12 × 12) to (11 × 11), as shown in Table 7.4.

We can use this to solve the case of a 1000 kN load applied by a 2.5 m wide wall placed centrally on a 2.5 m × 2.5 m × 0.75 m foundation, for which the moment of inertia, $I = 0.0879$ m^4. Divide the base into ten sections, so $a = 0.25$ m and $c = 0.156$. Assuming $E = 28 \times 10^6$ kN/m^2 and the coefficient of subgrade reaction $q = 10\,000$ kN/m^3, then

Table 7.4 Half matrix for footing with central load

Equation	0 p	1 p	M	2 p	M	3 p	M	4 p	M	5 p	M	$\overline{W}$
E.0	$\frac{3}{2}$c	$\frac{c}{2}$	−4									0
A.1	g	−2g	4	g	1							0
E.1	c	2c	8	c	−4							0
A.2		g	1	−2g	4	g	1					0
E.2		c	−4	2c	8	c	−4					0
A.3				g	1	−2g	4	g	1			0
E.3				c	−4	2c	8	c	−4			0
A.4						g	1	−2g	4	g	1	0
E.4						c	−4	2c	8	c	−4	0
A.5								g	1	−g	2	0
E.5								c	−4	c	4	2aW$_5$

$$g = \frac{6 \times 28 \times 10^6 \times 0.0879}{10\,000 \times 0.25 \times 0.25} = 23\,654$$

and $\quad 2aW = 0.5 \times 1000 = 500$

These values are used to construct the data table set out in Table 7.5. The results obtained with this data using the MATIN program are shown in the bottom line. From this it will be seen that the maximum anticipated settlement equals $1000 \times (160.4/23\,654) = 44$ mm.

The foundation in this example is quite thick and the ground is firm, so the results do not differ very much from those using the conventional formulae for a rigid foundation,

$$p = \frac{W}{A} = \frac{1000}{2.5 \times 2.5} = 160 \text{ kN/m}^2$$

$$M = p\frac{bL^2}{8} = \frac{160 \times 2.5^3}{8} = 312.5 \text{ kNm per m width.}$$

However, considerably different results can be given by the two methods under other conditions, as described in the next section.

Concentrated load at different positions

To see the effect of varying the point of application of a load on a foundation, let us assume a foundation beam, 10 m long × 1 m wide × 400 mm thick with $E = 28 \times 10^6$ kN/m², resting on a soil for which $q = 5000$ kN/m³. The moment of inertia $I = 0.00533$ m⁴. If we divide the beam into ten sections, then $a = 1$ m, $c = 1$ and

76   FOUNDATION AND STRUCTURAL PROBLEMS

Table 7.5 Data table for central load example

Equation	0 p	1 p	1 M	2 p	2 M	3 p	3 M	4 p	4 M	5 p	5 M
E.0	0.234	0.078	−4								
A.1	23 654	−47 308	4								
E.1	0.156	0.312	8								
A.2		23 654	1	23 654	1						
E.2		0.156	−4	0.156	−4						
A.3				−47 308	4	23 654	1				
E.3				0.312	8	0.156	−4				
A.4				23 654	1	−47 308	4	23 654	1		
E.4				0.156	−4	0.312	8	0.156	−4		
A.5						23 654	1	−47 308	4	23 654	1
						0.156	−4	23 654	1	−23 654	2
E.5								0.156	−4	0.156	4
$\overline{W}$											500
MATIN N = 11	160.0	160.1	12.5	160.2	49.9	160.3	112.4	160.4	199.8	160.4	312.3

Table 7.6 Data table for examples in Fig. 7.6

Equation	0 P	0 M	1 P	1 M	2 P	2 M	3 P	3 M	4 P	4 M	5 P	5 M
E.0	1.5											
A.1	179.1	−4	0.5	−4								
E.1	1		−358.2	4								
A.2			2	8	179.1	1						
E.2			179.1	1	1	−4						
A.3			1	−4	−358.2	4	179.1	1				
E.3					2	8	1	−4				
A.4					179.1	1	−358.2	4	179.1	1		
E.4					1	−4	2	8	1	−4		
A.5							179.1	1	−358.2	4	−179.1	2
E.5									2	8	1	4
$\bar{W}$, Case 1	2000											
$\bar{W}$, Case 2					2000							
$\bar{W}$, Case 3											2000	
MATIN: Case 1	7.7		48.2	8.9	88.3	65.9	125.6	210.5	155.2	478.8	167.9	898.0
Case 2	108.6		108.3	54.3	105.6	216.2	97.9	−17.4	90.2	−153.2	87.2	−197.5
Case 3	266.6		180.6	−377.5	106.3	−571.4	50.5	−654.5	16.3	−681.7	4.8	−687.0

$$g = \frac{6 \times 28 \times 10^6 \times 0.00533}{5000 \times 1 \times 1} = 179.1$$

These data are used to construct the coefficient matrix for half the foundation beam, included in Table 7.6. Load vectors, $\overline{W}$, are shown for three different load cases:

Case 1 a concentrated load of 1000 kN at the centre
Case 2 concentrated loads at 0.3L from the centre of 500 kN each
Case 3 concentrated loads of 500 kN each at each end of the footing

The results at the foot of Table 7.6 have been used to construct the pressure distribution and bending moment diagrams in Fig. 7.6. Note how

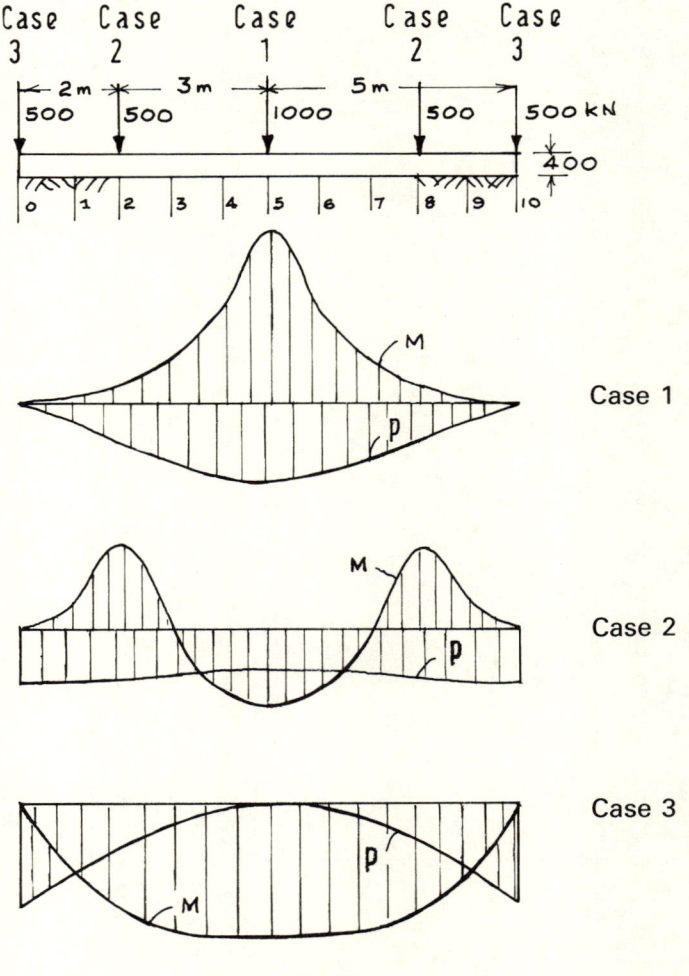

Fig. 7.6.

the maximum value of p increases considerably as the load moves towards the edge of the footing, increasing the danger of differential settlement. It is essential, of course, that a check be made on the value of p to ensure that it does not exceed the allowable bearing pressure for the soil type.

It should be observed that the rigid foundation method of analysis would indicate a uniform subgrade reaction pressure for all three cases equal to 100 kN/m^2. The variations in pressure demonstrated in the flexible method of analysis are much more in accordance with results obtained from practical tests by Terzaghi and others. This illustrates the serious discrepancies that can arise in using the rigid foundation hypothesis, especially where loads are applied near the edges of the foundation.

In a previous example we have seen the effect of different q values on the results. Varying the thickness of the foundation, d, will be found to have an even more significant effect. If the foundation is made too flexible or too long, then negative pressures will appear in the output. The dimensions should be altered to eliminate these or else the area showing negative pressures should be omitted from the calculations. It is advisable to check also that the pressure distribution diagram is in conformity with the settlement pattern assumed in the deformation diagram in each case. A negative bending moment, on the other hand, is an indicator of tension in the top surface of the foundation and reinforcement should be provided to cater for this.

8 Continuous foundations

Twin column loads

In this chapter, the theory of elastic beams on a continuous elastic support is used to examine continuous footings transmitting a number of column loads to the underlying soil stratum. As before, the release-deformation method provides the data for the computer analysis using a program such as MATIN, listed in Appendix B.

First, we will look at the simple case of two concentrated loads applied equally about the centre-line of the foundation beam (Fig. 8.1). Assume a foundation 1 m wide × 600 mm deep ($I = 0.018$ m^4) × 20 m long on a soil for which $q = 5000$. Dividing the length into ten strips each 2 m long, then

$$a = 2, c = 4 \text{ and } g = 151.2 \text{ (for } E = 28 \times 10^6 \text{ kN/m}^2\text{)}$$

These values will enable the data table for the half beam to be constructed. The point loads are taken to be applied at 1/5 the beam length from each end and each equal to 100 kN. As this example is similar to case 2 in the previous chapter, it is left to the reader to prepare the data table for the values to be used with MATIN. If the correct data statements are inserted you should get a printout like that shown below, although with a greater number of decimal places in the results. As usual, the printout incorporates the supplied data for checking purposes.

MATRIX SIZE?   11

DATA:-

6	2	−4	0	0	0	0	0	0	0	−0
151.2	−302.4	4	151.2	1	0	0	0	0	0	0
4	8	8	4	−4	0	0	0	0	0	0
0	151.2	1	−302.4	4	151.2	1	0	0	0	0
0	4	−4	8	8	4	−4	0	0	0	0
0	0	0	151.2	1	−302.4	4	151.2	1	0	0
0	0	0	4	−4	8	8	4	−4	0	0
0	0	0	0	0	151.2	1	−302.4	4	151.2	1
0	0	0	0	0	4	−4	8	8	4	−4
0	0	0	0	0	0	0	151.2	1	−151.2	2
0	0	0	0	0	4	0	4	−4	4	4

CONTINUOUS FOUNDATIONS 81

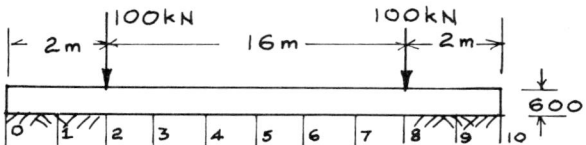

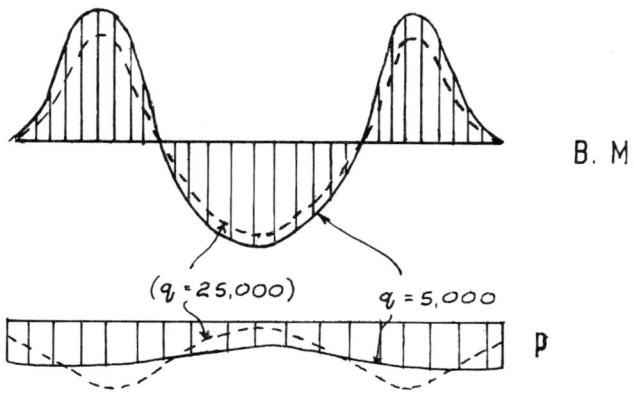

Fig. 8.1.

0
0
0
0
800
0
0
0
0
0
0

RESULTS:-
1   10.31
2   11.85
3   21.40
4   12.24
5   89.05
6   10.10
7   3.12
8   7.58
9   −42.79
10   6.54
11   −56.91

Results follow the same sequence as the data table headings so that items 1,2,4,6,8 and 10 are subgrade reaction pressures in kN/m² (which may be converted directly to settlements in mm) and items 3,5,7,9 and 11 are bending moments in kNm.

For the same foundation on a much firmer stratum for which, say, q = 25 000, results are obtained by substituting g = 30.24 in the matrix. These results have been used to plot the curves in Fig. 8.1. It will be seen that, as in the previous example, the firmer soil produces a greater variation in subgrade reaction pressure and a reduction in the peak bending moments. If the foundation depth is reduced from 600 mm to 350 mm on the original soil material, then g = 30.01, so similar results to the above will be obtained. Since g is more sensitive to variations in dimension, d, than coefficient, q, the change in flexibility has a more pronounced effect in percentage terms.

If the load system is altered so as to move the applied loads towards the centre of the foundation, negative pressures may appear in the results. This indicates that the end sections of the foundation are ineffective and should be ignored in the analysis. In practice, the foundation would be shortened in such cases to give a more economical support structure.

Multiple column loads

The methods used previously for analysing foundations carrying single or twin column loads can be applied to continuous footings supporting a series of column loads. The equations in Chapter 7, as summarised in Tables 7.2 and 7.4, can be used, but it will usually be necessary to divide the foundation beam into many more sections in order to deal with the loads applied along its length. The greater the number of sections, the more precise the results are likely to be, but there is a limit to the degree of precision worth attaining in view of uncertainties associated with the value of q, for instance. It is more important to ensure that sections (or nodes) are located at critical points where peak values or large variations can occur, such as changes in cross-section or points of load application.

As an aid to design of continuous footings supporting multiple column loads, a (20 × 20) tabular matrix is reproduced in Table 8.1 for a beam divided into ten equal sections. The full length of the beam has been taken to allow asymmetrical loading to be considered. The load vector, $\overline{W}$, assumes concentrated loads, W_0, W_1, W_2, etc. applied at nodes 0, 1, 2, etc. In practice, these would be altered as necessary to accord with the actual loading.

To demonstrate its practical application, consider the problem in Fig. 8.2, in which the foundation is 1 m wide by 300 mm thick and the coefficient q = 4000 kN/m³. The values required for the computer data, assuming $E = 28 \times 10^6$ kN/m², are

$$g = \frac{6 \times 28 \times 10^6 \times 0.00225}{4000 \times 1.5 \times 1.5} = 42; \quad c = 2.25$$

CONTINUOUS FOUNDATIONS

Table 8.1 Tabular matrix for continuous beam foundation (full matrix)

Equation	0 p	1 p	1 M	2 p	2 M	3 p	3 M	4 p	4 M	5 p	5 M	6 p	6 M	7 p	7 M	8 p	8 M	9 p	9 M	10 p	$\bar{W}$
E.0	$\tfrac{3}{2}c$	$\tfrac{c}{2}$	-4																		$4aW_0$
A.1	g	$-2g$	4	g	1																0
E.1	c	$2c$	8	c	-4																$4aW_1$
A.2		g	1	$-2g$	4	g	1														0
E.2		c	-4	$2c$	8	c	-4														$4aW_2$
A.3				g	1	$-2g$	4	g	1												0
E.3				c	-4	$2c$	8	c	-4												$4aW_2$
A.4						g	1	$-2g$	4	g	1										0
E.4						c	-4	$2c$	8	c	-4										$4aW_2$
A.5								g	1	$-2g$	4	g	1								0
E.5								c	-4	$2c$	8	c	-4								$4aW_2$
A.6										g	1	$-2g$	4	g	1						0
E.6										c	-4	$2c$	8	c	-4						$4aW_2$
A.7												g	1	$-2g$	4	g	1				0
E.7												c	-4	$2c$	8	c	-4				$4aW_2$
A.8														g	1	$-2g$	4	g	1		0
E.8														c	-4	$2c$	8	c	-4		$4aW_2$
A.9																g	1	$-2g$	4	g	0
E.9																c	-4	$2c$	8	c	$4aW_9$
E.10																		$\tfrac{c}{2}$	-4	$\tfrac{3}{2}c$	$4aW_{10}$

Note In this and subsequent tables 'boxes' are used to indicate repeated sets of coefficients.

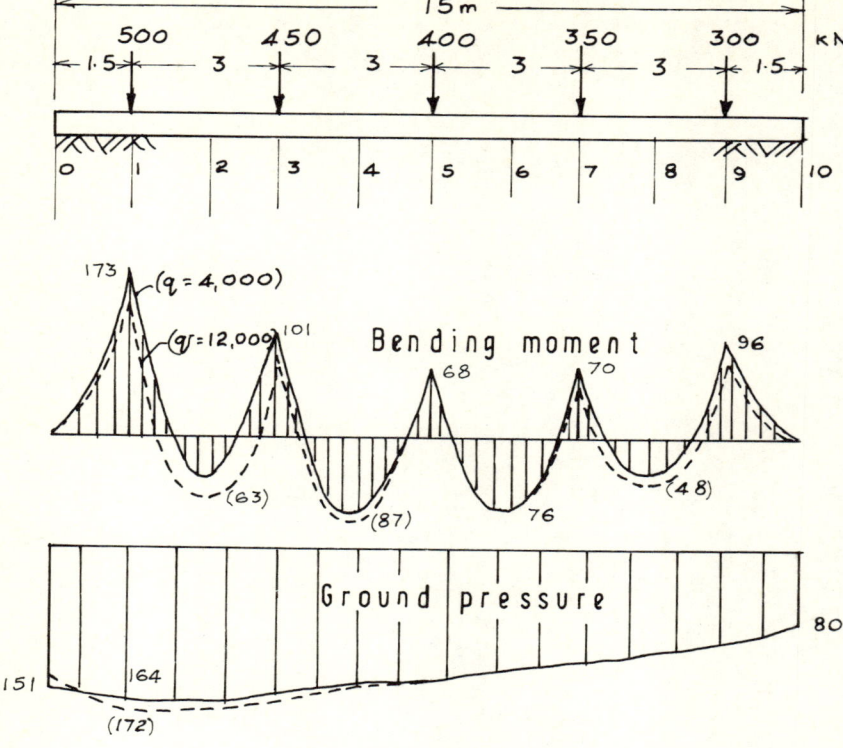

Fig. 8.2.

$4aW_1 = 3000$; $4aW_3 = 2700$; $4aW_5 = 2400$; $4aW_7 = 2100$; $4aW_9 = 1800$

By entering the appropriate values in Table 8.1 and running the MATIN program with these as data, the bending moments and ground pressures can be obtained at each nodal point. Results obtained by this means have been used to construct the diagrams in Fig. 8.2. The maximum anticipated settlement is equal to $(1 \times 164)/4 = 41$ mm and occurs under the 500 kN load. As this is on the high side, some improvement in ground conditions might be undertaken by compaction, ground drainage, soil mix or other means. If this succeeds in an increase in q to 12 000 kN/m^3, then by rerunning the program with g = 14 it will be seen that the maximum settlement is reduced to $172/12 = 14$ mm, which is more acceptable. The maximum bending moment is reduced also, but by a much smaller margin — from 173 kNm to 153 kNm, in fact. The general tendency is for the support moments to be reduced while the span moments are increased.

It is not unusual in practice for one or more of the columns to transmit bending moments in addition to axial loads to the continuous foundation. In the next chapter will be found a procedure for dealing with this situation.

Data processing

One problem highlighted in the last example is the time-consuming activity of preparing and keying in data where a number of nodes are involved. Even for the fairly elementary problems dealt with in these chapters, this can be a tedious operation. There are a few steps that can be taken to relieve matters, however.

Squared paper can save a lot of effort in presenting tabular data. Blank spaces are sufficient to indicate where zeroes should be entered into data statements. Once an initial set of figures has been written into the table, any repetition can be indicated very simply by framing the relevant area of the matrix in rectangular boxes, as in Table 8.1.

Square matrices, as used in MATIN, may involve keying in numerous zeroes. Once one or more sets of zeroes have been typed in they can be repeated at any stage merely by overtyping the line numbers. The same technique can be used to repeat specific sets of figures, such as those box-framed in the table. Full use should be made of repeat action facilities in keying in data. These measures will reduce the work load, but for problems involving the processing of large amounts of data, more sophisticated software is called for than that described in this book.

9 Foundations: applied moments and varying section

Applied moments

Assume we have a loaded foundation subject to settlement with a moment, M', applied at node 2 as shown in Fig. 9.1. If we examine the moments at node 2, it will be seen that M_{21} acts in opposition to moments M_{23} and M' (see deformation diagram in Fig. 9.1). The bending moment diagram will show a change of value at node 2 of M', that is

$$M_{23} = M_{21} - M'$$

Angular equations for nodes 0 and 1 will be the same as in the previous chapter, but we now have

A.2: $\dfrac{M_1}{6EI} a + \dfrac{M_{21}}{3EI} a + \dfrac{M_{23}}{3EI} a + \dfrac{M_3}{6EI} a = \dfrac{\Delta_2 - \Delta_1}{a} - \dfrac{\Delta_3 - \Delta_2}{a}$

$M_1 + 2M_{21} + 2M_{23} + M_3 = -\dfrac{P}{a}(\Delta_1 - 2\Delta_2 + \Delta_3)$

$M_1 + 2M_{21} + 2M_{21} - 2M' + M_3 + gp_1 - 2gp_2 + gp_3 = 0$

$M_1 + 4M_{21} + M_3 + gp_1 - 2gp_2 + gp_3 = 2M'$, where $g = \dfrac{6EI}{qa^2}$

Similarly,

A.3: $\dfrac{M_{23}}{6EI} a + \dfrac{2M_3}{3EI} a + \dfrac{M_4}{6EI} a = \dfrac{\Delta_3 - \Delta_2}{a} - \dfrac{\Delta_4 - \Delta_3}{a}$

$M_{21} + 4M_3 + M_4 + gp_2 - 2gp_3 + gp_4 = M'$

The equilibrium equations for nodes 0 and 1 are the same as before, but for nodes 2 and 3 we have:

E.2: $\dfrac{M_{21} - M_1}{a} + \dfrac{M_{23} - M_3}{a} + P_2 = W_2$

But $P_2 = \dfrac{ab}{4}(p_1 + 2p_2 + p_3)$

$M_{21} - M_1 + M_{23} - M_3 + \dfrac{a^2 b}{4}(p_1 + 2p_2 + p_3) = W_2 a$

$-4M_1 + 4M_{21} + 4M_{21} - 4M' - 4M_3 + cp_1 + 2cp_2 + cp_3 = 4aW_2$, where $c = a^2 b$

FOUNDATIONS: APPLIED MOMENTS AND VARYING SECTION

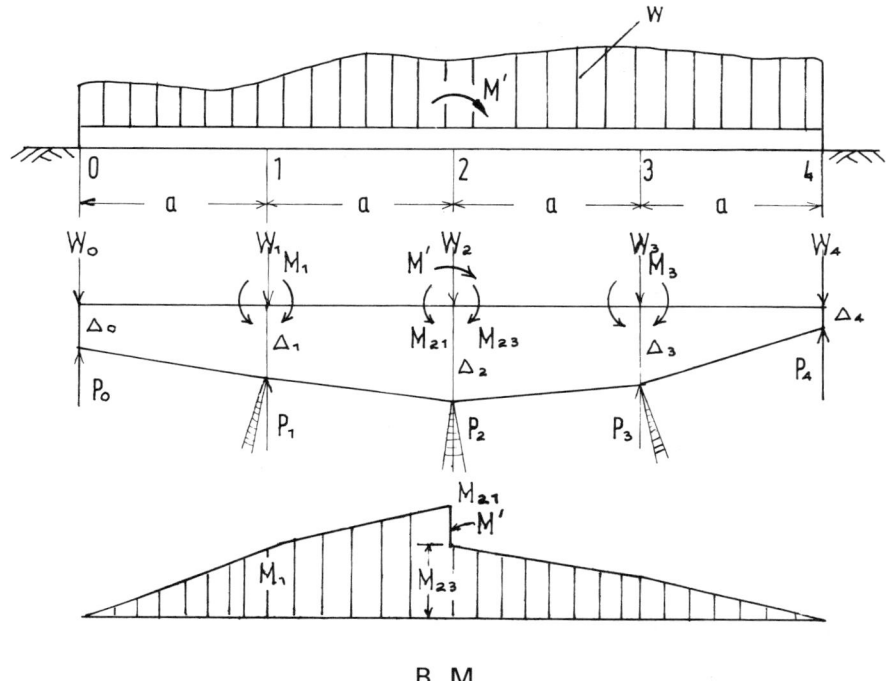

Fig. 9.1.

E.3:
$$-4M_1 + 8M_{21} - 4M_3 + cp_1 + 2cp_2 + cp_3 = 4aW_2 + 4M'$$

$$\frac{M_3 - M_{23}}{a} + \frac{M_3 - M_4}{a} + P_3 = W_3$$

$$P_3 = \frac{ab}{4}(p_2 + 2p_3 + p_4)$$

$$M_3 - M_{23} + M_3 - M_4 + \frac{a^2 b}{4}(p_2 + 2p_3 + p_4) = W_3 a$$

$$-4M_{21} + 4M' + 8M_3 - 4M_4 + cp_2 + 2cp_3 + cp_4 = 4aW_3$$

$$-4M_{21} + 8M_3 - 4M_4 + cp_2 + 2cp_3 + cp_4 = 4aW_3 - 4M'$$

Therefore, the normal coefficient matrix can be used, similar to that in Table 8.1, but the load vector $\overline{W}$ should have additional terms $2M'$, $4M'$, M' and $-4M'$ in equations for two nodes — the node of application and one adjoining node. Table 9.1 shows the general tabular terms for a foundation beam with four divisions, where the moment is applied at node 2 in the direction indicated in Fig. 9.1. Note that if the moment, M', is applied in the opposite direction to that in Fig. 9.1, then the signs of the four M' terms in the load vector should be reversed.

FOUNDATION AND STRUCTURAL PROBLEMS

Table 9.1 Foundation with applied moment

Equation	0 p	1 p	1 M	2 p	2 M	3 p	3 M	4 p	$\overline{W}$
E.0	$\frac{3}{2}c$	$\frac{c}{2}$	-4						$4aW_0$
A.1	g	$-2g$	4	g	1				0
E.1	c	$2c$	8	c	-4				$4aW_1$
A.2		g	1	$-2g$	4	g	1		$2M'$
E.2		c	-4	$2c$	8	c	-4		$4aW_2 + 4M'$
A.3				g	1	$-2g$	4	g	M'
E.3				c	-4	$2c$	8	c	$4aW - 4M'$
E.4						$\frac{c}{2}$	-4	$\frac{3}{2}c$	0

Combination of loads and moments

Where the foundation carries a combination of distributed or concentrated loading and applied moments, the values to be entered in the data table can be compiled in stages in a separate table, as described in the following example.

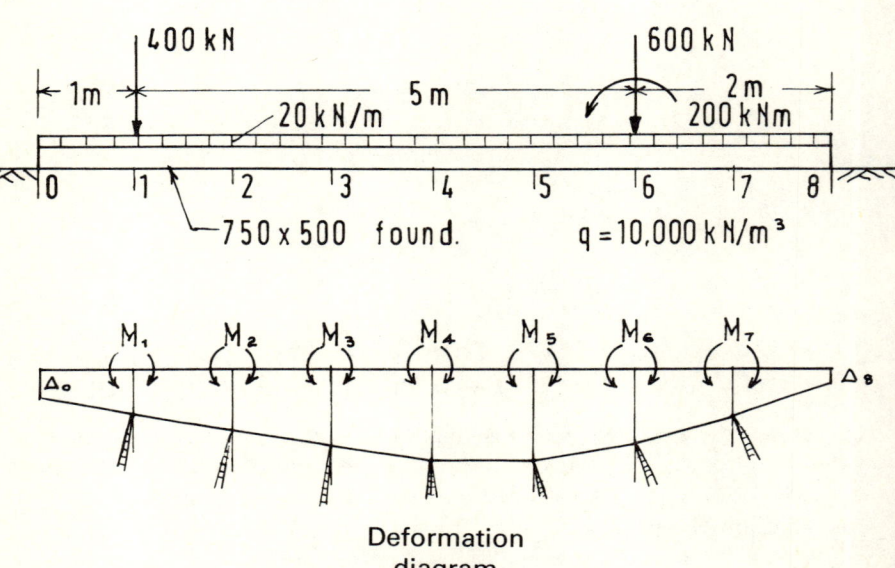

Deformation diagram

Fig. 9.2.

FOUNDATIONS: APPLIED MOMENTS AND VARYING SECTION

Figure 9.2 shows a foundation divided into eight strips carrying a combination of loads and a moment at node 6. The foundation size is 750 mm × 500 mm in depth, the coefficient of subgrade reaction $q = 10\,000$ kN/m³ and E for concrete $= 30 \times 10^6$ kN/mm². Thus

$$I_B = \frac{0.75 \times 0.5^3}{12} = 0.0078 \text{ m}^4; \quad c = 1 \times 0.75 = 0.75$$

$$g = \frac{6 \times 30 \times 10^6 \times 0.0078}{10\,000 \times 1} = 140.4$$

Load vector terms incorporate both nodal loads and moments and are given in Table 9.2. The column headed $\Sigma (4a\overline{W} + 4M')$ gives the values to be entered in the equilibrium equations E.0, E.1, etc. The final column, headed M' or $2M'$, gives the values to be entered in the angular compatibility equations A.1, A.2, etc. The bottom line in the table shows a check on the total applied load, which is equal to $400 + 600 + 160 = 1160$ kN. Note that as the moment is applied in the opposite direction to that in Fig. 9.1, the signs for M' are reversed, since in this case

$$M_{67} = M_{65} + M'$$

The tabular matrix for this problem is shown in Table 9.3. By entering these values in data statements with the MATIN program, the results shown in the bottom line should be displayed and printed out. These have been used to construct the BM diagram and pressure diagram shown in Fig. 9.3.

Table 9.2 Calculations for vector terms

			Equations, E			Equations, A
		$\overline{W}$				
Node	UDL	Pt loads	$4M'$	$4a\overline{W}$	$\Sigma(4a\overline{W} + 4M')$	M' or $2M'$
0	10	0	0	40	40	—
1	20	400	0	1680	1680	0
2	20	0	0	80	80	0
3	20	0	0	80	80	0
4	20	0	0	80	80	0
5	20	0	0	80	80	0
6	20	600	−800	2480	1680	−400
7	20	0	800	80	880	−200
8	10	0	0	40	40	—
		Check:	$\Sigma W =$	$\dfrac{4640}{4}$	$= 1160$ kN	

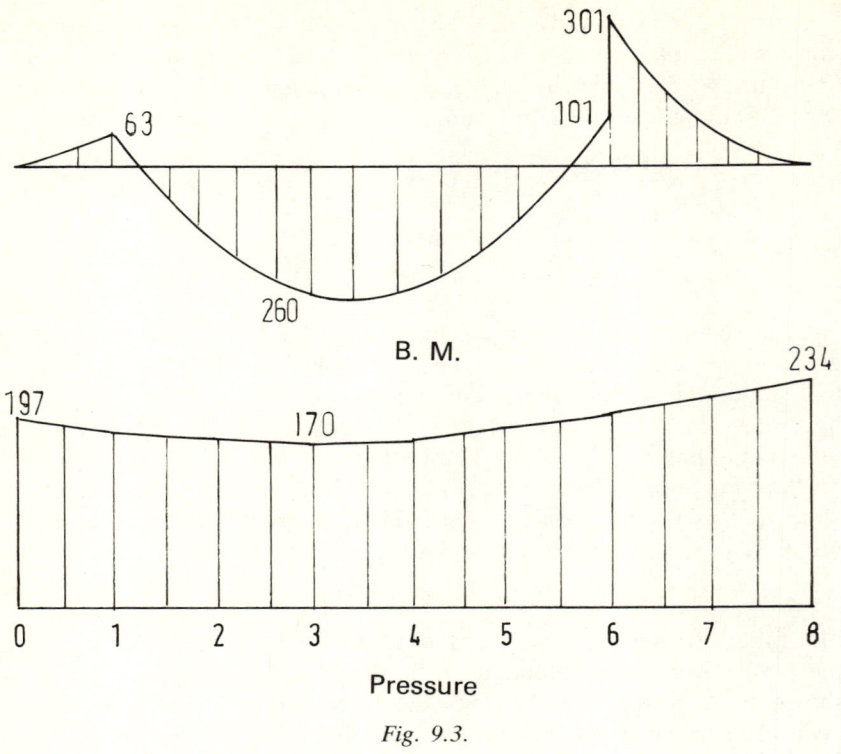

Fig. 9.3.

Comparison with rigid foundation

We can compare these results with those obtainable from a rigid foundation analysis, as follows.

Taking moments about node 6,

$$1160 \, \bar{x} = 400 \times 5 + 120 \times 3 - 200$$
$$\bar{x} = 1.86 \text{ m}$$

eccentricity, $e = 2 - 1.86 = 0.14$

$$\text{pressure, } p = \frac{1160}{8 \times 0.7} \left(1 \pm \frac{6 \times 0.14}{8}\right) = 230 \text{ or } 183 \text{ kN/m}^2$$

Therefore, the rigid analysis gives slightly lower maximum and minimum ground pressures than the flexibility analysis. The difference is not very marked in this case since $q = 10000$ represents a firm soil and the foundation slab is relatively stiff. It should be observed that the printout does not include the applied moment value at node 6 and so does not actually display the maximum bending moment. This is regarded as an external action and appears as a discontinuity at node 6 in the BM diagram in Fig. 9.3.

FOUNDATIONS: APPLIED MOMENTS AND VARYING SECTION

Table 9.3 Tabular matrix for Fig. 9.2

Equation	0 p	1 p	1 M	2 p	2 M	3 p	3 M	4 p	4 M	5 p	5 M	6 p	6 M	7 p	7 M	8 p	$\overline{W}$
E.0	1.125	0.375	−4														40
A.1	140.4	−280.8	4														0
E.1	0.75	1.5	8														1680
A.2		140.4	1	140.4	−4												0
E.2		0.75	−4	0.75	4												80
A.3				−280.8	8	140.4	1										0
E.3				1.5	1	0.75	−4										80
A.4				140.4	−4	−280.8	4	140.4	1								0
E.4				0.75		1.5	8	0.75	−4								80
A.5						140.4	1	−280.8	4	140.4	1						0
E.5						0.75	−4	1.5	8	0.75	−4						80
A.6								140.4	1	−280.8	4	140.4	1				−400
E.6								0.75	−4	1.5	8	0.75	−4				1680
A.7										140.4	1	−280.8	4	140.4	1		−200
E.7										0.75	−4	1.5	8	0.75	8	0.75	880
E.8												0.75	−4	0.375	−4	1.125	40
MATIN N = 16	197	186	63	175	−155	170	−260	174	−256	189	−139	209	101	224	77	234	Results
Δ	20	19		17.5		17		17		19		21		22		23	mm

Foundations of varying width

The width or depth of a foundation may be varied to suit the loading. The same method of analysis as before may be used, but the computation is complicated by the fact that each strip will have different loadings and physical properties.

Consider the foundation in Fig. 9.4, divided into small strips of length a. Variations in width require the introduction of a new factor, r, where

$$r_1 = \frac{b_1}{b_0}; \quad r_2 = \frac{b_2}{b_0}; \quad r_3 = \frac{b_3}{b_0}; \text{ etc.}$$

The first strip between nodes 0 and 1 is taken as the reference element, in which b_0 equals the width at the extreme end. Also, I_1 equals the moment of inertia of the reference strip, 01, based on its average width.

Flexibility of first strip $= \dfrac{a}{I_1} = k_1$, the reference flexibility.

Relative flexibility of strip $12 = \dfrac{a}{I_2} \times \dfrac{I_1}{a} = \dfrac{I_1}{I_2} = K_2$

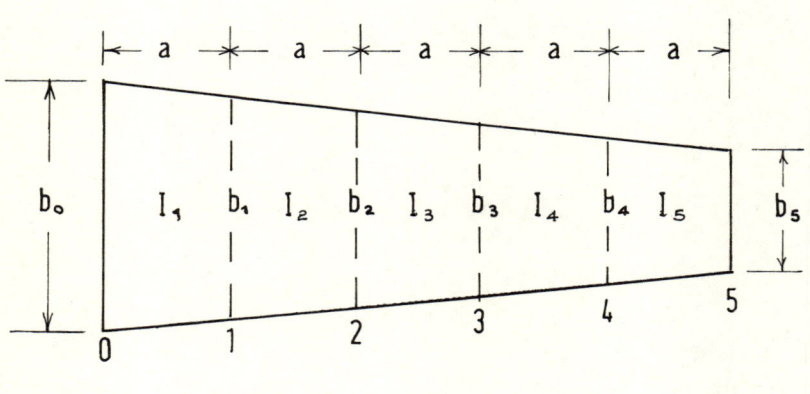

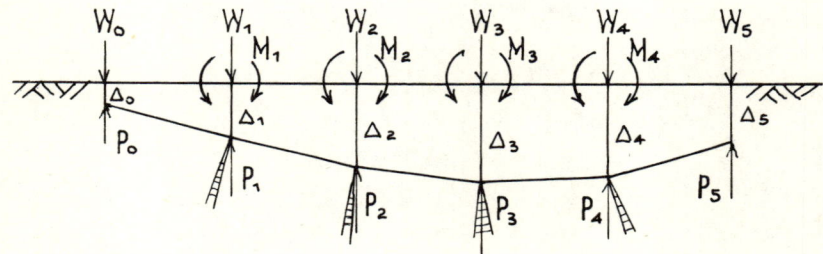

Fig. 9.4.

Relative flexibility of strip $23 = \dfrac{a}{I_3} \times \dfrac{I_1}{a} = \dfrac{I_1}{I_3} = K_3$ etc.

Let $g = \dfrac{6EI_1}{qa^2}$ and $c = a^2 b_0$

Equations of equilibrium are, therefore

E.0: $\quad \dfrac{M_1}{a} + W_0 = P_0,\qquad$ but $P_0 = \dfrac{a}{8}(3b_0 p_0 + b_1 p_1)$

$\dfrac{ab_0}{8}(3p_0 + r_1 p_1) - \dfrac{M_1}{a} = W_0$

$-4M_1 + \dfrac{3}{2} cp_0 + \dfrac{c}{2} r_1 p_1 = 4aW_0$

E.1: $\quad \dfrac{M_1}{a} + \dfrac{M_1 - M_2}{a} + P_1 = W_1$

$\dfrac{M_1}{a} + \dfrac{M_1}{a} - \dfrac{M_2}{a} + \dfrac{ab_0}{4}(p_0 + 2r_1 p_1 + r_2 p_2) = W_1$

$8M_1 - 4M_2 + cp_0 + 2cr_1 p_1 + cr_2 p_2 = 4aW_1$

E.2: $\quad \dfrac{M_2 - M_1}{a} + \dfrac{M_2 - M_3}{a} + P_2 = W_2$

$\dfrac{M_2}{a} - \dfrac{M_1}{a} + \dfrac{M_2}{a} - \dfrac{M_3}{a} + \dfrac{ab_0}{4}(r_1 p_1 + 2r_2 p_2 + r_3 p_3) = W_2$

$-4M_1 + 8M_2 - 4M_3 + cr_1 p_1 + 2cr_2 p_2 + cr_3 p_3 = 4aW_2$

Loads W_0, W_1, etc. may be due to concentrated loads applied at the nodal points or to a uniformly distributed load on the foundation beam. In the latter case, the nodal loads can be obtained in the same way as that used in determining p_0, p_1, etc. If w equals the intensity of the distributed load (kN/m^2), then

$$W_0 = \dfrac{wa}{8}(3b_0 + b_1) = \dfrac{wab_0}{4}\left(\dfrac{3}{2} + \dfrac{r_1}{2}\right)$$

$$W_1 = \dfrac{wab_0}{4}(1 + 2r_1 + r_2)$$

$$W_2 = \dfrac{wab_0}{4}(r_1 + 2r_2 + r_3)\quad \text{etc.}$$

The angular compatibility equations in turn are:

A.1: $\quad \dfrac{M_1}{3EI_1}a + \dfrac{M_1}{6EI_2}a + \dfrac{M_2}{6EI_2}a = \dfrac{\Delta_1 - \Delta_0}{a} - \dfrac{\Delta_2 - \Delta_1}{a}$

Multiplying by $\dfrac{6EI_1}{a}$,

$$2M_1 + K_2M_1 + K_2M_2 + \frac{6EI_1}{a}\left(\frac{\Delta_0 - 2\Delta_1 + \Delta_2}{a}\right) = 0$$

$$2(1 + K_2)M_1 + K_2M_2 + g(p_0 - 2p_1 + p_2) = 0$$

$$2(1 + K_2)M_1 + K_2M_2 + gp_0 - 2gp_1 + gp_2 = 0$$

A.2: $\dfrac{M_1}{6EI_2}a + \dfrac{M_2}{3EI_2}a + \dfrac{M_2}{3EI_3}a + \dfrac{M_3}{6EI_3}a = \dfrac{\Delta_2 - \Delta_1}{a} - \dfrac{\Delta_3 - \Delta_2}{a}$

$$K_2M_1 + 2(K_2 + K_3)M_2 + K_3M_3 + gp_1 - 2gp_2 + gp_3 = 0 \quad \text{etc.}$$

These are summarised in tabular form in Table 9.4. To demonstrate use of the table, consider the problem shown in Fig. 9.5 for a concrete foundation with $E = 30 \times 10^6$ kN/mm^2 supported by a soil with coefficient of subgrade reaction $q = 5000$ kN/m^3. The foundation is divided into six strips so that $a = 1$.

$$c = a^2b = 3; \quad I_1 = \frac{0.3^3}{12} \times 2.83 = 2.25 \times 10^{-3} \times 2.83 = 63.68 \times 10^{-4}$$

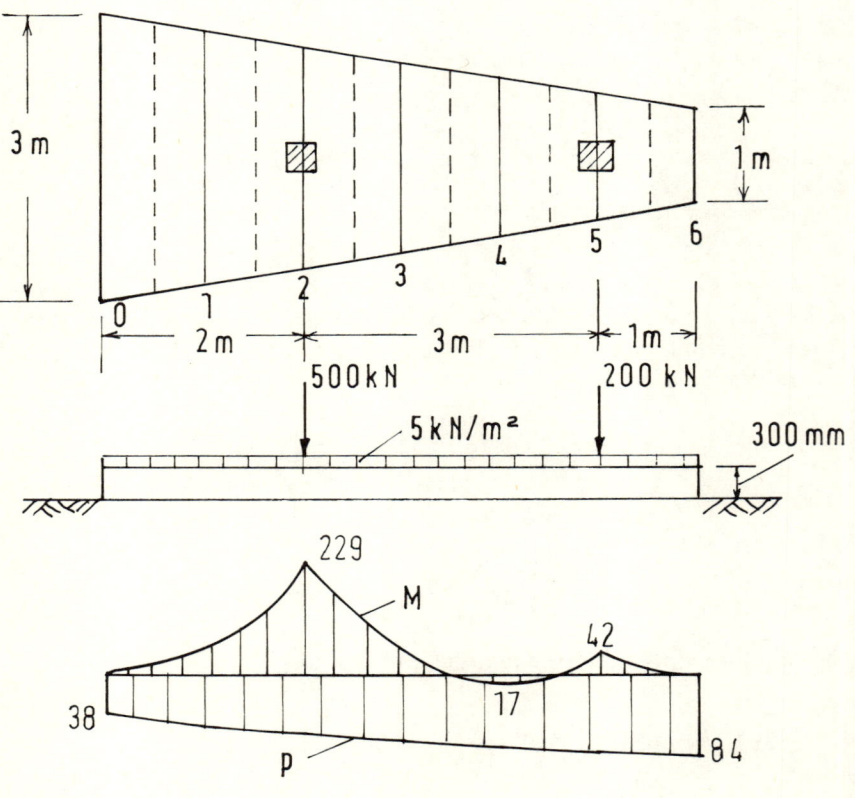

Fig. 9.5.

Table 9.4 Foundation of variable width

Equation	0		1		2		3		4		5		$\overline{W}$
	p	M	p	M	p	M	p	M	p	M	p	M	
E.0	$\tfrac{3}{2}c$	$\tfrac{c}{2}r_1$	-4										$4aW_0$
A.1	g	$-2g$	$2(1+K_2)$										0
E.1	c	$2cr_1$	8										$4aW_1$
A.2		g	K_2	g	K_2								0
E.2		cr_1	-4	cr_2	-4								$4aW_2$
A.3				$-2g$	$2(K_2+K_3)$	g	K_3						0
E.3				$2cr_2$	8	cr_3	-4						$4aW_3$
A.4					g	$-2g$	$2(K_3+K_4)$	g	K_4				0
E.4					cr_2	$2cr_3$	8	cr_4	-4			g		$4aW_4$
E.5							g	K_4	-4	$2(K_4+K_5)$		cr_5		$4aW_5$
							cr_3	-4	$\tfrac{c}{2}r_4$	-4	$\tfrac{3}{2}cr_5$		

$$g = \frac{6 \times 30 \times 10^6 \times 63.68 \times 10^{-4}}{5 \times 10^3 \times 1} = 229.2$$

$\frac{wab}{4} = 3.75$, so

$W_0 = 3.75 \times 1.945 = 7.29$

$W_1 = 3.75 \times 3.56 = 13.35$ etc.

Similar calculations are carried out for I_2, I_3, etc., enabling K_2, K_3, etc. to be ascertained. These are not reproduced because of their repetitious nature; for this reason they can be calculated readily on a programmable calculator. These figures enable the data table to be prepared, as shown in Table 9.5. The results in the bottom line have been used to construct the diagrams in Fig. 9.5. These indicate a substantial variation in soil pressure underneath the base. The base design could be improved by increasing the width at the narrower end and widening that at the wider end, so as to reduce the danger of differential settlement. This change is likely to result in some increase in the maximum bending moment under the 500 kN column load. Since the data are based on relative and not absolute values of moments of inertia, etc., the necessary adjustments can be made fairly quickly before the program is re-run.

Foundation beams of varying depth

In some foundations it may be economical to vary the depth to suit the bending moment values along its length. Usually this is done by reducing the depth uniformly from the point of application of the load, where the shears and bending moments are greatest, to the edge, as shown in Fig. 9.6. If we consider unit width of this foundation, then $b = 1$, $r = a$ and $c = a^2$, so these parameters may well be omitted by inserting a and a^2 where appropriate.

By writing down the equations as before, it will be found that these are similar to those derived in the previous section. The tabular matrix can be constructed exactly as before. This is reproduced in Table 9.6. The bottom lines indicate alternatives for (a) full base length and (b) half base length for symmetrical conditions, where the suffix L indicates the last node.

Special case

It may be of interest to observe that a version of the same tabular matrix can be used to find the moments and deflections in a beam of variable depth, as shown in Fig. 9.7. In this case, since the beam is free to deflect

FOUNDATIONS: APPLIED MOMENTS AND VARYING SECTION 97

Table 9.5 Data table for Fig. 9.5

Equation	0		1		2		3		4		5		6	
	p	p	p	M	p	M	p	M	p	M	p	M	p	
E.0	4.5	1.34	−4											
A.1	229.2	−458.4	4.26		229.2	1.13								
E.1	3	5.34	8		2.34	−4								
A.2		229.2	1.13		−458.4	5.72	229.2	1.55						
E.2		2.67	−4		4.68	8	2.01	−4						
A.3					229.2	1.31	−458.4	5.72	229.2	1.55				
E.3					2.34	−4	4.02	8	1.65	−4				
A.4							229.2	1.55	−458.4	6.88	229.2			
E.4							2.01	−4	3.30	8	1.32	−4		
A.5									229.2	1.89	−458.4	8.62	229.2	
E.5									1.65	−4	2.64	8	0.99	
E.6											0.66	−4	1.49	
W	29.16	0	53.4		0	2046.8	0	40.04	0	33.16	0	826.4		10.8
MATIN N = 12	38.1	52.4	53.1		64.7	229.3	71.5	40.4	76.2	−17.4	80.7	41.9	83.8	

Table 9.6 Foundation of varying depth

Equation	0 p	0 M	1 p	1 M	2 p	2 M	3 p	3 M	L(a) p	L(a) M	L(b) M	$\overline{W}$
E.0	$\tfrac{3}{2}a^2$	-4										$4aW_0$
A.1	$\dfrac{g}{a^2}$		$\dfrac{-2g}{2a^2}$	$\dfrac{2(1+K_2)}{8}$								0
E.1			$\dfrac{g}{a^2}$	$\dfrac{K_2}{-4}$								$4aW_1$
A.2			$\dfrac{g}{a^2}$		$\dfrac{-2g}{2a^2}$	$\dfrac{2(K_2+K_3)}{8}$	$\dfrac{g}{a^2}$	$\dfrac{K_3}{-4}$				0
E.2					$\dfrac{g}{a^2}$	$\dfrac{K_2}{-4}$						$4aW_2$
A.3							$\dfrac{g}{a^2}$		$\dfrac{-2g}{2a^2}$	$\dfrac{2(K_2+K_3)}{8}$		0
E.3							$\dfrac{g}{a^2}$	-4				$4aW_2$
E.L(a)									$\dfrac{a^2}{2}$	-4		$4aW_L$
E.L(b)									a^2	-4	4	$2aW_L$

Table 9.7 Beam of varying depth

Equation	1 $\frac{P}{a}\Delta$	1 M	2 $\frac{P}{a}\Delta$	2 M	3 $\frac{P}{a}\Delta$	3 M	L(a) $\frac{P}{a}\Delta$	L(a) M	L(b) M	$\overline{W}$
A.1	-2	$\dfrac{2(1+K_2)}{2}$								0
E.1	1	K_2		-1						Wa
A.2	1	K_2	-2	$\dfrac{2(K_2+K_3)}{2}$	1	K_3				0
E.2		-1	1	K_2	-2	$\dfrac{2(K_2+K_3)}{2}$	1	K_3		Wa
A.3				-1				-1		0
E.3								-1		Wa
E.L(a)										Wa
E.L(b)									1	$Wa/2$

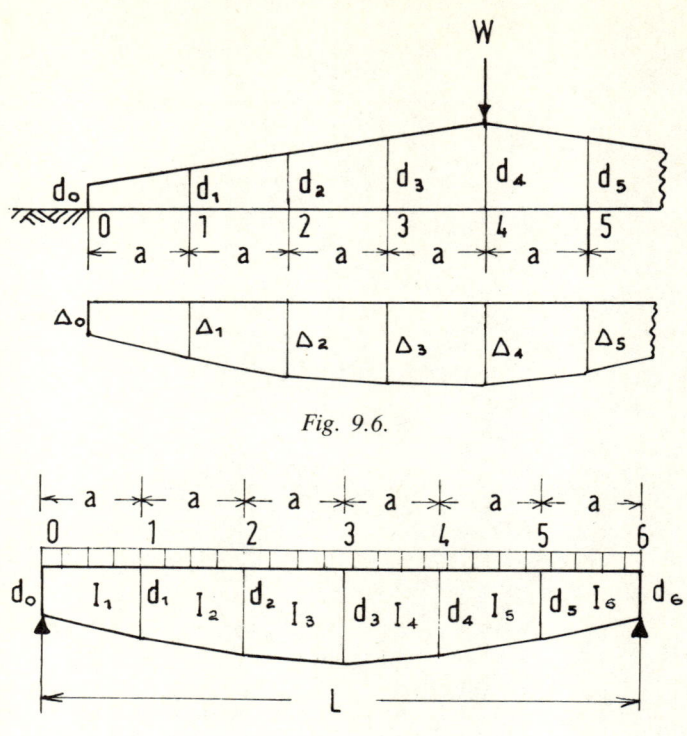

Fig. 9.6.

Fig. 9.7.

between end supports, the coefficient $q = 1$, so $p = \Delta$ and $g = 6EI/a^2 = P/a$. The equilibrium conditions no longer include nodal reactions p_0, p_1, etc., so these are omitted as well as equation E.0. The equilibrium equations are then divided across by 4, so the tabular matrix for this type of problem appears as shown in Table 9.7. This may be compared with Table 5.3. Since $I_n = D_n^3/12$, the relative flexibility factor, $K_n = (D_1/D_n)^3$, where D equals the average depth of the strip in each case. The output gives the moments at each node. To obtain the nodal deflections, multiply the other results by $a/P = a^2/6EI_1 = 2a^2/ED_1$.

10 Raft foundations

Transverse bending

In the previous chapters, consideration was given to the analysis of strip foundations supporting various column arrangements. If the width of the foundation is increased, bending in the transverse direction becomes significant. To consider conditions at different points, a grid system may be adopted to divide the foundation into a number of small elements. Ignoring torsional effects, we may consider the bending moments operating in two directions at right angles at the nodal intersections. By studying the effect of releases introduced at the nodes, equations can be formulated, as before, to represent the deformations and equilibrium conditions at these locations.

In analysing foundation slabs, the subgrade reaction theory is adopted once again, so that the soil reaction is taken to be proportional to the vertical deflection of the slab at any point,

i.e. $p = q\Delta$, where q = coefficient of subgrade reaction.

In the following section nodes are divided into three categories according to location:

(1) internal
(2) edge
(3) corner.

Node types

(1) Internal

Figure 10.1 shows a typical internal node, O, in a rectangular grid of size a × b (dimension a is assumed to be measured in the XX direction and b in the YY direction). Nodes to the left and right are designated A and B; those below and above it are designated C and D. A load, W, is applied at O. The foundation slab rests on an elastic medium for which the coefficient of subgrade reaction is q. The deformations are assumed to take the form shown in diagrams (a) and (b).

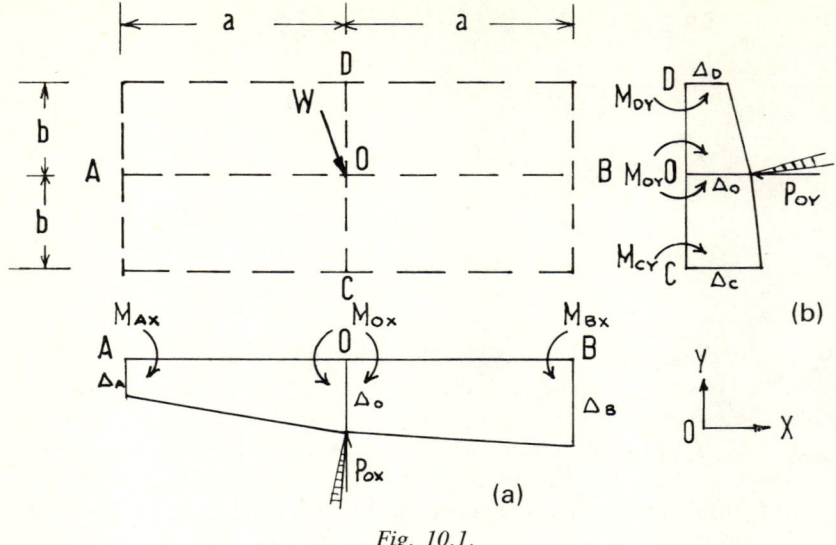

Fig. 10.1.

Diagram (a) indicates the deformation in the XX direction when node O is released to permit rotation in that direction. (Bending between nodal points is ignored as these are assumed close together. A change in sign of the settlement gradients at O will not invalidate the results, therefore.) Diagram (b) illustrates conditions in the YY direction when node O is released to permit rotation in that direction. The directions of the moments required to eliminate the angular changes are indicated by arrows in the diagrams, as well as the reactions holding the nodes in their displaced position when releases are introduced. Note that $P_O = P_{OX} + P_{OY}$ in this and the following cases.

Pressures p_O, p_A, p_B, etc. occur at nodes O, A, B, etc. The reactions are taken to be gathered to the nodal points, so the reaction at any node is equal to the area of the element about that node multiplied by the average pressure between it and the adjoining nodes. Thus,

$$P_O = ab \times \frac{1}{4} \left(\frac{p_O + p_A}{2} + \frac{p_O + p_B}{2} + \frac{p_O + p_C}{2} + \frac{p_O + p_D}{2} \right)$$
$$= \frac{ab}{8} (4p_O + p_A + p_B + p_C + p_D)$$

When node O is released the angular compatibility and equilibrium equations may be written in the usual way, but using suffixes to distinguish between terms that differ in the XX and YY directions.

A.Ox: $\quad \dfrac{M_{AX}}{6EI_X} a + \dfrac{2M_{OX}}{3EI_X} a + \dfrac{M_{BX}}{6EI_X} a = \dfrac{\Delta_O - \Delta_A}{a} - \dfrac{\Delta_B - \Delta_O}{a}$

$$M_{AX} + 4M_{OX} + M_{BX} = \frac{6EI_X}{a^2}(-\Delta_A + 2\Delta_O - \Delta_B)$$

$$= \frac{6EI_X}{qa^2}(-p_A + 2p_O - p_B)$$

Let $g_X = \dfrac{6EI_X}{qa^2}$

$$M_{AX} + 4M_{OX} + M_{BX} + g_X p_A - 2g_X p_O + g_X p_B = 0$$

A.Oy: $\dfrac{M_{CY}}{6EI_Y}b + \dfrac{2M_{OY}}{3EI_Y}b + \dfrac{M_{DY}}{6EI_Y}b = \dfrac{\Delta_O - \Delta_C}{b} - \dfrac{\Delta_D - \Delta_O}{b}$

$$M_{CY} + 4M_{OY} + M_{DY} = \frac{6EI_Y}{b^2}(-\Delta_C + 2\Delta_O - \Delta_D)$$

$$= \frac{6EI_Y}{qb^2}(-p_C + 2p_O - p_D)$$

Let $r = \dfrac{a}{b}$ and $g_Y = \dfrac{6EI_Y}{qb^2}$

$$g_Y = g_X \frac{a^2}{I_X} \frac{I_Y}{b^2} = r^2 g_X \frac{I_Y}{I_X}$$

For rectangular elements, such as that shown in Fig. 10.2,

$$\frac{I_Y}{I_X} = \frac{ad^3}{12} \cdot \frac{12}{bd^3} = \frac{a}{b} = r$$

(I_X used in considering bending along the XX axis refers to the moment of inertia of the section at right-angles to this; hence $I_X = bd^3/12$ and similarly $I_Y = ad^3/12$.)

We can now substitute for $g_Y = r^3 g_X$

$$M_{CY} + 4M_{OY} + M_{DY} = g_Y(-p_C + 2p_O - p_D)$$
$$= r^3 g_X(-p_C + 2p_O - p_D)$$

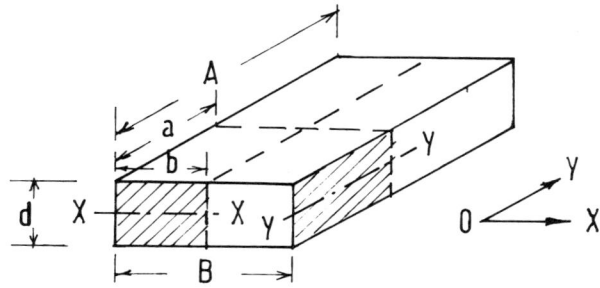

Fig. 10.2.

$$M_{CY} + 4M_{OY} + M_{DY} + r^3 g_X p_C - 2r^3 g_X p_O + r^3 g_X p_D = 0$$

E.O: $\dfrac{M_{OX} - M_{AX}}{a} + \dfrac{M_{OX} - M_{BX}}{a} + \dfrac{M_{OY} - M_{CY}}{b} + \dfrac{M_{CY} - M_{DY}}{b} + P_O = W$

But $P_O = \dfrac{ab}{8}(4p_O + p_A + p_B + p_C + p_D)$

$\therefore\ 2M_{OX} - M_{AX} - M_{BX} + 2\dfrac{a}{b}M_{OY} - \dfrac{a}{b}M_{CY} - \dfrac{a}{b}M_{DY} + \dfrac{a^2 b}{2}p_O$

$+ \dfrac{a^2 b}{8}(p_A + p_B + p_C + p_D) = Wa$

$2M_{OX} - M_{AX} - M_{BX} + 2rM_{OY} - rM_{CY} - rM_{DY} + \dfrac{a^3}{2r}p_O$

$+ \dfrac{a^3}{8r}(p_A + p_B + p_C + p_D) = Wa$

These equations are incorporated in the tabular matrix, shown in Table 10.1.

(2) Edge

Where the node occurs at the edge of the slab, angular change takes place on release in one direction only so either equation A.Ox or A.Oy alone is sufficient. For example, referring to diagram (a) in Fig. 10.3, bending occurs at node O on release in the XX direction only, since M_{OY} is zero. The relevant equations for the *top edge node*, therefore, are

A.Ox: $\dfrac{M_{AX}}{6EI_X}a + \dfrac{2M_{OX}}{3EI}a + \dfrac{M_{BX}}{6EI}a = \dfrac{\Delta_O - \Delta_A}{a} - \dfrac{\Delta_B - \Delta_O}{a}$

$M_{AX} + 4M_{OX} + M_{BX} + g_X p_A - 2g_X p_O + g_X p_B = 0$

E.O: $\dfrac{M_{OX} - M_{AX}}{a} + \dfrac{M_{OX} - M_{BX}}{a} - \dfrac{M_{CY}}{b} + P_O = W$

But $P_O = \dfrac{ab}{2}\dfrac{1}{3}\left(\dfrac{P_O + P_A}{2} + \dfrac{P_O + P_B}{2} + \dfrac{P_O + P_C}{2}\right)$

$= \dfrac{ab}{12}(3p_O + p_A + p_B + p_C)$

$\therefore\ 2M_{OX} - M_{AX} - M_{BX} - rM_{CY} + \dfrac{a^3}{4r}p_O + \dfrac{a^3}{12r}p_A + \dfrac{a^3}{12r}p_B$

$+ \dfrac{a^3}{12r}p_C = Wa$

Table 10.1 Raft foundation

Node	Equation	M_{OX}	M_{OY}	P_O	M_{AX}	M_{AY}	P_A	M_{BX}	M_{BY}	P_B	M_{CX}	M_{CY}	P_C	M_{DX}	M_{DY}	P_D	$\overline{W}$	Key figure
Internal	A.Ox	4		$-2g$	1		g	1		g		1			1	r^3g	0	A-[-O-]-B / D,C (dashed box)
	A.Oy		4	$-2r^3g$		1			1			1	$\dfrac{a^3}{8r}$		1		0	
	E.O	2	$2r$	$\dfrac{a^3}{2r}$	-1		$\dfrac{a^3}{8r}$	-1		$\dfrac{a^3}{8r}$		$-r$	$\dfrac{a^3}{8r}$		$-r$	$\dfrac{a^3}{8r}$	Wa	
Top edge	A.Ox	4		$-2g$	1		g	1		g							0	A — O — C
	E.O	2		$\dfrac{a^3}{4r}$	-1		$\dfrac{a^3}{12r}$	-1		$\dfrac{a^3}{12r}$		$-r$	$\dfrac{a^3}{12r}$				Wa	
LH edge	A.Oy		4	$-2r^3g$					1	$\dfrac{a^3}{12r}$		1	$\dfrac{a^3}{12r}$		1	r^3g	0	D,O,C — B
	E.O		$2r$	$\dfrac{a^3}{4r}$				-1		$\dfrac{a^3}{12r}$		$-r$	$\dfrac{a^3}{12r}$		$-r$	$\dfrac{a^3}{12r}$	Wa	
Corner	E.O			$\dfrac{a^3}{8r}$				-1		$\dfrac{a^3}{16r}$		$-r$	$\dfrac{a^3}{16r}$				Wa	O — B / C

Note: g_x is abbreviated to g in the table.

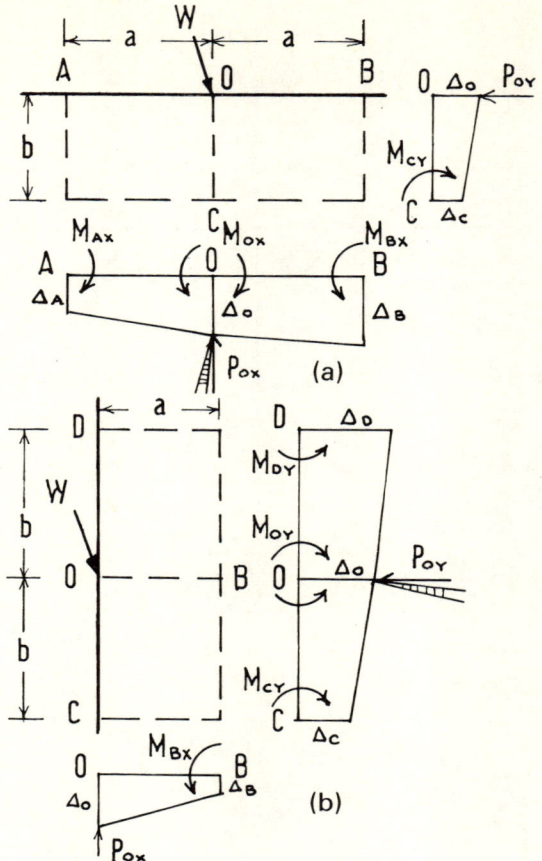

Fig. 10.3.

Similarly, referring to diagram (b) in Fig. 10.3, the relevant equations for the *left-hand edge node* are

A.Oy: $\dfrac{M_{CY}}{6EI_Y}b + \dfrac{2M_{OY}}{3EI_Y}b + \dfrac{M_{DY}}{6EI_Y}b = \dfrac{\Delta_O - \Delta_C}{b} - \dfrac{\Delta_D - \Delta_O}{b}$

$M_{CY} + 4M_{OY} + M_{DY} + r^3 g_X p_C - 2r^3 g_X p_O + r^3 g_X p_D = 0$

E.O: $\dfrac{M_{OY} - M_{CY}}{b} + \dfrac{M_{OY} - M_{DY}}{b} - \dfrac{M_{BX}}{a} + P_O = W$

But $P_O = \dfrac{ab}{12}(3p_O + p_B + p_C + p_D)$

∴ $2rM_{OY} - rM_{CY} - rM_{DY} - M_{BX} + \dfrac{a^3}{4r}p_O + \dfrac{a^3}{12r}p_B + \dfrac{a^3}{12r}p_C$

$+ \dfrac{a^3}{12r}p_D = Wa$

(Similar equations for right-hand edge nodes may be written merely by changing the suffixes.)

Equations for the two types of edge nodes are summarised in Table 10.1.

(3) *Corner*

When the node is located at the corner of the slab, as in Fig. 10.4, no angular changes occur on release, i.e. $M_{OX} = M_{OY} = 0$. Therefore, equilibrium conditions only need be considered. Diagrams in Fig. 10.4 indicate the moments effecting the reaction at node O, which is assumed to be held in position by 'props' on release. The equation for equilibrium is:

E.O: $\quad \dfrac{M_{OX} - M_{BX}}{a} + \dfrac{M_{OY} - M_{CY}}{b} + P_O = W$

$$P_O = \frac{a}{2} \frac{b}{2} \frac{1}{2} \left(\frac{p_O + p_B}{2} + \frac{p_O + p_C}{2} \right)$$

$$= \frac{ab}{16} (2p_O + p_B + p_C)$$

also, $M_{OX} = M_{OY} = 0$

$\therefore \quad -M_{BX} - rM_{CY} + \dfrac{a^3}{8r} p_O + \dfrac{a^3}{16r} p_B + \dfrac{a^3}{16r} p_C = Wa$

This is reproduced in the tabular matrix, Table 10.1.

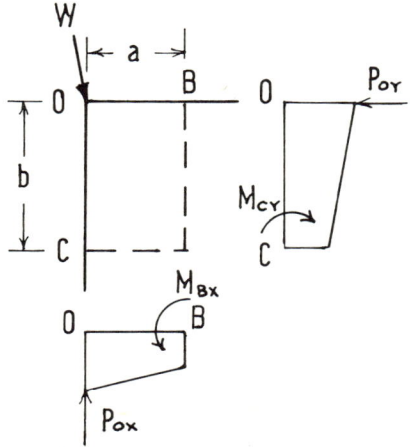

Fig. 10.4.

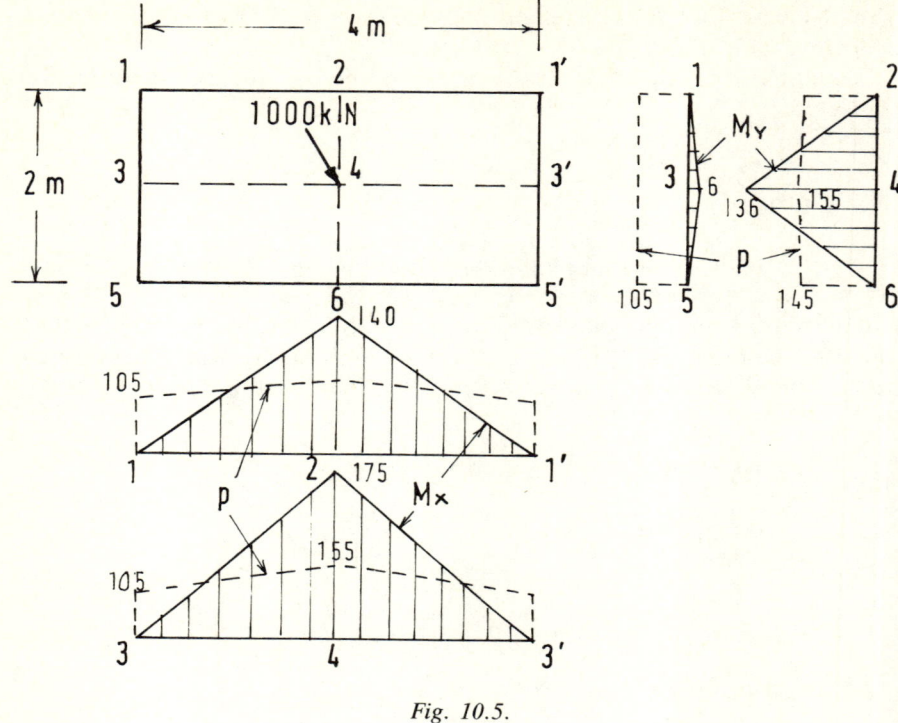

Fig. 10.5.

Column on raft foundation

To illustrate the use of Table 10.1, we can take the simple case of a rectangular raft foundation, 4 m × 2 m, supporting a centre column load of 1000 kN. The raft is divided into a grid, as shown in Fig. 10.5, in which a = 2 and b = 1. The raft is assumed to be 200 mm thick and constructed of concrete with E = 28 × 10^6 kN/mm². The coefficient of subgrade reaction of the soil q = 4000 kN/m³. From these values,

$$r = 2, \quad I = \frac{1 \times 0.2^3}{12} = 666.7 \times 10^{-6}$$

$$\text{and } g = \frac{6 \times 28 \times 10^6 \times 666.67 \times 10^{-6}}{4000 \times 2 \times 2} = 7$$

For symmetrical conditions we need only consider one quarter of the slab. The general matrix for nodes 1, 2, 3 and 4 is set out in Table 10.2. The sequence of equations as shown avoids a blank space occurring on the main diagonal. Terms shown under p_1, p_2, etc. have been doubled where appropriate to take account of reflected nodes p'_1, p'_2, etc. Thus the answer gives the full joint moments, assuming the full nodal loads, W, are entered in the $\overline{W}$ vector. Using this general matrix, the values shown in

Table 10.2 Quarter matrix for raft foundation

Equation	1 p	M_X	2 p	M_Y	3 p	4 M_X	M_Y	p	$\overline{W}$
E.1	$\dfrac{a^3}{8r}$	-1	$\dfrac{a^3}{16r}$	$-r$	$\dfrac{a^3}{16r}$				Wa
A.2x	2g	4	$-2g$						0
E.2	$\dfrac{a^3}{6r}$	2	$\dfrac{a^3}{4r}$				$-r$	$\dfrac{a^3}{12r}$	Wa
A.3y	$2r^3g$			4	$-2r^3g$				0
E.3	$\dfrac{a^3}{6r}$			2r	$\dfrac{a^3}{4r}$	-1		$\dfrac{a^3}{12r}$	Wa
A.4x					2g	4		$-2g$	0
A.4y			$2r^3g$				4	$-2r^3g$	0
E.4			$\dfrac{a^3}{4r}$		$\dfrac{a^3}{4r}$	2	2r	$\dfrac{a^3}{2r}$	Wa

the data table, Table 10.3, have been derived for solving the problem. Results obtained from running the MATIN program with the data are shown near the bottom of the table. The bending moments are those on strips of widths a and b. For design purposes, we usually require the moments per unit width or length of slab. These are shown in the bottom line indicated as $\dfrac{M_X}{b}$ or $\dfrac{M_Y}{a}$. These values were used to plot the BM diagrams in Fig. 10.5. These appear as a series of straight lines but, by

Table 10.3 Data table for Fig. 10.5

Equation	1 p	M_X	2 p	M_Y	3 p	4 M_X	M_Y	p
E.1	0.5	-1	0.25	-2	0.25			
A.2	14	4	-14					
E.2	0.667	2	1				-2	0.333
A.3	112			4	-112			
E.3	0.667			4	1	-1		0.333
A.4x					14	4		-14
A.4y			112				4	-112
E.4			1		1	2	4	2
$\overline{W}$								2000
MATIN N = 8	104.8	139.8	144.7	-12.6	104.3	175.4	272.8	154.5
$\dfrac{M_X}{b}$ or $\dfrac{M_Y}{a}$		139.8		-6.3		175.4	136.4	kNm

taking a finer grid (even one extra subdivision would suffice), it becomes evident that the moment diagrams are curved, in fact, with peak values as shown at node 4. Positive bending moments, in accordance with the arrow convention in Fig. 10.1, indicate tension at the bottom surface. Positive moments are drawn above the baseline, as usual, and therefore on the compression face of the foundation.

General matrix and examples

In most problems it will be found necessary to divide the slab into more sections than in the previous example. A general matrix has been tabulated in Table 10.4 for a (4a × 4b) grid of unit size (a × b) for solving loaded raft foundations. The (21 × 21) size matrix covers one quarter of the slab only, but this is sufficient for symmetrical conditions. The coefficient terms can be derived from a study of Table 10.1 for the various node types. Figures to be entered in the $\overline{W}$ vector column will vary with the load conditions, of course.

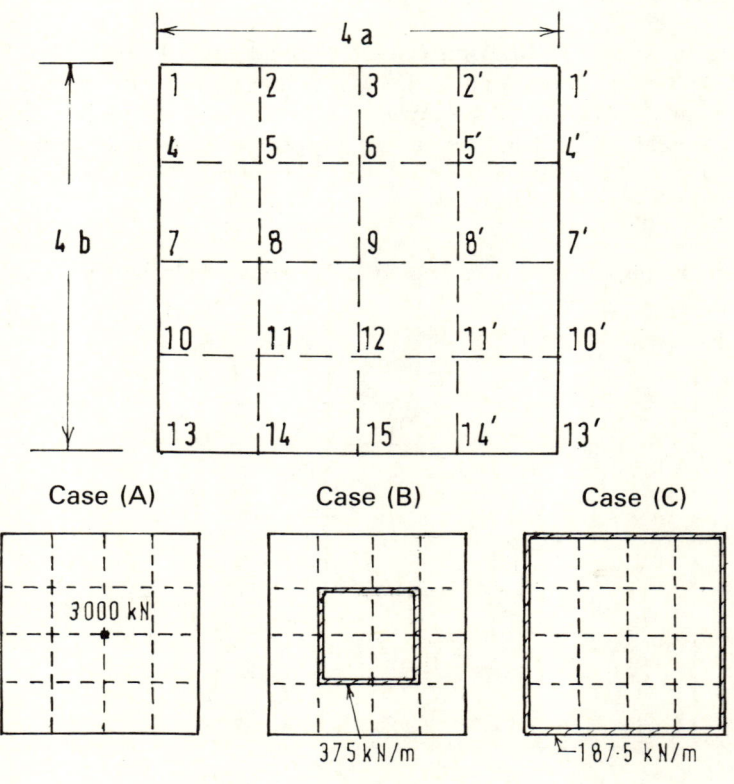

Fig. 10.6.

RAFT FOUNDATIONS 111

Table 10.4 Quarter matrix for raft foundation slab

Equation	1 p	1 M_X	2 p	3 M_X	3 p	4 M_Y	4 p	5 M_X	5 M_Y	5 p	6 M_X	6 M_Y	6 p	7 M_Y	7 p	8 M_X	8 M_Y	8 p	9 M_X	9 M_Y	9 p	$\overline{W}$
E.1	$\dfrac{a^3}{8r}$	-1	$\dfrac{a^3}{16r}$			$-r$	$\dfrac{a^3}{16r}$															(Wa)
A.2x	$\dfrac{g}{a^3}$	4	$-\dfrac{2g}{a^3}$	1	$\dfrac{g}{a^3}$																	0
E.2	$\dfrac{a^3}{12r}$	2	$\dfrac{a^3}{4r}$	-1	$\dfrac{a^3}{12r}$																	(Wa)
A.3x		2	$\dfrac{2g}{a^3}$	4	$-\dfrac{2g}{a^3}$																	0
E.3		-2	$\dfrac{a^3}{6r}$	2	$\dfrac{a^3}{4r}$						$-r$		$\dfrac{a^3}{12r}$									(Wa)
A.4y	$\dfrac{r^3 g}{a^3}$					4	$-\dfrac{2r^3 g}{a^3}$	-1	4	$-\dfrac{2g}{a^3}$												0
E.4	$\dfrac{a^3}{12r}$					2r	$\dfrac{a^3}{4r}$	4		g	1		g									0
A.5x A.5y			$\dfrac{r^3 g}{a^3}$					2	2r	$\dfrac{a^3}{2r}$	-1	4	$\dfrac{a^3}{8r}$	1	$\dfrac{r^3 g}{a^3}$							0 0
E.5			$\dfrac{a^3}{8r}$					-2		$\dfrac{a^3}{4r}$	2	2r	$\dfrac{a^3}{2r}$	$-r$	$\dfrac{a^3}{8r}$							0
A.6x A.6y							$\dfrac{2r^3 g}{a^3}$			$-\dfrac{2g}{a^3}$ -$\dfrac{2r^3 g}{a^3}$	4	4	$-\dfrac{2g}{a^3}$ -$\dfrac{2r^3 g}{a^3}$	1	$-\dfrac{2r^3 g}{a^3}$	-1	1	$\dfrac{a^3}{12r}$				0 0
E.6						2	$\dfrac{2r^3 g}{a^3}$		2	$\dfrac{a^3}{2r}$	2	4	$\dfrac{a^3}{2r}$	4	$\dfrac{a^3}{4r}$	4	4	$\dfrac{a^3}{2r}$				0
A.7y						$-2r$	$\dfrac{a^3}{6r}$		$-2r$	$\dfrac{a^3}{4r}$		2r	$\dfrac{a^3}{4r}$	2r	$-2r^3 g$/a^3	2	2r	$\dfrac{a^3}{2r}$				0
E.7														4	$\dfrac{a^3}{4r}$			$\dfrac{a^3}{2r}$				(Wa)
A.8x A.8y									2	$\dfrac{2r^3 g}{a^3}$			$\dfrac{2r^3 g}{a^3}$	2r	g $\dfrac{a^3}{2r}$	4	4	$-\dfrac{2g}{a^3}$ $\dfrac{a^3}{2r}$	1		g $\dfrac{a^3}{2r}$	0 0
E.8									$-2r$	$\dfrac{a^3}{4r}$		$-2r$	$\dfrac{a^3}{4r}$		$\dfrac{a^3}{8r}$	2	2r	$\dfrac{a^3}{2r}$	-1		$\dfrac{a^3}{8r}$	0
A.9x A.9y																-2	4	2g	4	4	$-\dfrac{2g}{a^3}$ $-2r^3 g$/a^3	0 0
E.9																2	2r	$\dfrac{a^3}{4r}$	2	2r	$\dfrac{a^3}{2r}$	(Wa)

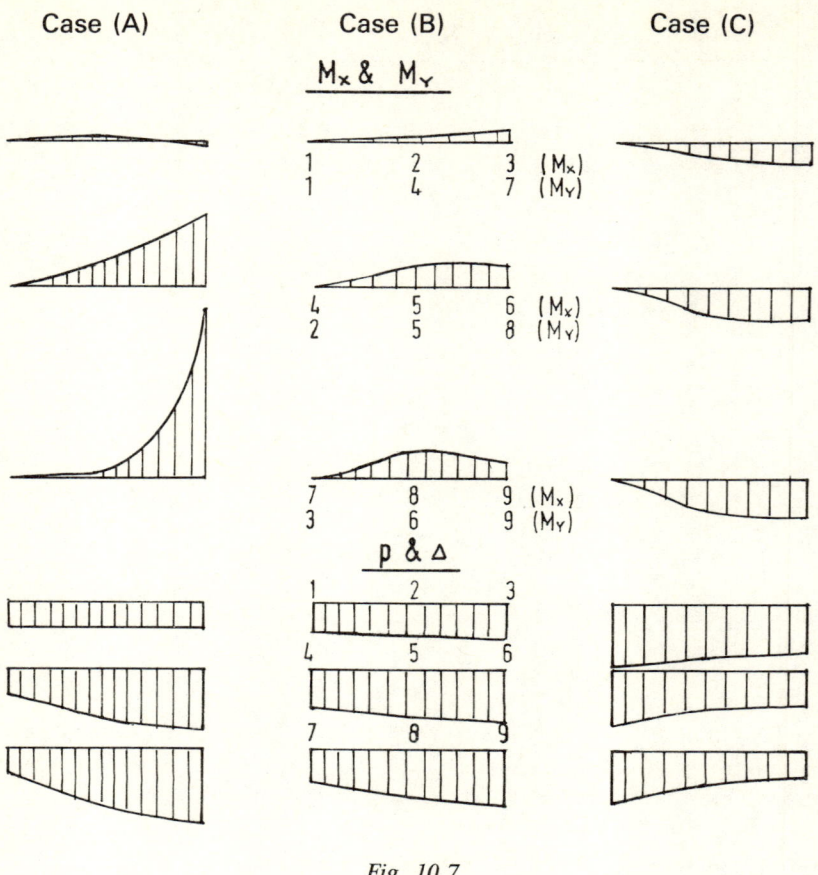

Fig. 10.7.

The (4a × 4b) grid is reproduced in the problem shown in Fig. 10.6. The 200 mm thick concrete slab, square on plan, for which $E_C = 28 \times 10^6$ kN/mm², is supported on a soil for which q = 8000 kN/m³. Therefore,

$$a = b = r = 1; \quad I = \frac{1 \times 0.2^3}{12} = 666.67 \times 10^{-6} \text{m}^4$$

$$g = \frac{6 \times 28 \times 10^6 \times 666.67 \times 10^{-6}}{8000 \times 1} = 14 \text{ kN/m}^2$$

These values may be used to construct the (21 × 21) coefficient matrix. Three different load conditions are shown in Fig. 10.6. A total load of 3000 kN is applied to the slab in each case (A) on a central column, (B) on a core wall inside the perimeter of the slab and (C) on a core wall on the slab perimeter. The DATA statements to reproduce these load systems successively are:

(A) DATA 0,0,0,0,0,0,0,0,0,0
    DATA 0,0,0,0,0,0,0,0,0,0,3000
(B) DATA 0,0,0,0,0,0,0,0,0,0,375
    DATA 0,0,375,0,0,0,0,375,0,0,0
(C) DATA 187.5,0,187.5,0,187.5,0,187.5,0,0,0
    DATA 0,0,0,0,187.5,0,0,0,0,0,0

Results from the printouts obtained by running MATIN with the above data were used to construct the diagrams for M_X, M_Y and p in Fig. 10.7. The important figures for design purposes are:

Case (A) Central column load, W = 3000 kN
          BM max = 689.5 kNm; p (max) = 365.1 kN/m²
Case (B) Inner core wall, W = 3000 kN
          BM max = 120.1 kNm; p (max) = 237.5 kN/m²
Case (C) Outer core wall, W = 3000 kN
          BM max = −160.1 kNm; p (max) = 274.7 kN/m²

This summary shows clearly the considerable difference in bending moments where the same size load is applied but arranged in a different manner on the foundation slab.

It will be observed from Fig. 10.7 that the bending moment diagrams in the XX and YY directions are identical in all cases, for example $M_{2X} = M_{4Y}$, $M_{3X} = M_{7Y}$, etc. This could have been deduced from considerations of symmetry, so the work load might have been reduced by omitting the M_Y terms and corresponding equations from the tabular data, thereby reducing the matrix from (21 × 21) to (15 × 15) size. This can only be done, of course, where loading, support and physical conditions are all symmetrical.

11 Lateral loads on piles

Piles in an elastic medium

Piles are usually designed to take vertical or direct loads only. There are occasions, however, where they will be required to resist lateral loads and bending moments. The subgrade reaction theory may be applied to the case of a pile embedded in an elastic medium and subject to lateral loading. The equation $p = q\Delta$ applies as before, except that the term Δ denotes lateral displacement instead of vertical settlement, and q is described as the coefficient of lateral reaction.

In Fig. 11.1, an elastic pile of length L and width b is considered to be divided into a number of small strips of length a below ground level. A lateral load, W, is applied to the head of the pile which is at a height h above ground level. Lateral soil reactions are assumed gathered to the nodal points, resulting in the same equations for nodal reactions as before, namely,

$$P_0 = \frac{ab}{8}(3p_0 + p_1)$$

$$P_1 = \frac{ab}{4}(p_0 + 2p_1 + p_2)$$

$$P_2 = \frac{ab}{4}(p_1 + 2p_2 + p_3) \quad \text{etc.}$$

The pile is assumed to take up the shape indicated in the deformation diagram in Fig. 11.1(b) with lateral displacements Δ_0, Δ_1, Δ_2, etc. at the nodes at intervals a apart. Also shown are the soil reactions at the nodes, designated P_0, P_1, P_2, etc. (The diagram has been rotated into the horizontal to suit construction of the BM and pressure diagrams.) The angular change at each node is indicated by the shaded angles in the diagram. Thus,

$$\text{Angular change in strip 0A} = \frac{\Delta_A - \Delta_0}{h}$$

$$\text{in strip 01} = \frac{\Delta_0 - \Delta_1}{a}$$

LATERAL LOADS ON PILES

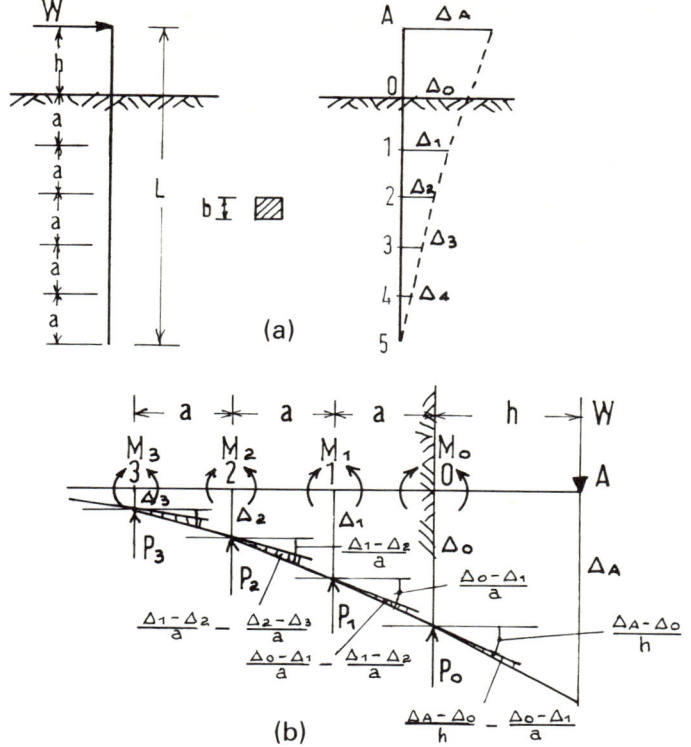

Fig. 11.1.

$$\text{Angular change} \quad \text{at node 0} = \frac{\Delta_A - \Delta_0}{h} - \frac{\Delta_0 - \Delta_1}{a}$$

$$\text{in strip 12} = \frac{\Delta_1 - \Delta_2}{a}$$

$$\text{at node 1} = \frac{\Delta_0 - \Delta_1}{a} - \frac{\Delta_1 - \Delta_2}{a} \quad \text{etc.}$$

Release moments are to be applied in turn so as to close the angular discontinuities and restore the strips to their original alignment. Angular change is measured relative to the reference element, i.e. to the angle of strip 01 measured from the origin, 0. Assuming node 0 to be 'free', the release moment applied at node 1 to strip 01 should act so as to close the discontinuity indicated by the shaded angle in diagram (b). This determines its direction and that of the moments applied at subsequent nodes.

It will be seen from Fig. 11.1(b) that the release moments act in the same direction as that adopted for continuous beams in Chapter 6. Therefore, positive moments indicate tension at the top surface, i.e. *tension on the left-hand side of the pile.* Also, since positive values are plotted above the baseline, the BM diagrams will appear on the upper or, when the latter is plotted vertically, left-hand side of the baseline. Ground pressures will be measured downwards as in previous foundation examples, to permit them to function also as displacement diagrams. This means that, when the baseline is plotted vertically, positive values indicate *displacements to the right of the pile* from the vertical position.

In Fig. 11.1, the load, W, produces a bending moment in the pile above ground level equal to Wh,

$$\text{i.e.} \quad M_0 = Wh$$

The compatibility equations for angular displacement at the nodes may be written in the usual way as:

A.0: $\dfrac{M_0}{3EI} h + \dfrac{M_0}{3EI} a + \dfrac{M_1}{6EI} a = \dfrac{\Delta_A - \Delta_0}{h} - \dfrac{\Delta_0 - \Delta_1}{a}$

$2M_0 \dfrac{h}{a} + 2M_0 + M_1 = \dfrac{6EI}{a} \left(\dfrac{\Delta_A}{h} - \dfrac{\Delta_0}{h} + \dfrac{\Delta_1}{a} - \dfrac{\Delta_0}{a} \right)$

$2M_0 \left(1 + \dfrac{h}{a} \right) + M_1 - \dfrac{6EI}{ah} \Delta_A = - \dfrac{6EI}{qa} \left(\dfrac{p_0}{h} + \dfrac{p_0}{a} - \dfrac{p_1}{a} \right)$

$M_0 = Wh; \quad g = \dfrac{6EI}{qa^2}; \quad c = a^2 b$

$M_1 - g \dfrac{qa}{h} \Delta_A + gp_0 \dfrac{a}{h} + gp_0 - gp_1 = -2Wh \left(1 + \dfrac{h}{a} \right)$

$M_1 + gp_0 \left(1 + \dfrac{a}{h} \right) - gp_1 - \dfrac{gaq}{h} \Delta_A = -2Wh \left(1 + \dfrac{h}{a} \right)$

A.1: $\dfrac{M_0}{6EI} a + \dfrac{2M_1}{3EI} a + \dfrac{M_2}{6EI} a = \dfrac{\Delta_0 - \Delta_1}{a} - \dfrac{\Delta_1 - \Delta_2}{a}$

$M_0 + 4M_1 + M_2 = \dfrac{6EI}{a^2} (\Delta_0 - 2\Delta_1 + \Delta_2)$

$4M_1 + M_2 - gp_0 + 2gp_1 - gp_2 = -Wh$

A.2: $M_1 + 4M_2 + M_3 - gp_1 + 2gp_2 - gp_3 = 0 \quad$ etc.

From consideration of equilibrium conditions in Fig. 11.1(b), the following equations may be derived:

E.0: $\dfrac{M_0 - M_1}{a} = P_0 - W = \dfrac{ab}{8} (3p_0 + p_1) - W$

$M_0 = Wh; \quad c = a^2 b$

$Wh + Wa - M_1 = \dfrac{3}{8} cp_0 + \dfrac{c}{8} p_1$

LATERAL LOADS ON PILES

E.1:
$$4M_1 + \frac{3}{2}cp_0 + \frac{c}{2}p_1 = 4W(a+h)$$
$$\frac{M_1 - M_0}{a} + \frac{M_1 - M_2}{a} = P_1 = \frac{ab}{4}(p_0 + 2p_1 + p_2)$$
$$-M_0 + 2M_1 - M_2 = \frac{c}{4}(p_0 + 2p_1 + p_2)$$

E.2:
$$-8M_1 + 4M_2 + cp_0 + 2cp_1 + cp_2 = -4Wh$$
$$4M_1 - 8M_2 + 4M_3 + cp_1 + 2cp_2 + cp_3 = 0 \quad \text{etc.}$$

These equations enable Table 11.1 to be constructed. The coefficient matrix, apart from the extra equation A.0, is identical to that derived for the continuous beam foundation in Chapter 7, except that some signs are altered because of the different way in which pile deflections are measured.

The table can be used to prepare data statements that, when incorporated in the MATIN program, will solve problems involving the use of flexible piling subject to lateral loads applied above ground level. An example is given in the next section.

Laterally loaded pile

In this example, a circular concrete pile, 500 mm diameter, is assumed to be driven 10 m below ground. A lateral load of 100 kN is applied at a height of 1 m above ground level (Fig. 11.2). If the soil has a coefficient of lateral restraint $q = 5000$ kN/m^3 and $E_c = 28 \times 10^6$ kN/m^2, and assuming nodes at 1 m intervals from ground level to the toe of the pile, then

$$I = \frac{3.1416 \times 0.5^4}{64} = 0.0031 \text{ m}^4; \quad g = \frac{6 \times 28 \times 10^6 \times 0.0031}{5 \times 10^3 \times 1} = 104.2$$

Table 11.1 Pile with horizontal load at head

Equation	Δ	A 0 p	1 p	M	2 p	M	3 p	M	p	$\overline{W}$
A.0	$-\frac{aq}{h}g$	$\left(1+\frac{a}{h}\right)g$	$-g$	1						$-2Wh\left(1+\frac{h}{a}\right)$
E.0		$\frac{3}{2}c$	$\frac{c}{2}$	4						$4W(a+h)$
A.1		$-g$	$2g$	4	$-g$	1				$-Wh$
E.1		c	$2c$	-8	c	4				$-4Wh$
A.2			$-g$	1	$2g$	4	$-g$	1		0
E.2			c	4	$2c$	-8	c	4		0
E.L							$\frac{c}{2}$	4	$\frac{3}{2}c$	0

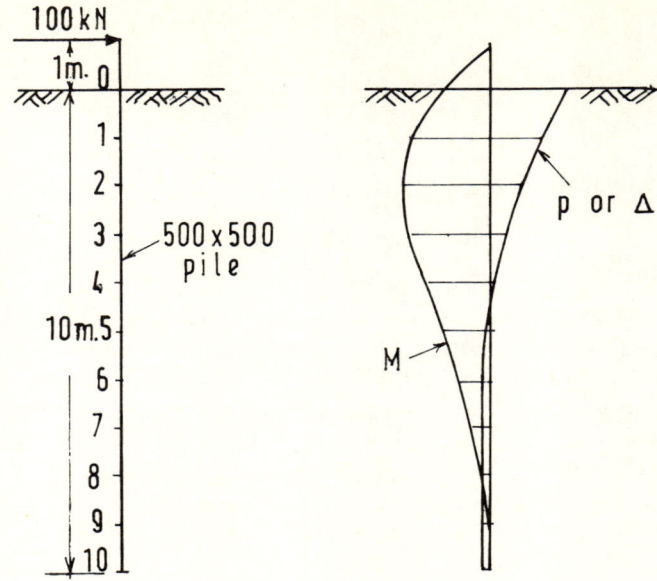

Fig. 11.2.

$$a = 1;\ c = a^2\ b = 0.5;\ 1 + \frac{a}{h} = 2 = 1 + \frac{h}{a};\ \frac{aq}{h} = 5000$$

A (21 × 21) tabular matrix can be prepared on the lines indicated in Table 11.1 and incorporating the above values. This is reproduced in Table 11.2. By inserting these figures in data statements before MATIN is run, we can obtain the pile moments and ground pressures. (Hints on keying in the (21 × 21) matrix are given in the next section.) Pile deflections in millimetres are obtained by multiplying the p values by one-fifth, since $\Delta = p/q$. Results have been used to plot the BM and deflection diagrams in Fig. 11.2.

The bending moment at the foot of the pile is zero, as we have assumed that the pile is free to bend at the toe. The maximum bending moment occurs at node 2. However, when the pile is designed, a check should be made to ensure that the maximum value of the pressure p (in our case, 148 kN/m^2) does not exceed the allowable bearing pressure for the soil. This needs to be interpreted in a common sense fashion because of the rapid reduction in pressure with depth, as can be seen from the results. As well as this, weathering of the soil at or immediately below ground level may seriously affect its resistance to lateral pressure. To counteract this, it may be advisable to locate the node for initial soil resistance at some distance (commonly 1 m) below actual ground level. This may be checked using a modified (19 × 19) matrix.

Table 11.2 Lateral load on pile in Fig. 11.2

Equation	A	0	1		2		3		4		5		6		7		8		9		10	
	Δ	p	p	M	p	M	p	M	p	M	p	M	p	M	p	M	p	M	p	M	p	M
A.0	−520 800	208.4	−104.2	1																		
E.0		0.75		4																		
A.1		−104.2	208.4	4		−8																
E.1		0.5	1																			
A.2			−104.2	1	208.4	4	−104.2	1														
E.2				0.5	1	−8																
A.3					−104.2	1	208.4	4	−104.2	1												
E.3						0.5	4	−8														
A.4							−104.2	1	208.4	4	−104.2	1										
E.4								0.5	4	−8												
A.5									−104.2	1	208.4	4	−104.2	1								
E.5										0.5	4	−8										
A.6											−104.2	1	208.4	4	−104.2	1						
E.6												0.5	4	−8								
A.7													−104.2	1	208.4	4	−104.2	1				
E.7														0.5	4	−8						
A.8															−104.2	1	208.4	4	−104.2	1		
E.8																0.5		−8				
A.9																	−104.2	1	208.4	4	−104.2	0.5
E.9																			0.25	4		0.75
E.10																						
$\overline{W}$	−400	800	−100	−400	58.7	181	28.6	166	7.9	135.3	−5	99.6	−12	65.7	−15.6	37.4	−16.7	16.7	−16.8	4.2	−16.6	
MATIN N = 21 Δ / mm	0.04	40	30		20		12		6		1		−2		−3		−3		−3		−3	

It will be seen from the pressure diagram, Fig. 11.2, that a point of rotation, indicated by a sign change in ground pressure, occurs at node 4. A study of the diagram suggests that the embedded pile length could be reduced to about 7 m in this case for economy.

Keying in a large matrix

Keying in a (21 × 21) matrix can be a tedious task, so any short cuts that will reduce the work load, such as overtyping line numbers or use of the 'repeat' facility, is worth considering. Overtyping is also a means of reducing keyboard errors, and so is to be strongly recommended. The matrix should be displayed symmetrically to facilitate visual checking of the data. In the above example each row might occupy two statements, the first containing 10 coefficients and the second 11 in each case.

It may help to key in initially just part of the top quarter of the matrix with half the last row and load vector. The repetitive 'boxed' values in the table should be incorporated in two statements as well as, say, two rows of zeroes for overtyping purposes. In the example, the data statements initially might be as follows:

```
1000 DATA −520800, 208.4, −104.2,1,0,0,0,0,0
1005 DATA 0,0.75,0.25,4,0,0,0,0,0,0
1010 DATA 0,−104.2,208.4,4,−104.2,1,0,0,0,0
1015 DATA 0,0.5,1,−8,0.5,4,0,0,0,0
1100 DATA 0,0,0,0,0,0,0,0,0.25,4,0.75
1150 DATA −400,800,−100,−400,0,0,0,0,0,0
1200 DATA −104.2,1,208.4,−104.2,1
1205 DATA 0.5,4,1,−8,0.5,4
1210 DATA 0,0,0,0,0,0,0,0,0,0
1211 DATA 0,0,0,0,0,0,0,0,0,0,0
```

The first four statements need no alteration, so to complete the tabulated rows overtype line 1211 with new line numbers, e.g. 1001, 1006, 1011 and 1016. Statements 1200 and 1205 need to be altered by inserting zeroes or truncating the last items as necessary, e.g.

```
1020 DATA 0,0,−104.2,1,208.4,4,−104.2,2,1,0,0
1025 DATA 0,0,0.5,4,1,−8,0.5,4,0,0
```

These two matrix rows are completed by overtyping line 1211 as before, in this case with line numbers 1021 and 1026.

By proceeding in this way the complete matrix can be keyed in merely by inserting zeroes and typing in truncated terms in following statements, where necessary. For instance, the whole of the lower left-hand quarter of the matrix may be keyed in by overtyping line 1210 with new line numbers, e.g. 1059, 1064, 1069, 1074, 1079, 1084, 1089, 1094 and 1099.

Use of overtyping reduces the danger of keying in numbers incorrectly but, of course, it is still important to check that the tabular data are correctly interpreted in the data statements when the program is ready to run.

Pile groups

Lateral loads applied over a number of vertical piles may be analysed in a similar fashion to that used for a single pile. It is usual to distribute the lateral load to the piles in proportion to their relative stiffnesses. This type of problem is commonly encountered in the design of piled jetties, dolphins or quay walls where horizontal forces caused during berthing of vessels have to be considered. In bridges constructed on piles, horizontal forces due to braking or acceleration of vehicles and lateral wind loads have to be catered for in the same way.

It must be pointed out that the analysis only applies to piles embedded in a uniform soil with elastic properties in accordance with the subgrade reaction theory. Where soils of different q values are encountered at different depths, these may be dealt with simply by modifying the value of g in the matrix at the appropriate depth. Nodes should be selected so as to occur at or near each change in soil stratum.

Pile caps

In the above analysis it was assumed that each end of the pile was free to rotate. For the special case where the lateral load is applied to the pile directly at ground level $\Delta_A = 0$ and since the head of the pile is 'free', the equation A.0 is no longer relevant. This simplifies the tabular matrix, the first three rows of which are reproduced in Table 11.3. The terms for the rest of the matrix are exactly the same as before. The $\overline{W}$ terms take into account the fact that h is now equal to zero.

Table 11.3 Part tabular matrix for pile with lateral load, W, applied directly at ground level

Equation	0	1		2		$\overline{W}$
	p	p	M	p	M	
E.0	$\frac{3}{2}c$	$\frac{c}{2}$	4			4aW
A.1	$-g$	$2g$	4	$-g$	1	0
E.1	c	$2c$	-8	c	4	0
A.2		$-g$	etc.			
E.2		c				

If the pile is constructed with a rigid pile cap at ground level, as frequently done in practice (Fig. 11.3), rotation of the pile head is prevented and so a bending moment is introduced at the head of the pile. As mentioned before, great care has to be taken to ensure the use of the correct sign where bending and displacement effects are combined. The deformation diagram due to lateral displacement alone is shown in diagram (b) and that due to a bending moment applied at node 0 in diagram (c); these combine to give the outline shown in diagram (a). The individual diagrams are rotated into the horizontal and shown in greater detail in Fig. 11.4.

Since node 0 is now fixed, a release moment is introduced there, as shown. It will be seen that the release moment and the applied moment M_0 act in opposite directions. Therefore, in accordance with the rule for combining settlement and bending effects described in Chapter 6, we should subtract the lesser settlement from the larger one and insert a negative sign in front of the difference divided by the span in compiling equation A.0, thus

A.0: $\dfrac{M_0}{3EI} a + \dfrac{M_1}{6EI} a = -\left(\dfrac{\Delta_0 - \Delta_1}{a}\right)$

$2M_0 + gp_0 - gp_1 + M_1 = 0$

Other equations are derived as before:

A.1: $\dfrac{M_0}{6EI} a + \dfrac{2M_1}{3EI} a + \dfrac{M_2}{6EI} a = \dfrac{\Delta_0 - \Delta_1}{a} - \dfrac{\Delta_1 - \Delta_2}{a}$

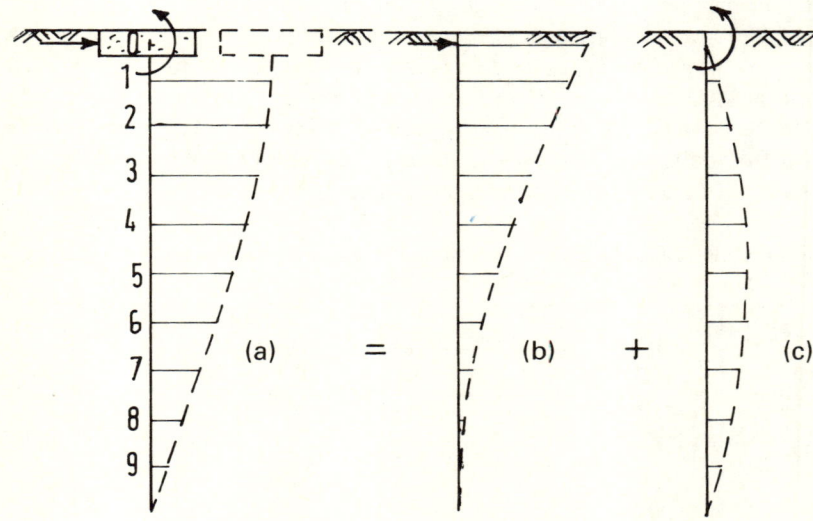

Fig. 11.3.

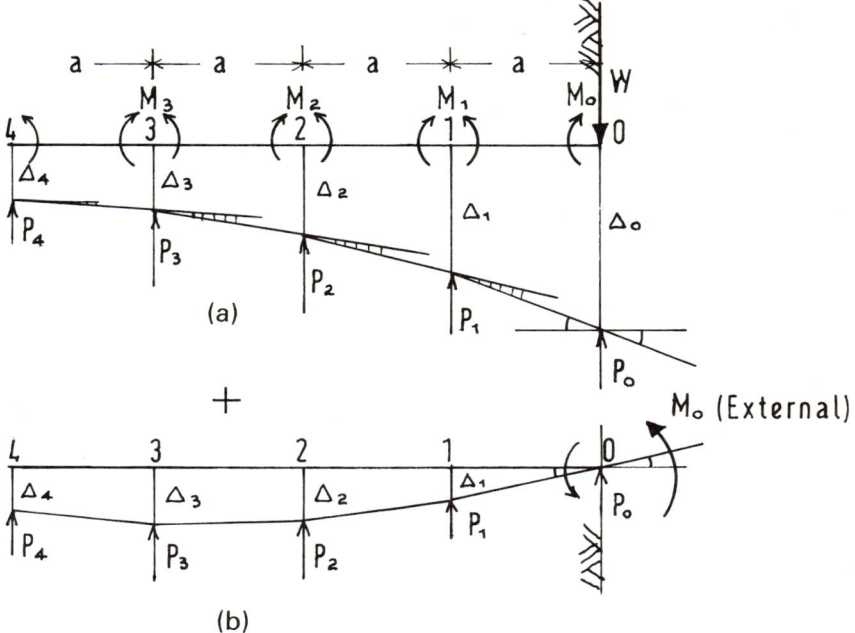

Fig. 11.4.

$$M_0 - gp_0 + 2gp_1 + 4M_1 - gp_2 + M_2 = 0$$

E.0: $\quad \dfrac{M_0 - M_1}{a} = P_0 - W$

$$Wa = M_1 - M_0 + aP_0 = M_1 - M_0 + \dfrac{a^2 b}{8}(3p_0 + p_1)$$

Table 11.4 Part tabular matrix for pile subject to lateral load, W, at ground level with rigid pile cap

Equation	0		1		2		$\overline{W}$
	p	M	p	M	p	M	
A.0	g	2	−g	1			0
E.0	$\frac{3}{2}$c	−4	$\frac{c}{2}$	4			4aW
A.1	−g	1	2g	4	−g	1	0
E.1	c	4	2c	−8	c	4	0
A.2			−g	etc.			0
E.2			c				0

$$-4M_0 + \frac{3}{2}cp_0 + \frac{c}{2}p_1 + 4M_1 = 4aW$$

E.1: $\dfrac{M_1 - M_0}{a} + \dfrac{M_1 - M_2}{a} = P_1 = \dfrac{ab}{4}(p_0 + 2p_1 + p_2)$

$$4M_0 - 8M_1 + 4M_2 + cp_0 + 2cp_1 + cp_2 = 0$$

The first few rows of the general matrix for this case are shown in Table 11.4. The rest will be identical to the terms in Table 11.1.

Piles subject to bending

Although bending moments are normally resisted by piles acting in groups, the situation can arise, especially with sheet piling or eccentrically loaded piles, where moments are applied directly to the head of the pile from the superstructure. Assuming the toe of the pile is free to rotate, the deformation induced in the pile will be similar to that in Fig. 11.5 where, in addition to the lateral load W, a moment M_A is applied at the head. As

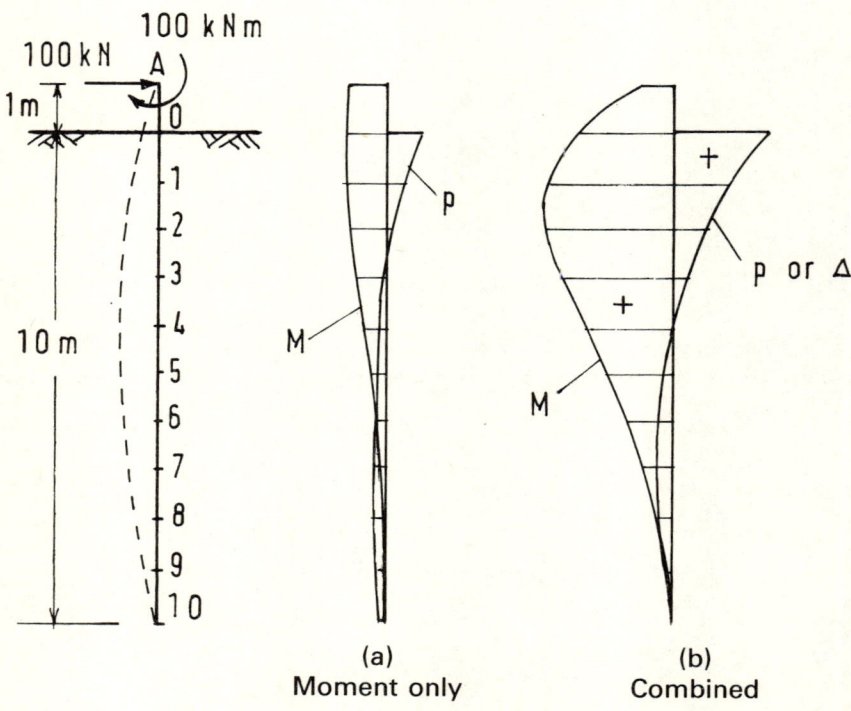

Fig. 11.5.

there is no restraint above ground, $M_0 = Wh + M_A$, when M_A acts in the direction shown. By making this substitution in the equations obtained previously, the $\overline{W}$ vector in Table 11.1 will require adjustment to include M_A as shown in Table 11.5.

Table 11.5 $\overline{W}$ terms to take account of the moment M_A at the head

Equation	$\overline{W}$
A.0	$-2(Wh + M_A)\left(1 + \dfrac{h}{a}\right)$
E.0	$4(Wa + Wh + M_A)$
A.1	$-(Wh + M_A)$
E.1	$-4(Wh + M_A)$

Assuming the same pile section and soil properties as before, but with a moment of 100 kNm applied at A in addition to the lateral load of 100 kN, as shown in Fig. 11.5, the values in Table 11.2 may be used but with the following revised $\overline{W}$ terms:

$$-800, \ 1200, \ -200, \ -800.$$

Results obtained with MATIN have been used to construct the bending moment and pressure diagrams in Fig. 11.5. Diagrams show the effect of applied moment only and combined moment and lateral loading. Note that, since both of these actions produce tension on the left-hand side of the pile, the bending moment diagrams are of the same sign.

12 Sheet piling and strutted abutments

Cantilever walls

In the design of sheet piled retaining walls, the weight of the wall is assumed negligible and stability is dependent on the bending and shear resistance of the sheet piling. The cantilever type wall, which is not restrained by tie rods, is sufficiently flexible to yield and allow the full active earth pressure to develop behind the wall. Below ground level, the pressure is assumed to develop in accordance with the subgrade reaction theory. The toe of the sheet pile is considered to be free to bend in this method of analysis so the bending moment there is zero. Active pressure is taken to be in accordance with the Rankine distribution, so that the intensity of pressure is given by

$$p = \gamma\, wh$$

where w = unit weight of soil, $\gamma = (1 - \sin \phi)/(1 + \sin \phi)$ for granular soil material with an internal angle of friction ϕ and h = depth below the surface of the retained soil.

As the pressure varies uniformly with depth below ground level, the total horizontal force per unit length of wall is given by

$$W = \frac{ph}{2} = \gamma\, \frac{wh^2}{2} \qquad (12.1)$$

This acts at a height $h/3$ above the base 0 (Fig. 12.1).

Therefore, the bending moment at $0 = \dfrac{Wh}{3} = \gamma\, \dfrac{wh^3}{6} \qquad (12.2)$

The effect of this force is to produce a deflection, Δ_A, at the head of the sheet piling; in addition, because of the triangular distributed load on the length 0A, an angular rotation θ_{0A} occurs at 0. This is shown in the deformation diagram in Fig. 12.1, rotated into the horizontal. On this are shown also the moments applied at the nodes on release to counteract the angular changes, including that due to the active earth pressure on 0A represented by the loading diagram 0AK.

Below ground, pressure is exerted on the sheet piling in accordance with the subgrade reaction convention; this is shown in the diagram as a series of reactions, P_0, P_1, P_2, etc., applied at the nodes. The conditions

SHEET PILING AND STRUTTED ABUTMENTS 127

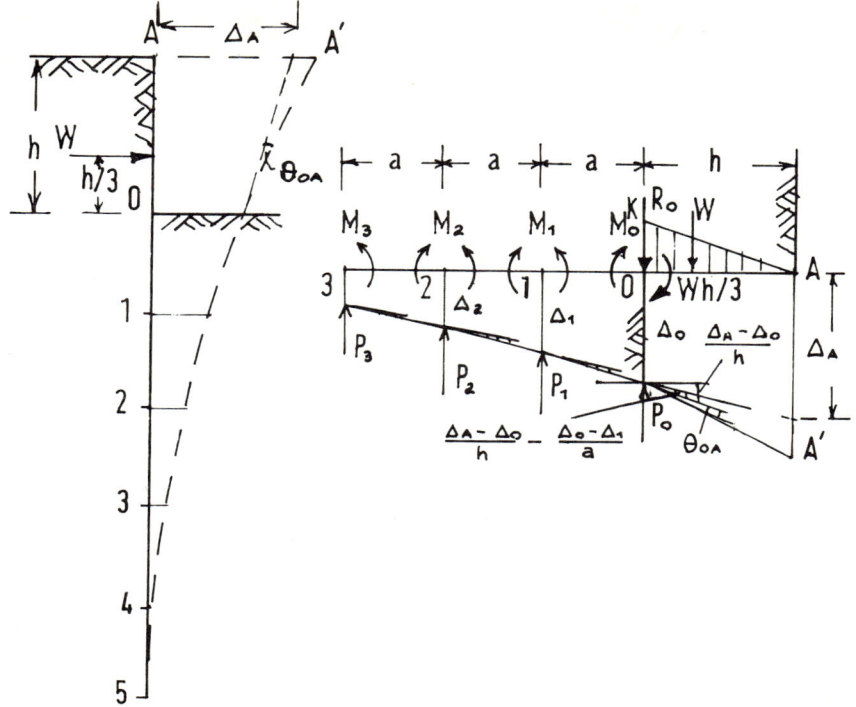

Fig. 12.1.

depicted in the deformation diagram are similar to those in Fig. 11.2 for a laterally loaded pile with the addition of an angular change at 0 due to active earth pressure. Much the same equations may be used to construct the matrix, therefore, as long as we include a provision for θ_{0A} at node 0. From first principles, the angular change on release at 0 with a triangular load on $0A = \theta_{0A} = \dfrac{2}{45} \dfrac{WL^2}{EI}$ (see Table 2.1 and equation 2.9c).

Since we are considering unit length of wall, then $b = 1$ and $c = a^2$. Applying moments M_0, M_1, etc. to eliminate the angular discontinuities we obtain equations at the nodes as follows:

A.0: $\dfrac{M_0}{3EI} h + \dfrac{M_0}{3EI} a + \dfrac{M_1}{6EI} a = \dfrac{\Delta_A - \Delta_0}{h} - \dfrac{\Delta_0 - \Delta_1}{a} + \theta_{0A}$

$2M_0 \left(\dfrac{h}{a} + 1 \right) + M_1 = \dfrac{6EI}{a} \left(\dfrac{\Delta_A}{h} - \dfrac{\Delta_0}{h} - \dfrac{\Delta_0}{a} + \dfrac{\Delta_1}{a} \right)$

$+ \dfrac{6EI}{a} \dfrac{2}{45} \dfrac{Wh^2}{EI}$

$$M_1 - \frac{6EI}{ah}\Delta_A + \frac{6EI}{ah}\Delta_0 + \frac{6EI}{a^2}\Delta_0 - \frac{6EI}{a^2}\Delta_1$$
$$= \frac{4Wh^2}{15a} - 2M_0\left(1 + \frac{h}{a}\right)$$

But $p = q\Delta$; $M_0 = \frac{Wh}{3}$; $g = \frac{6EI}{qa^2}$

$$M_1 - \frac{gaq}{h}\Delta_A + gp_0\left(1 + \frac{a}{h}\right) - gp_1 = \frac{4Wh^2}{15a} - \frac{2Wh}{3}\left(1 + \frac{h}{a}\right)$$

$$M_1 + gp_0\left(1 + \frac{a}{h}\right) - gp_1 - \frac{gaq}{h}\Delta_A = -\frac{2Wh}{15a}(5a + 3h)$$

A.1: $\dfrac{M_0}{6EI}a + \dfrac{2M_1}{3EI}a + \dfrac{M_2}{6EI}a = \dfrac{\Delta_0 - \Delta_1}{a} - \dfrac{\Delta_1 - \Delta_2}{a}$

$$4M_1 + M_2 - gp_0 + 2gp_1 - gp_2 = -M_0 = -\frac{Wh}{3}$$

A.2: $M_1 + 4M_2 + M_3 - gp_1 + 2gp_2 - gp_3 = 0$

E.0: $\dfrac{M_0 - M_1}{a} = P_0 - R_0 = \dfrac{a}{8}(3p_0 + p_1) - \dfrac{2}{3}W$

$$M_0 + \frac{2}{3}W - M_1 = \frac{3a}{8}p_0 + \frac{a}{8}p_1$$

$$4M_1 + \frac{3}{2}cp_0 + \frac{c}{2}p_1 = \frac{4W}{3}(h + 2)$$

E.1: $\dfrac{M_1 - M_0}{a} + \dfrac{M_1 - M_2}{a} = P_1 = \dfrac{a}{4}(p_0 + 2p_1 + p_2)$

$$2M_1 - M_2 - \frac{a^2}{4}p_0 - \frac{a^2}{2}p_1 - \frac{a^2}{4}p_2 = M_0 = \frac{Wh}{3}$$

$$cp_0 - 8M_1 + 2cp_1 + 4M_2 + cp_2 = -\frac{4}{3}Wh$$

E.2: $4M_1 - 8M_2 + 4M_3 + cp_1 + 2cp_2 + cp_3 = 0$, etc.

These equations are used to construct the matrix in Table 12.1. It will be seen, as might be expected from the deformation diagrams, that the coefficient matrix is exactly the same as that derived in the previous chapter for a pile subject to lateral loading; the load vector in this case reflects the varying intensity of loading over ground level.

Embedded length

In order to carry out the analysis of the sheet piled retaining wall, an embedded length has to be assumed or calculated initially. For preliminary

SHEET PILING AND STRUTTED ABUTMENTS

Table 12.1 Cantilevered sheet piling wall matrix

Equation	A — A	0 — p	0 — M	1 — p	1 — M	2 — p	2 — M	3 — p	3 — M	4 — p	4 — M	n−1 — p	n−1 — M	n — p	$\overline{W}$
A.0	$-\dfrac{qag}{h}$	$\left(1+\dfrac{a}{h}\right)g$		$-g$	1										$-\dfrac{2Wh}{15a}(3h+5a)$
E.0		$\dfrac{3}{2}c$		$\dfrac{c}{2}$	4										$\dfrac{4W}{3}(h+2)$
A.1		$-g$		$2g$	4	$-g$	1								$-\dfrac{Wh}{3}$
E.1		c		$2c$	-8	c	4								$-\dfrac{4Wh}{3}$
A.2				$-g$	1	$2g$	4	$-g$	1						0
E.2				c	4	$2c$	-8	c	4						0
A.3						$-g$	1	$2g$	4	$-g$	1				
E.3						c	4	$2c$	-8	c	4				
E.n												$\dfrac{c}{2}$	4	$\dfrac{3}{2}c$	0

design purposes, the factors given in Table 12.2 may be used as a guide. These give the amount by which the effective height of the retained soil, H, is to be multiplied to obtain the penetration length. Where appropriate, the effective height should include an allowance, H', for surcharge equal to w_s/w_e, where w_s = surcharge loading in kN/m² and w_e = unit weight of soil in kN/m³.

Table 12.2 Factors for embedded length

Angle of internal friction of soil (degrees)	Condition at head*		
	Free	Hinged	Fixed
30	1.4	1.0	0.8
35	1.2	0.8	0.6
40	0.85	0.6	0.58
45	0.8	0.56	0.55

* Free = cantilevered wall.
 Hinged = tie at top of wall.
 Fixed = tie below top of wall.

Table 12.2 applies to cohesionless soils only; where ground water is present, weepholes are assumed to prevent build-up of pressure behind the wall. The length of embedment may also be calculated by assuming a simplified pressure distribution below ground and a specified safety factor, as described in standard textbooks on the subject; this provides a rigid-body statics solution. Having decided on the pile length, we then have to select a pile section for which E and I values are known in order to determine the lateral pressures and moments.

The bending moments obtained in the computer run must then be checked against the assumed section modulus. Even if the stresses are satisfactory, a heavier pile section may be needed to suit adverse driving conditions, for instance. Also the computed lateral pressure should not exceed the permissible value for the soil in question. For equilibrium, the active pressure on the back of the wall must be balanced by the passive pressure, which may act both in front of and behind the wall. This should not exceed the theoretical value for the soil divided by a factor of safety, usually taken as 2.0.

Where there is doubt about the ability of the soil to mobilise the necessary passive resistance at ground level due to weathering, lack of compaction etc., it will be advisable to locate node 0 at, say, 0.5 m to 1 m below this level and to increase h accordingly. Alternatively, one should ensure that the calculated soil pressure is well below the estimated passive resistance of the soil by adopting a suitable safety factor.

Cantilever example

To illustrate a practical use of the method, let us assume a steel sheet piling section for which $I_x = 50 \times 10^{-6}$ m^4 per m run retaining a granular soil with properties $w_e = 18$ kN/m^3, coefficient of lateral reaction $q_H = 15\,000$ kN/m^3 and angle of internal friction $\phi = 30°$. If the height of the retained soil $h = 3$ m and the length of pile, $L = 11$ m (i.e. 8 m depth of penetration) then, assuming $E_S = 200 \times 10^6$ kN/m^2 and nodes at 1 m intervals below ground level, we have

$$g = \frac{6 \times 200 \times 50}{15\,000 \times 1} = 4; \quad \gamma = \frac{1}{3} \text{ so } W = \frac{18 \times 3^2}{3 \times 2} = 27$$

$$M_0 = \frac{Wh}{3} = 27 \text{ kNm}; \quad c = a^2 = 1; \quad \frac{gaq}{h} = 20\,000$$

$$\frac{2Wh}{15a}(3h + 5a) = \frac{2 \times 27}{5} 14 = 151.2$$

These values are reproduced in Table 12.3; results obtained by entering the figures as data statements in the MATIN program before running it are shown in the bottom line of the table and plotted in Fig. 12.2. It will be seen from the pressure distribution diagram that, for the given set of conditions, the depth of penetration could safely be reduced to about 5 m; a lighter and more economical pile section could be selected also.

Anchored sheet piling

The bending effects in the sheet piling can be reduced markedly by the use of tie rods. These are embedded at intervals in the retained soil at some distance, d, above the lower ground level and anchored at a distance B behind the sheet piling (Fig. 12.3). The anchorage is generally located beyond the potential slip plane generated by a line drawn from the foot of the retained soil at the angle of internal friction, ϕ, as shown. In cohesive soils, the anchorage should be located outside the critical slip circle and at a sufficient distance behind the sheet piling to develop a shear resistance equal to the ultimate design capacity of the anchorage.

The active earth pressure, W, is now resisted partly by the tension, T, in the tie bar and partly by the passive resistance of the soil as well, of course, as by bending in the sheet piling. The proportions resisted by each can be ascertained on a trial-and-error basis by assuming a certain force in the tie rod and consequently a given displacement in the sheet piling at that level. From this we can determine the pressure distribution below lower ground level and moments in the piling. The total passive

Table 12.3 Cantilevered sheet piling example

Equation	A Δ	0 p	1 p	1 M	2 p	2 M	3 p	3 M	4 p	4 M	5 p	5 M	6 p	6 M	7 p	7 M	8 p	8 M
A.0	−200000	5.333	−4	1														
E.0		1.5	0.5	4														
A.1		−4	8	4	−4	1												
E.1		1	2	−8	1	4												
A.2			−4	1	8	4	−4	1										
E.2			1	4	2	−8	1	4										
A.3					−4	1	8	4	−4	1								
E.3					1	4	2	−8	1	4								
A.4							−4	1	8	4	−4	1						
E.4							1	4	2	−8	1	4						
A.5									−4	1	8	4	−4	1				
E.5									1	4	2	−8	1	4				
A.6											−4	1	8	4	−4	1		
E.6											1	4	2	−8	1	4		
A.7													−4	1	8	4	−4	1
E.7													1	4	2	−8	1	4
E.8																0.5	4	1.5
W̄	−151.2	180	−27	−108	−9.7	9.9	−7.4	0.58	−2.55	−1.98	0.11	−1.44	0.71	−0.5	0.44	−0.05		0
MATIN N = 17	0.002	52.32	4.27	24.85														

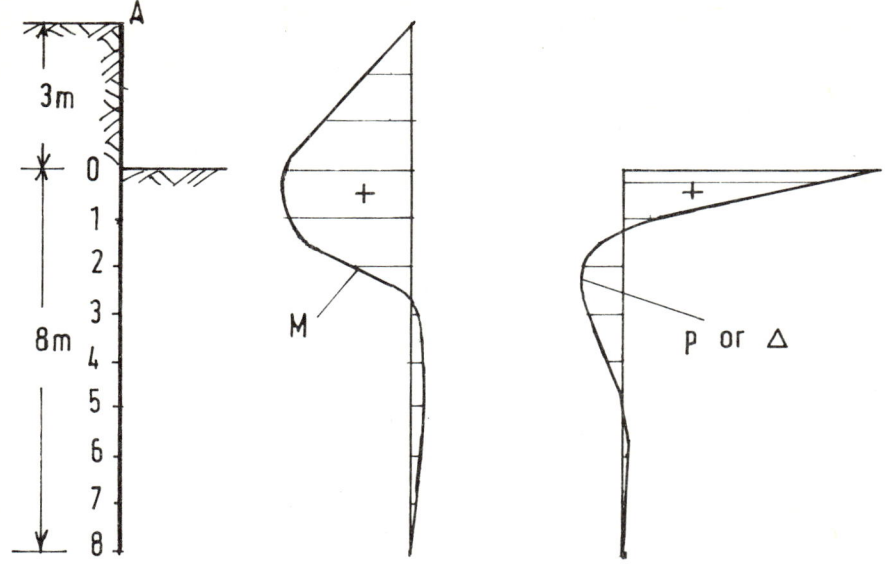

Fig. 12.2.

resistance can be calculated and checked against the assumed value. If necessary, a new tie-bar tension is selected that will meet the statical conditions and the process repeated until satisfactory agreement is achieved.

The tabular matrix for dealing with this problem can be derived in the usual way by considering the equations for angular compatibility and equilibrium of forces. Referring to Fig. 12.3, node A is unrestrained, so $M_A = 0$. Node B is restrained by the tie bar and may be treated as a simple end support with an applied cantilever moment from pile length AB equal to M_B. From equation (12.2),

$$M_B = \gamma \frac{w(h-d)^3}{6} \text{ per m length of wall}$$

At node 0, the angular change on release comprises the net horizontal displacement equal to $(\Delta_B - \Delta_0)/d - (\Delta_0 - \Delta_1)/a$ plus the rotational displacement due to the ground pressure on OB.

The average pressure on strip OB $= \gamma w \left(h - \dfrac{d}{2}\right)$

Therefore the total load on OB, say $P' = \gamma wd \left(h - \dfrac{d}{2}\right)$

$$= \frac{\gamma wd}{2}(2h - d)$$

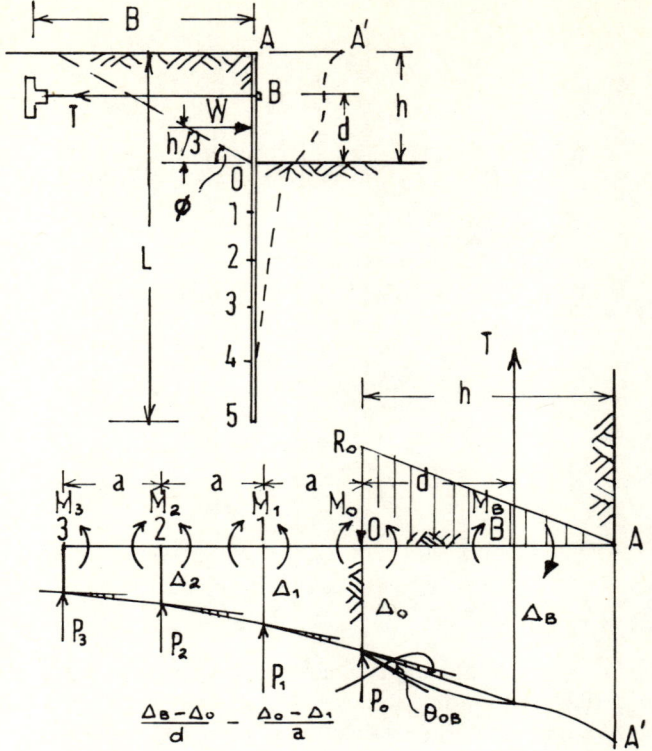

Fig. 12.3.

The angular change at 0 on release due to the distributed load on OB

$= \theta_{0B} = \dfrac{P^1 d^2}{24EI}$ (compare with equation (2.9a)).

Thus, the total angular change $= \dfrac{\Delta_B - \Delta_0}{d} - \dfrac{\Delta_0 - \Delta_1}{a} + \dfrac{P'd^2}{24EI}$

The relevant equations at nodes 0 and 1 are, therefore:

A.0: $\quad \dfrac{M_0}{3EI} d + \dfrac{M_0}{3EI} a + \dfrac{M_1}{6EI} a + \dfrac{M_B}{6EI} d = \dfrac{\Delta_B - \Delta_0}{d} - \dfrac{\Delta_0 - \Delta_1}{a}$

$$+ \dfrac{P'd^2}{24EI}$$

$$2M_0 \dfrac{d}{a} + 2M_0 + M_1 + M_B \dfrac{d}{a} = \dfrac{6EI}{a}\left(\dfrac{\Delta_B}{d} - \dfrac{\Delta_0}{d} - \dfrac{\Delta_0}{a} + \dfrac{\Delta_1}{a}\right)$$

$$+ \dfrac{P'd^2}{4a}$$

SHEET PILING AND STRUTTED ABUTMENTS

$$2M_0 \left(1 + \frac{d}{a}\right) + M_1 + gp_0 \left(1 + \frac{a}{d}\right) - gp_1 = \frac{gaq}{d} \Delta_B +$$
$$\frac{P'd^2}{4a} - M_B \frac{d}{a}$$

A.1: $\quad M_0 + 4M_1 + M_2 = \dfrac{6EI}{a^2} (\Delta_0 - 2\Delta_1 + \Delta_2)$

$\quad\quad M_0 - gp_0 + 4M_1 + 2gp_1 + M_2 - gp_2 = 0$

E.0: $\quad \dfrac{M_0 - M_B}{d} + \dfrac{M_0 - M_1}{a} = P_0 - R_0 = P_0 - \dfrac{P'}{2}$

$$= \frac{a}{8} (3p_0 + p_1) - \frac{P'}{2}$$

$\quad M_0 \dfrac{a}{d} - M_B \dfrac{a}{d} + M_0 - M_1 = \dfrac{a^2}{8} (3p_0 + p_1) - \dfrac{P'a}{2}$

$\quad 4M_0 \left(1 + \dfrac{a}{d}\right) - 4 \dfrac{a}{d} M_B - 4M_1 = \dfrac{3}{2} a^2 p_0 + \dfrac{a^2}{2} p_1 = \dfrac{3}{2} cp_0$
$$+ \frac{c}{2} p_1 + 2P'a$$

$\quad -4M_0 \left(1 + \dfrac{a}{d}\right) + 4M_1 + \dfrac{3}{2} cp_0 + cp_1 = -\dfrac{4a}{d} M_B + 2P'a$

E.1: $\quad \dfrac{M_1 - M_0}{a} + \dfrac{M_1 - M_2}{a} = P_1 = \dfrac{a}{4} (p_0 + 2p_1 + p_2)$

$\quad\quad 4M_0 + cp_0 - 8M_1 + 2cp_1 + 4M_2 + cp_2 = 0$

Other equations are identical with those derived in the previous chapter for a pile subject to bending at the head. The equations are summarised in Table 12.4. It will be seen that the coefficients are the same as before, but additional terms appear in the $\overline{W}$ vector because of the extra restraint at B.

Anchored pile example

For simplicity, assume the same sheet piled wall as before, but with a lighter steel section for which $I_X = 20 \times 10^{-6}$ m^4 per m wall, with tie bars 8 m long at 6 m intervals located 1 m below top ground level, i.e. $d = 2$ m (Fig. 12.4). Soil properties, etc. have the same values as before.

The centre of passive resistance will occur at some depth x, below lower ground level. For a first trial we might assume, from an inspection of Fig. 12.2, an initial value of $x = 1$ m. Taking moments for a 6 m length of wall, where T = force in the tie bar,

Table 12.4 Anchored sheet pile

Equation	0		1		2		3		N−1		N	$\overline{W}$
	p	M	p	M	p	M	p	M	p	M	p	
A.0	$\left(1+\dfrac{a}{d}\right)g$	$2\left(1+\dfrac{d}{a}\right)$	$-g$	1								$\dfrac{gaq}{d}\Delta_B + \dfrac{P'd^2}{4a}$ $-M_B\dfrac{a}{d}$
E.0	$\dfrac{3}{2}c$	$-4\left(1+\dfrac{a}{d}\right)$	$\dfrac{c}{2}$	4								$-4M_B\dfrac{a}{d}+2P'a$
A.1	$-g$	1	$2g$	4	$-g$	1						0
E.1	c	4	$2c$	-8	c	4						0
A.2			$-g$	1	$2g$	4	$-g$	1				0
E.2			c	4	$2c$	-8	c	4				0
A.n−1								1	$2g$	4	$-g$	0
E.n−1								4	$2c$	-8	c	0
E.n									$\dfrac{c}{2}$	4	$\dfrac{3}{2}c$	0

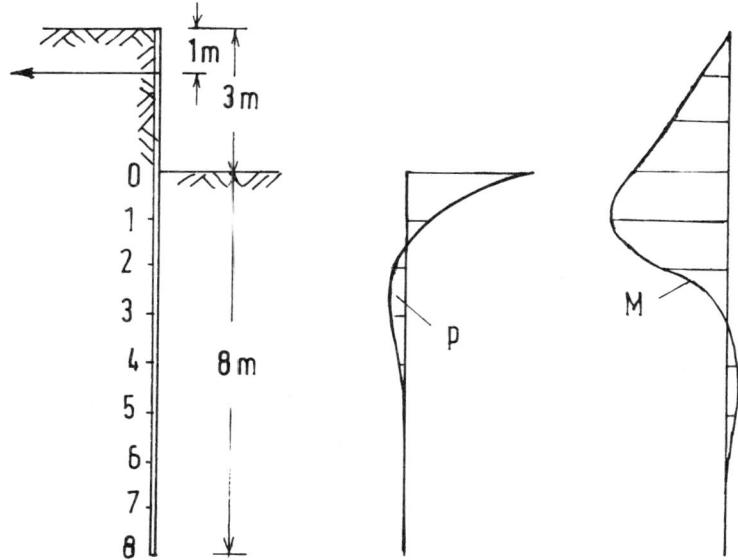

Fig. 12.4.

$$T \times 3 = 27 \times 6 \times 2$$
$$T = 108 \text{ kN}$$

For this load select a 16 mm diameter HYS tie bar, for which the extension will be equal to $(108 \times 8)/(200 \times 10^6 \times 202 \times 10^{-6}) = 0.0214$ m. The tie-bar force is transferred to the pile through a stiff horizontal waler, so we can take $\Delta_B = 0.0214$ m. The moment, M_B, due to active pressure on AB is given by

$$M_B = \frac{18 \times (3-2)^3}{3 \times 6} = 1 \text{ kNm/m}$$

The load on OB,

$$P' = \frac{18 \times 2}{3 \times 2}(6 - 2) = 24 \text{ kN/m}$$

As before, $M_0 = 27$ kNm, $g = 4$ and $a = 1 = c$. We can now evaluate the pressures and moments in the sheet piling resulting from a horizontal displacement of 21.4 mm at 2 m above 0, using the terms in Table 12.4. In this case,

$$\frac{gaq}{d}\Delta_B = 10\,000 \times 0.0214 = 214; \quad \frac{P'd^2}{4a} = 24; \quad 2P'_a = 48$$

and $\quad M_B \dfrac{d}{a} = 2 = 4M_B \dfrac{a}{d}$

Table 12.5 Anchored sheet piling example

Equation	0 p	0 M	1 p	1 M	2 p	2 M	3 p	3 M	4 p	4 M	5 p	5 M	6 p	6 M	7 p	7 M	8 p	8 M
A.0	6	6	−4	1														
E.0	1.5	−6	0.5	4														
A.1	−4	1	8	4	−4	1												
E.1	1	4	2	−8	1	4												
A.2			−4	1	8	4	−4	1										
E.2			1	4	2	−8	1	4										
A.3					−4	1	8	4	−4	1								
E.3					1	4	2	−8	1	4								
A.4							−4	1	8	4	−4	1						
E.4							1	4	2	−8	1	4						
A.5									−4	1	8	4	−4	1				
E.5									1	4	2	−8	1	4				
A.6											−4	1	8	4	−4	1		
E.6											1	4	2	−8	1	4		
A.7													−4	1	8	4	−4	1
E.7													1	4	2	−8	1	
E.8															0.5	4	1.5	
$\overline{W}$	236	46																
MATIN N=17	31.9	9.21	5.82	12.63	−3.79	6.11	−3.89	1.0	−1.65	−0.8	−0.15	−0.76	0.31	−0.32	0.25	−0.05	0.05	−0.05

Results obtained by running the MATIN program with the data in Table 12.5 are shown in the bottom line of the table. These were used to construct the bending moment and pressure diagrams in Fig. 12.4. The total passive ground resistance is estimated from the pressure diagram, by 'averaging' the nodal values, at about 9 kN/m. To balance the horizontal forces, the tie-bar load should be 6 (27 − 9) = 6 × 18 = 108 kN. This agrees with the assumed force; if not, another computer run would have been required using a different tension in the anchorage.

Stresses in the sheet piling should be checked for maximum bending moment to see if a heavier section is needed. The maximum ground pressure needs to be compared with the allowable soil pressure. One might observe from the pressure diagram that the centre of pressure is about 1 m lower than assumed, so the value of T requires some adjustment. However, it will be found that a change of this order does not materially affect the tension in the tie rod.

Varying soil conditions

Although a homogeneous soil has been assumed in deriving the equations and tables in this chapter, varying soil or groundwater conditions can be dealt with simply by inserting the appropriate values for g and W where the changes occur. A nodal point should be located at or near the position where the soil properties alter or where the groundwater causes a change in conditions.

For cohesionless soils, the Coulomb distribution, taking wall friction into account, may be preferred to the Rankine method. Cohesive soils may be analysed using Bell's equations, but uncertainties associated with clays demand considerable care. Tension cracks may occur at the surface in which hydrostatic water pressure can develop. Similarly, shrinkage of clay from the wall or sheet piling can permit ingress of water with similar effect. To avoid these contingencies, it is usual to backfill against the sheet piling with granular material. In this event, the Rankine distribution assumed earlier is quite suitable.

Load factors or factors of safety should be included in the analysis to conform with current code requirements. An alternative approach is to increase the depth of penetration by some arbitrary amount for safety. This may be advisable in any case to guard against future lowering or disturbance of the dredge line (the lower ground level). Any surcharge on the retained soil should be included in the analysis, of course (see note on effective height, under *Embedded Length*).

Diaphragm walls

Diaphragm retaining walls are formed by driving contiguous concrete piles or by placing in situ concrete in deep trenches, often stabilised by bentonite slurry after excavation. Cantilever diaphragm walls are usually sufficiently slender to yield under pressure and permit full active earth pressure to develop behind the wall. The same can be said of most conventional concrete cantilever walls, indeed. Backfill to cantilever walls is usually subject to vibration during construction; provided the fill is compacted carefully in layers, the active pressure is considered appropriate in checking ultimate stability.

Diaphragm walls may be analysed in the same way as described previously for sheet piling. Likewise, diaphragm walls restrained by steel bars or stressed cables may be investigated on the same lines as described for anchored sheet piling. Note that for walls of rectangular cross section and thickness t, the factor $g = Et^3/2a^2 q$.

Alternative analytical methods

Design of anchored sheet piling has been carried out for many years based either on the 'free earth' or 'fixed earth' theories. The former assumes that the piling is rigid and rotates about the anchor rod level. Penetration is assumed to be sufficient for the passive resistance to prevent forward movement of the toe but not sufficient to prevent rotation. The design is considered satisfactory when sufficient depth of penetration is provided to give an adequate safety margin against failure by rotation. Passive resistance is supplied by the soil in front of the piling below dredge level. The active pressure on the back of the sheet piling is usually taken to act over its full height.

Flexure of the sheeting can result in considerable redistribution of pressure and the free earth design values tend to be unduly conservative on this account. A moment reduction method was proposed by Rowe to secure more realistic values, but a similar effect can be obtained by assuming fixity to occur at the base of the sheet piling with a point of contraflexure at some distance below the dredge level. In this case, penetration is assumed sufficient to prevent both lateral movement and rotation at the toe. This 'fixed earth' approach has largely been superseded, however, by finite element methods using much the same principles as those described in this chapter. The properties of the sheet piling must be known or assumed, of course, when undertaking the analysis, so the classical methods, for which this is not a requirement, are commonly used to obtain an initial trial section.

Strutted abutments

Strutted or vertical beam abutments have been used for many small bridges in which the deck acts as a strut to support the top of the abutment. Since the deck has to react against each abutment in the span, no expansion joint is permissible. This limits the span to about 15 m; in addition, the span should not have a skew in excess of about 20°.

As the abutment is supported by the deck at the top and by the foundation at the bottom, it can be made narrower and so becomes more economical than the equivalent free-standing cantilever wall. The lateral support is particularly useful where conditions render installation of raking piles difficult. However, problems can arise during construction since backfilling cannot be placed until the deck is in position, unless temporary strutting is used. Backfilling has to be carried out carefully and done simultaneously behind both abutments. Mainly on this account, the popularity of vertical beam abutments has declined in recent years, although still used occasionally in small bridges with precast deck units and where sufficient working space is available on site.

Since the abutment is propped top and bottom, very little movement can take place, so either 'at rest' earth pressure or the trapezoidal pressure distribution recommended for strutted excavations in Chapter 13 is appropriate for design purposes. The wall itself may be treated as a propped cantilever, since a stiff foundation is normally provided for stability. Apart from retaining the fill and transmitting vertical reactions from the bridge deck to the foundations, vertical beam abutments have to cope with horizontal loads caused by the braking or acceleration of vehicles on the deck. If the deck is simply supported then these longitudinal forces

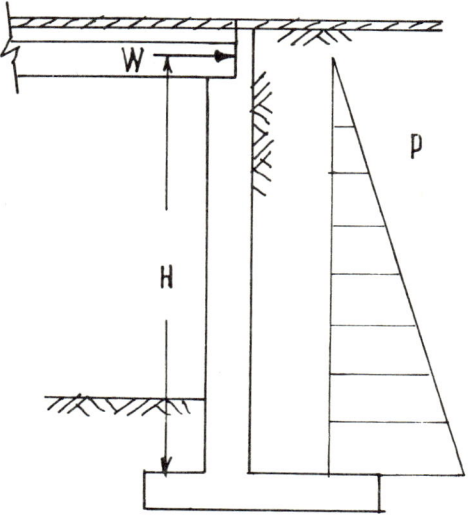

Fig. 12.5.

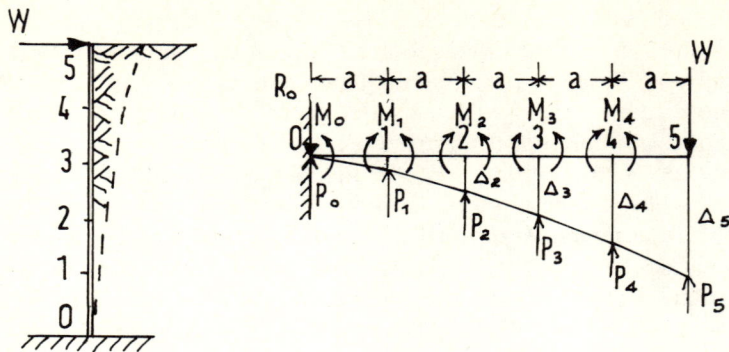

Fig. 12.6.

will have to be absorbed by the abutments reacting against the backfilling, as shown in Fig. 12.5. This is the equivalent of a cantilever supported on an elastic medium and subject to a line load at its extremity, as shown diagrammatically in Fig. 12.6. The deformation diagram is drawn after rotation through 90°. Release moments are applied at node 0 (considered fixed) but not at the end node, as the wall is treated as 'free ended' at the top. From a study of the deformation diagram the following equations may be formulated:

A.0: $\dfrac{M_0}{3EI} a + \dfrac{M_1}{6EI} a = \dfrac{\Delta_1}{a}$

$2M_0 + M_1 - gp_1 = 0$

A.1: $\dfrac{M_0}{6EI} a + \dfrac{2M_1}{3EI} a + \dfrac{M_2}{6EI} a = \dfrac{\Delta_2 - \Delta_1}{a} - \dfrac{\Delta_1}{a}$

$M_0 + 4M_1 + M_2 + 2gp_1 - gp_2 = 0$

A.2: $M_1 + 4M_2 + M_3 - gp_1 + 2gp_2 - gp_3 = 0$

E.0: $\dfrac{M_0 - M_1}{a} = R_0 - P_0 = W - (P_0 + P_1 + P_2 + P_3 + P_4 + P_5)$

$= W - \left(p_0 \dfrac{a}{2} + p_1 a + p_2 a + p_3 a + p_4 a + p_5 \dfrac{a}{2} \right)$

$M_0 - M_1 + \dfrac{c}{2} p_0 + cp_1 + cp_2 + cp_3 + cp_4 + \dfrac{c}{2} p_5 = Wa$

E.1: $\dfrac{M_1 - M_0}{a} + \dfrac{M_1 - M_2}{a} = P_1 = \dfrac{a}{4} (p_0 + 2p_1 + p_2)$

$4M_0 - 8M_1 + 4M_2 + cp_0 + 2cp_1 + cp_2 = 0$, etc.

The equations are summarised in Table 12.6. The way in which this table is used may be illustrated by the example in Fig. 12.7. The soil behind the strutted abutment is assumed to have the properties $w_e = 20$ kN/m³, $\phi = 30°$, $q_H = 12\,000$ kN/m³ and concrete to have $E_c = 30 \times 10^6$ kN/m². The wall is divided into five sections, so $a = 1$ and $g = (30 \times 10^6 \times 0.25^3)/2 \times 1 \times 12\,000 = 19.53$. The analysis shows the effect of a line load of 100 kN/m applied to the head of the wall.

Table 12.6 Strutted abutment

Equation	0 p	0 M	1 p	1 M	2 p	2 M	3 p	3 M	4 p	4 M	5 p	$\overline{W}$
E.0	$\dfrac{c}{2}$	1	c	−1	c		c		c		$\dfrac{c}{2}$	Wa
A.0		2	−g	1								0
E.1	c	4	2c	−8	c	4						0
A.1		1	2g	4	−g	1						0
E.2			c	4	2c	−8	c	4				0
A.2			−g	1	2g	4	−g	1				0
E.3					c	4	2c	−8	c	4		0
A.3					−g	1	2g	4	−g	1		0
E.4							c	4	2c	−8	c	0
A.4							−g	1	2g	4	−g	0
E.5							$\dfrac{c}{2}$	4	$\dfrac{3}{2}c$			4Wa

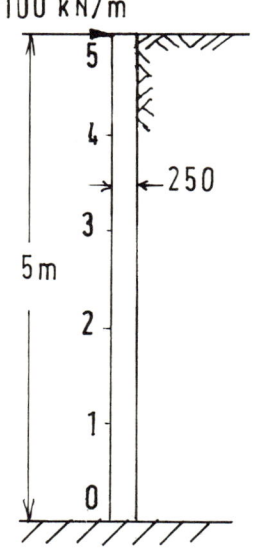
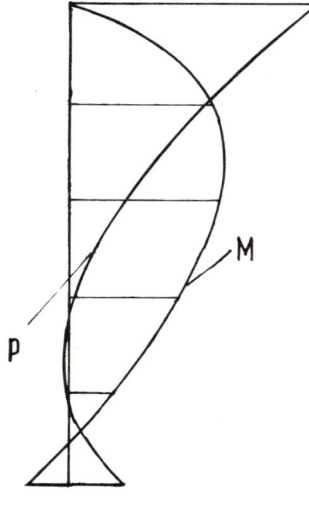

Fig. 12.7.

Appropriate values from Table 12.6 enable the data statements to be entered in MATIN which, when run, provides the results shown in the printout below. These have been used to plot the diagrams in Fig. 12.7 (results are shown to one or two decimal places only).

MATRIX SIZE? 11

DATA:−

0.5	1	1	−1	1	0	1	0	1	0	0.5
0	2	−19.53	1	0	0	0	0	0	0	0
1	4	2	−8	1	4	0	0	0	0	0
0	1	39.06	4	−19.53	1	0	0	0	0	0
0	0	1	4	2	−8	1	4	0	0	0
0	0	−19.53	1	39.06	4	−19.53	1	0	0	0
0	0	0	0	1	4	2	−8	1	4	0
0	0	0	0	−19.53	1	39.06	4	−19.53	1	0
0	0	0	0	0	0	1	4	2	−8	1
0	0	0	0	0	0	−19.53	1	39.06	4	−19.53
0	0	0	0	0	0	0	0	0.5	4	1.5

100
0
0
0
0
0
0
0
0
0
400

RESULTS:−

1	24.58
2	−16.32
3	−0.9
4	15.06
5	2.5
6	40.11
7	17.94
8	59.65
9	50.58
10	56.96
11	97.93

13 Shoring and braced excavations

Shoring and strutting are two operations commonly associated with excavation work and the construction of foundations. Underpinning or the reconstruction of buildings often demands use of temporary works including raking shores, flying shores and horizontal struts to prevent lateral movement or tilt during reconstruction. In the same way, vertical struts and props are used to avoid settlements taking place.

Temporary structures of this kind were usually designed by empirical methods in the past. However, with greater emphasis being placed nowadays on checking procedures and development of novel forms of construction, analytical assessments of stresses and strains are becoming more prevalent. The release-deformation approach can be used to investigate the behaviour of such temporary works, as described below.

Raking and flying shores

Raking shores provide support to defective or dangerous buildings and are also commonly installed when structural alterations are being carried out that might affect the stability of a structure. The purpose of the shores is to counteract any tendency of the wall to overturn or bulge outwards. The thrust likely to be imposed on a raking shore cannot be ascertained with any certainty, so empirical methods are commonly adopted. Nevertheless, analysis of the forces and displacements in the members based on realistic but conservative assumptions is to be encouraged. In regard to loads, the Code of Practice on Falsework (BS 5975) stipulates that raking and flying shores supporting masonry or brickwork structures should be capable of taking a horizontal thrust equal to 10% of the all-up weight above the head of the shore.

Figure 13.1 shows a system of raking shores. The base 0A is assumed fixed in position and the displacements of the bearing points 1, 2 and 3 are marked on the diagram assuming releases are inserted there. Equations for equilibrium of forces are designated E.1X, E.2X, etc. and those for linear displacement L.A1, L.A2, etc., as described previously in Chapter 2.

The braced structure (usually a masonry or concrete wall) will be extremely stiff in the vertical plane. If the building moves bodily as shown in Fig. 13.1 by tilting about the base, then the lateral displacement at any

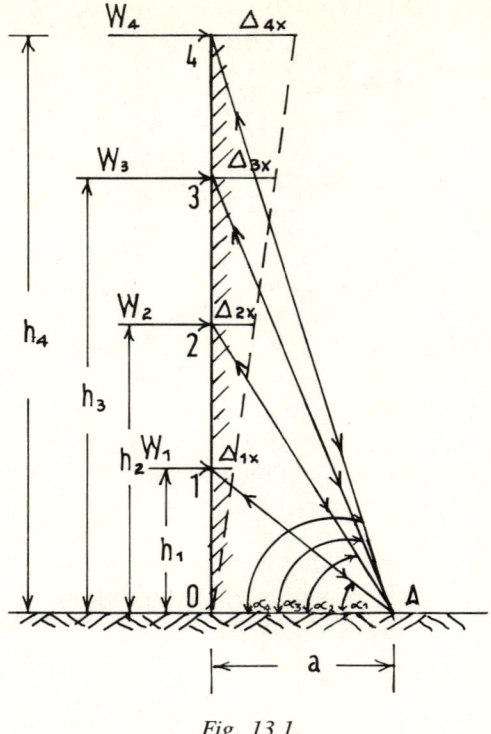

Fig. 13.1.

floor is proportional to the height over ground level. If Δ_{4X} is the horizontal displacement at roof level, then

$$\Delta_{3X} = \frac{h_3}{h_4}\Delta_{4X}, \quad \Delta_{2X} = \frac{h_2}{h_4}\Delta_{4X} \quad \text{and} \quad \Delta_{1X} = \frac{h_1}{h_4}\Delta_{4X}$$

The raking angles α_1, α_2, etc. are measured from the horizontal. As there is no horizontal member in the bracing system, the use of a reference element and relative flexibility factors is impractical. By putting $r = 1$ and using actual flexibility values for each member, the general compatibility equation for linear displacement derived in Chapter 2 becomes:

$$(\Delta_{1X} - \Delta_{2X}) \cos \alpha + (\Delta_{2Y} - \Delta_{1Y}) \sin \alpha - F_{12} S_{12} = 0$$

$$\text{where } S_{12} = \frac{L_{12}}{E_{12} A_{12}} \qquad \text{(compare with equation (2.3))}$$

For horizontal displacements only in a member with the opposite end fixed in position this reduces to

$$\pm \Delta_X \cos \alpha - F_{12} S_{12} = 0$$

The direction of the forces in the members is indicated by arrows in Fig.

Table 13.1 Raking shores

Equation	F				Δ	$\overline{W}$
	1A	2A	3A	4A	4X	
E.1X	cos α					W_1
E.2X		cos α				W_2
E.3X			cos α			W_3
E.4X				cos α		W_4
L.4A				−S	cos α	0

13.1. These help in determining the sign for the displacement equations, as explained in Chapter 2. In this case we have merely to find Δ_{4X}, the other lateral displacements being determinable from the equations above. Therefore, we need just one displacement equation, namely

L.4A: $\qquad \Delta_{4X} \cos \alpha_4 - F_{4A} S_{4A} = 0$

The equilibrium equations obtained by simple resolution of the forces are:

E.1X: $\qquad F_{1A} \cos \alpha_1 = W_1$
E.2X: $\qquad F_{2A} \cos \alpha_2 = W_2$
E.3X: $\qquad F_{3A} \cos \alpha_3 = W_3$
E.4X: $\qquad F_{4A} \cos \alpha_4 = W_4$

The equations are expressed in tabular form in Table 13.1, from which the forces and displacements may be determined.

Flying shores are used to form a bracing system between two buildings, usually when an intervening building is demolished. Figure 13.2 shows a very common arrangement comprising a horizontal strut with shorter inclined struts wedged at one end against the horizontal member and at the other against vertical wall plates. This construction is assumed to be loaded symmetrically so that we can solve the displacements and forces by considering one quarter of the frame only as depicted in the deformation diagram, Fig. 13.2(a). Bulging of the wall and vertical movement at the joints is provided for in the diagram; with symmetrical conditions, node A will remain fixed in position, so we can select 2A as the reference member for convenience. The two sets of equations for the quarter-frame are:

E.1X: $\qquad F_{1A} \cos \alpha = W_1$
E.2X: $\qquad F_{2A} = W_2$
E.1Y: $\qquad F_{1A} \sin \alpha - F_{12} = 0$
L.1A: $\qquad r\Delta_{1X} \cos \alpha - r\Delta_{1Y} \sin \alpha - F_{1A} S_{1A} = 0$
L.12: $\qquad r\Delta_{1Y} - F_{12} S_{12} = 0$
L.2A: $\qquad r\Delta_{2X} - F_{2A} S_{2A} = 0$

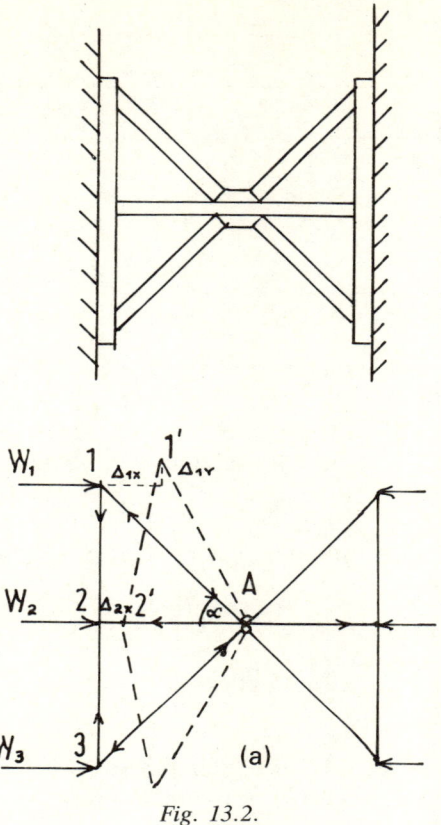

Fig. 13.2.

It will be seen that, in order to solve for both forces and displacements, we need to know at least the E and A values for the reference element and the comparative flexibility factors for the other members as well, of course, as the geometry of the frame. To see how the tables work in practice, consider the problem of a flying shore of the type shown in Fig. 13.2, with bracing walls 6 m apart, in which the timber members are of uniform 150 × 50 mm size with $E_T = 10$ kN/mm^2. Inclined members are at 45° and the applied loads are $W_1 = W_3 = 100$ kN and $W_2 = 200$ kN. Thus, with 2A as the reference member,

$$r = \frac{10 \times 150 \times 50}{3000} = 25 \text{ and}$$

$$S_{1A} = \frac{4243}{3000} = 1.414$$

Although this problem can be solved manually without much difficulty, we shall assume use of the MATIN program, for which the following data statements will apply:

Table 13.2 Quarter matrix for flying shore

Equation	F			Δ			$\overline{W}$
	1A	2A	12	1X	1Y	2X	
E.1X	cos α						W_1
E.2X		1					W_2
E.1Y	sin α		−1				0
L.1A	−S			r cos α	−r sin α		0
L.12			−S		r		0
L.2A		−S				r	0

```
1000 DATA 0.7071,0,0,0,0,0
1005 DATA 0,1,0,0,0,0
1010 DATA 0.7071,0,-1,0,0,0
1015 DATA -1.414,0,0,17.68,-17.68,0
1020 DATA 0,0,-1,0,25,0
1025 DATA 0,-1,0,0,0,25
1030 DATA 100,200,0,0,0,0
```

Answers obtained when the program is run are:

> Forces: 141.42, 200, 100 kN
> Displacements: 15.3, 4, 8 mm

As usual, these are printed out in the same order as the headings to the tabular matrix, Table 13.2.

Trench sheeting and cofferdams

Figure 13.3 shows a cutting supported by vertical sheeting, horizontal walings and transverse struts. The pressure on the struts is affected by the method of excavation and type of soil. In a clay soil, the normal practice is to excavate to a certain safe depth, d, whereupon the first strut is inserted and wedged tightly in position. The pressure distribution in the trench sheeting over this depth is triangular, as shown in Fig. 13.3. As further excavation proceeds, sheeting and strutting is done in stages; however, before the struts can be wedged tightly, the soil yields. The amount of yield is only slightly affected by the depth below ground level. As a result, struts are fairly uniformly loaded. From the lowest strut to the bottom of the excavation, the horizontal pressure reduces gradually to zero, aided by arching action in the soil. The overall result is the type of trapezoidal distribution on the sheeting shown in Fig. 13.3. This is a somewhat idealised version of actual pressures measured in field tests.

Terzaghi and Peck (1967) suggested from test results the pressure values shown in Fig. 13.4(a) as applicable to stiff or fissured clays. For

150 FOUNDATION AND STRUCTURAL PROBLEMS

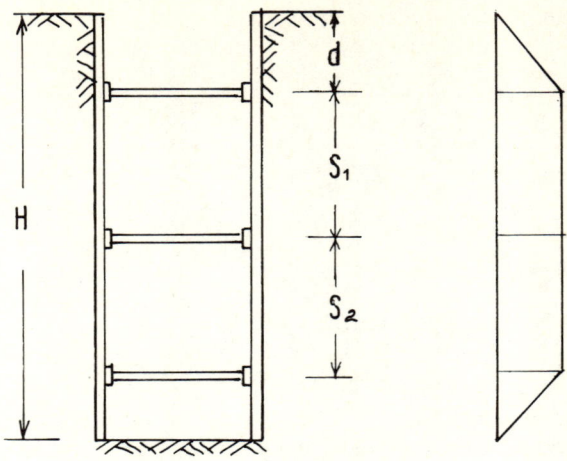

Fig. 13.3.

soft to medium clays, where arching action is less likely to occur, the distribution in Fig. 13.4(b) is recommended; for sands the pressure is assumed to be uniform, as shown in Fig. 13.4(c). These diagrams appear also in BS 6031 Code of Practice for Earthworks, where they are described as the envelopes for maximum load on the struts.

In the case of a sheeted trench or cofferdam in water, the normal arrangement is that the excavation is dewatered to just below the first

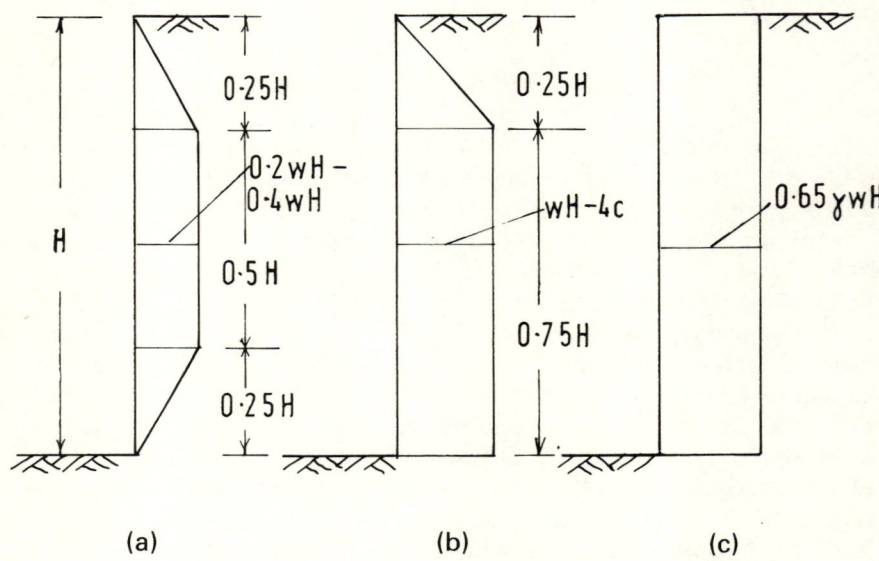

Fig. 13.4.

strut level and the strut and frame installed. This procedure is repeated at each successive strut level. Thus, the pressure distribution diagram is similar to the previous case, assuming the trench is kept dewatered by pumping when the excavation is complete. An allowance for seepage pressure may be desirable, however. If no dewatering is carried out, the lateral pressure at any depth will be equal to the hydrostatic pressure.

In clays, a useful guide is that the first brace should be located at a depth h, not exceeding that of the potential tension crack. This may be determined from the formula,

$$h = \frac{2c}{w \sqrt{(1 - \sin \phi)/(1 + \sin \phi)}}$$

where c = soil cohesion
       w = unit weight of soil
       ϕ = internal angle of friction of soil

Let us consider the equations for the case of a trench excavated in sand using braced sheeting, as shown in Fig. 13.5(a). The loading diagram for one half of the trench is shown in Fig. 13.5(b) and the assumed deformation diagram in Fig. 13.5(c). The cantilever load on AB produces a moment

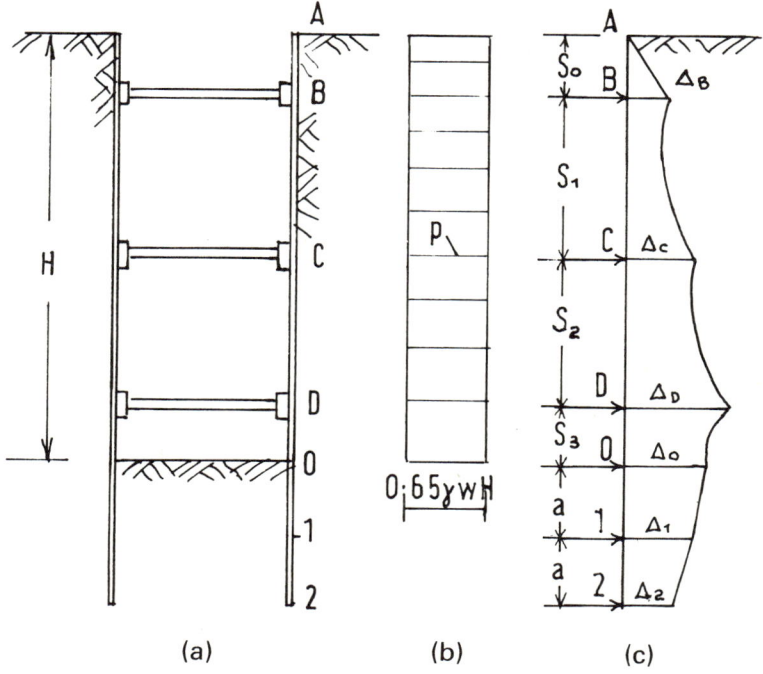

Fig. 13.5.

M_B at B, which is assumed pinned. Bending is induced in the sheeting between struts, B, C, D and node 0. Below excavation level, nodal deflections only are assumed with the ground pressure collected to each node in turn.

Moments are applied to eliminate the angular changes on release of joints C, D, 0, etc. as shown. Clearly the conditions for spans BC, CD and D0 are similar to those described in Chapter 6 for a continuous beam with support settlements. The conditions for spans 01, 12, etc. are more akin to those described in the previous chapter for sheet piling, so we can expect the equations to include both these cases. This is reflected in the deformation diagram divided into two separate figures in Fig. 13.6. Figure 13.6(a) shows the deformations due to bending only and Fig. 13.6(b) the angular changes due to the assumed horizontal displacement pattern. The arrows in Fig. 13.6(b) acting in an opposite direction to those in Fig. 13.6(a) are marked with a negative sign. The corresponding $(P/L)\Delta$ vector terms for these moments are given a negative sign in the angular compatibility equations, A.C and A.D.

$$\text{At C, angular change} = -\frac{\Delta_C - \Delta_B}{S_1} + \frac{\Delta_D - \Delta_C}{S_2}$$

$$\text{At D, angular change} = -\frac{\Delta_D - \Delta_C}{S_2} - \frac{\Delta_D - \Delta_0}{S_3}$$

$$\text{At 0, angular change} = \frac{\Delta_D - \Delta_0}{S_3} - \frac{\Delta_0 - \Delta_1}{a}$$

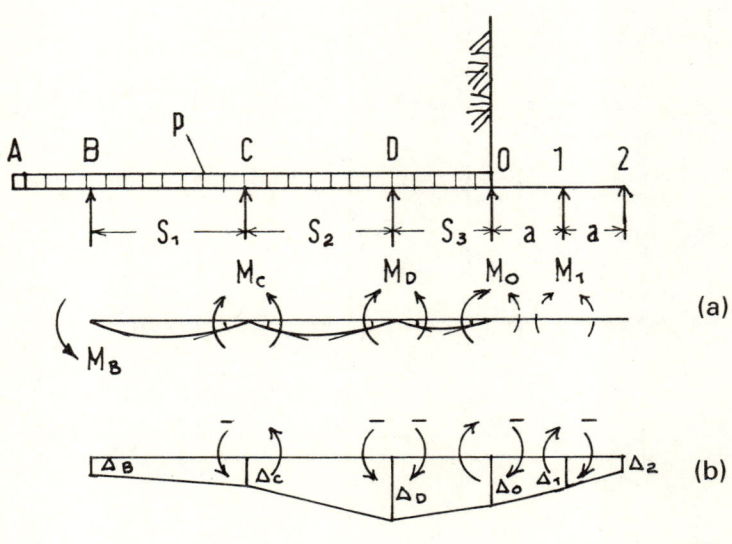

Fig. 13.6.

The terms S_1, S_2, etc. denote the strut spacings, as shown in Fig. 13.6. For a sandy soil, the uniform pressure distribution shown in Fig. 13.4(c) is assumed in which $p = 0.65 \gamma wH$. The earth pressure is transmitted via the struts spaced at intervals b along the trench to the walings and sheeting on the opposite face of the excavation. Strut loads are calculated directly from the pressure distribution diagram (Terzaghi and Peck, 1967). Struts and walings are considered as completely stiff and the sheeting as flexible and subject to bending and lateral displacement. These conditions are reproduced in the deformation diagram, Fig. 13.5(c), for unit length of sheeting. Below ground level the soil yields under pressure causing lateral displacement of the sheeting, as shown.

At the head of the sheeting, a cantilever moment M_B occurs at B in which $M_B = p_0 S_0^2/2 = 0.65 \gamma wH S_0^2/2$. The reference element is taken to be the trench sheeting length 01. This has a length a, elastic modulus E and moment of inertia, $I = I_a$. The relevant equations are:

A.C: $$\frac{M_B}{6EI} S_1 + \frac{M_C}{3EI} S_1 + \frac{M_C}{3EI} S_2 + \frac{M_D}{6EI} S_2 = \theta_{CB} + \theta_{CD} - \frac{\Delta_C - \Delta_B}{S_1}$$
$$+ \frac{\Delta_D - \Delta_C}{S_2}$$

$$K_1 M_B + (2K_1 + 2K_2) M_C + K_2 M_D = \frac{1}{4} K_1 W_1 S_1 + \frac{1}{4} K_2 W_2 S_2 +$$
$$P \left(\frac{\Delta_B}{S_1} - \frac{\Delta_C}{S_2} - \frac{\Delta_C}{S_1} + \frac{\Delta_D}{S_2} \right)$$

where $K_1 = \frac{I_a S_1}{aI_1}$; $K_2 = \frac{I_a S_2}{aI_2}$, etc; $P = \frac{6EI_a}{a}$

W_1, W_2, etc. = total loads on spans S_1, S_2, etc.

A.D: $$\frac{M_C}{6EI} S_2 + \frac{M_D}{3EI} S_2 + \frac{M_D}{3EI} S_3 + \frac{M_0}{6EI} S_3 = \theta_{DC} + \theta_{D0} - \frac{\Delta_D - \Delta_C}{S_2}$$
$$- \frac{\Delta_D - \Delta_0}{S_3}$$

$$K_2 M_C + (2K_2 + 2K_3) M_D + K_3 M_0 = \frac{1}{4} K_2 W_2 S_2 + \frac{1}{4} K_3 W_3 S_3 +$$
$$P \left(\frac{\Delta_C}{S_2} - \frac{\Delta_D}{S_3} - \frac{\Delta_D}{S_2} + \frac{\Delta_0}{S_3} \right)$$

$$K_2 M_C + 2(K_2 + K_3) M_D + K_3 M_0 = \frac{1}{4} (K_2 W_2 S_2 + K_3 W_3 S_3) +$$
$$P \left(\frac{\Delta_C}{S_2} - \frac{\Delta_D}{S_3} - \frac{\Delta_D}{S_2} \right) +$$
$$\frac{ga}{S_3} p_0$$

where $g = \frac{6EI}{qa^2}$

A.0: $\dfrac{M_D}{6EI} S_3 + \dfrac{M_0}{3EI} S_3 + \dfrac{M_0}{3EI} a + \dfrac{M_1}{6EI} a = \theta_{0D} + \dfrac{\Delta_D - \Delta_0}{S_3} - \dfrac{\Delta_0 - \Delta_1}{a}$

$K_3 M_D + 2(1 + K_3) M_0 + M_1 = \dfrac{1}{4} K_3 W_3 S_3 + P \left(\dfrac{\Delta_D}{S_3} - \dfrac{\Delta_0}{a} - \dfrac{\Delta_0}{S_3} + \dfrac{\Delta_1}{a} \right)$

$K_3 M_D + 2(1 + K_3) M_0 + M_1 = \dfrac{1}{4} K_3 W_3 S_3 + P \dfrac{\Delta_D}{S_3} - g \left(1 + \dfrac{a}{S_3}\right) p_0 + g p_1$

A.1: $\dfrac{M_0}{6EI} a + \dfrac{2 M_1}{3EI} a = \dfrac{\Delta_1 - \Delta_0}{a} - \dfrac{\Delta_2 - \Delta_1}{a}$

$M_0 + 4 M_1 = -g p_0 + 2 g p_1 - g p_2$

E.0: $\dfrac{M_0 - M_D}{S_3} + \dfrac{M_0 - M_1}{a} = P_0 = \dfrac{a}{8}(3 p_0 + p_1)$

$M_0 \left(1 + \dfrac{a}{S_3}\right) - M_D \dfrac{a}{S_3} - M_1 = \dfrac{3 a^2}{8} p_0 + \dfrac{a^2}{8} p_1$

E.1: $\dfrac{M_1 - M_0}{a} + \dfrac{M_1}{a} = P_1 = \dfrac{a}{4}(p_0 + 2 p_1 + p_2)$

$2 M_1 - M_0 = \dfrac{a^2}{4} p_0 + \dfrac{a^2}{2} p_1 + \dfrac{a^2}{4} p_2$

E.2: $-\dfrac{M_1}{a} = P_2 = \dfrac{a}{8}(p_1 + 3 p_2)$

$-M_1 = \dfrac{a^2}{8} p_1 + \dfrac{3 a^2}{8} p_2$

These equations are embodied in Table 13.3.

Cofferdam example

Consider the example in Fig. 13.7 in which $S_0 = 1.5$, $S_1 = 3$, $S_2 = 3$ and $S_3 = 2$; i.e. depth of excavation $= 9.5$ m. If the sheeting is driven 2 m below excavation level and nodes 0, 1 and 2 are at 1 m intervals, then $a = 1$. Soil properties for sand are taken as $w = 16.5$ kN/m^3, $\phi = 30°$ and $q = 10\,000$ kN/m^3. Steel sheeting is selected for which $E = 200 \times 10^6$ kN/m^2 and $I = 22 \times 10^{-6}$ m^4. Bracings are to be inserted in the trench at 4 m intervals.

Table 13.3 Trench sheeting

Equation	M				p			$\overline{W}$	$\overline{\Delta}/P$
	C	D	0	1	0	1	2		
A.C	$2(K_1 + K_2)$	K_2						$\frac{1}{4}(K_1W_1S_1 + K_2W_2S_2) - K_1M_B$	$\frac{\Delta_B}{S_1} + \frac{\Delta_D}{S_2} - \Delta_C\left(\frac{1}{S_1} + \frac{1}{S_2}\right)$
A.D	K_2	$2(K_2 + K_3)$	K_3		$-\frac{ga}{S_3}$			$\frac{1}{4}(K_2W_2S_2 + K_3W_3S_3)$	$\frac{\Delta_C}{S_2} - \Delta_D\left(\frac{1}{S_2} + \frac{1}{S_3}\right)$
A.0		K_3	$2(1+K_3)$	1	$g\left(1+\frac{a}{S_3}\right)$	$-g$		$\frac{1}{4}K_3W_3S_3$	$\frac{\Delta_D}{S_3}$
A.1			1	4	g	$-2g$	g		
E.0		$-\frac{a}{S_3}$	$1+\frac{a}{S_3}$	-1	$\frac{3a^2}{8}$	$\frac{a^2}{8}$			
E.1			-1	2	$\frac{a^2}{4}$	$\frac{a^2}{2}$	$\frac{a^2}{4}$		
E.2				-1	$\frac{a^2}{8}$	$\frac{a^2}{8}$	$\frac{3a^2}{8}$		

156   FOUNDATION AND STRUCTURAL PROBLEMS

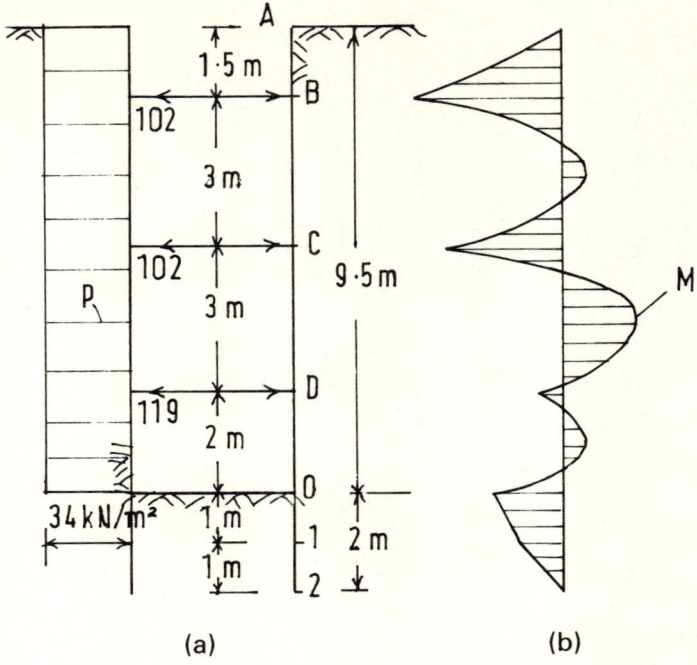

Fig. 13.7.

Horizontal soil pressure $= 0.65 \times 16.5 \times 9.5 \times \dfrac{1 - \sin \phi}{1 + \sin \phi} = 34$ kN/m²

$$P = \frac{6EI}{a} = \frac{6 \times 200 \times 10^6 \times 22 \times 10^{-6}}{1} = 26\,400; \quad g = \frac{6EI}{qa^2} = 2.64$$

$K_1 = K_2 = 3; \quad K_3 = 2; \quad M_B = \dfrac{pS_0^2}{2} = 38.25; \quad K_1 M_B = 114.75$

From the pressure diagram, Fig. 13.7(a), the strut loads from 1 m length of wall can be determined as 102, 102 and 119 respectively. Using the relationship $\Delta =$ Strut load/q, the lateral displacements are $\Delta_B = \Delta_C = 0.0102$ and $\Delta_D = 0.0119$. These values may now be used to obtain the Δ vector terms. The data statements used to solve this problem using the SBEAM program are shown below with the answers. Only a single value has to be input when the program is run, namely N = 7.

```
1000 DATA 12,3,0,0,0,0,0
1005 DATA 3,10,2,0,−1.32,0,0
1010 DATA 0,2,6,1,3.96,−2.64,0
1015 DATA 0,0,1,4,2.64,−5.28,2.64
1020 DATA 0,−0.5,1.5,−1,−0.375,−0.125,0
1025 DATA 0,0,−1,2,−0.25,−0.5,−0.25
```

1030 DATA 0,0,0,−1,0,−0.125,−0.375
1035 DATA 344.25,297.5,68,0,0,0,0
1040 DATA 15.05,−171.6,157.08,0,0,0,0

Results:	23.51	20.71	6.62	1.71	−5.04	−1.93	−3.92
	5.30	−16.16	11.33	9.70	36.75	12.81	−30.12
	28.81	4.55	17.95	11.41	31.71	10.88	−34.04

These results give nominal values for the design of the cofferdam sheeting, the maximum bending moment being 38.25 kNm at B. The maximum strut load = 4 [102 + (28.81 − 38.25)/3 + (28.81 − 4.55)/3] = 428 kN. The maximum soil pressure at the toe of the sheeting is 34 kN/m². This is low for this situation but braced excavations and cofferdams in soft clay strata should be checked for possible 'heave' resulting from the flow of clay beneath the sheeting into the excavation. This can be analysed as an incipient bearing failure and the factor of safety against heave checked as described by Terzaghi and Peck (1967).

Bottom failure can also occur due to 'piping' caused by water flowing into the excavation beneath the sheeting under excess hydrostatic pressure. This may be remedied by driving the sheeting deeper or by use of a loaded filter layer, for example.

Strutted retaining walls

Figure 13.8 shows a vertical flexible retaining wall supported against lateral forces P_1, P_2, etc. by raking struts. Unlike the cantilever wall discussed in Chapter 12 the struts prevent yielding despite the flexibility

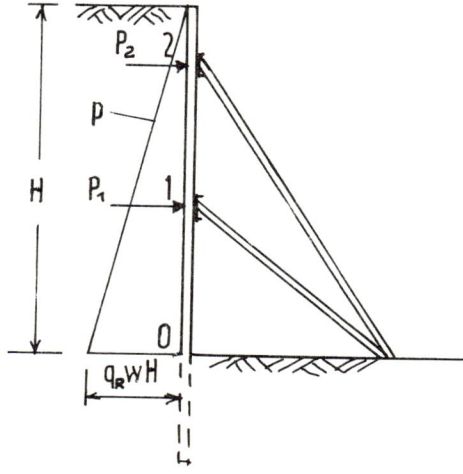

Fig. 13.8.

of the wall, so active earth pressure is not developed. Instead, the wall is acted upon by the 'at rest' earth pressure which is dependent on, among other factors, the method of placing the fill material. For compacted granular material

$$p = q_r wh$$

where q_r = coefficient of earth pressure at rest.

For dense sand, Terzaghi suggested a value of $q_r = 0.4$, which is a commonly adopted value (Reynolds and Steedman, 1981). The wall may be analysed in much the same way as the strutted abutment described in the previous chapter.

14 Formwork and falsework

Strutted formwork

The form of construction shown in Fig. 14.1 represents a typical framed construction of vertical posts and raking struts supporting formwork to a vertical concrete wall. The pressure exerted on the formwork during construction of the wall will depend on the rate of pour, constituents of the mix and temperature at the time of placing the concrete. These factors affect the time in which the concrete attains its initial set; this may vary from around one half-hour to less than 2 hours. Thus a fluid head of concrete will exist from the top to some distance, h, below. In this length

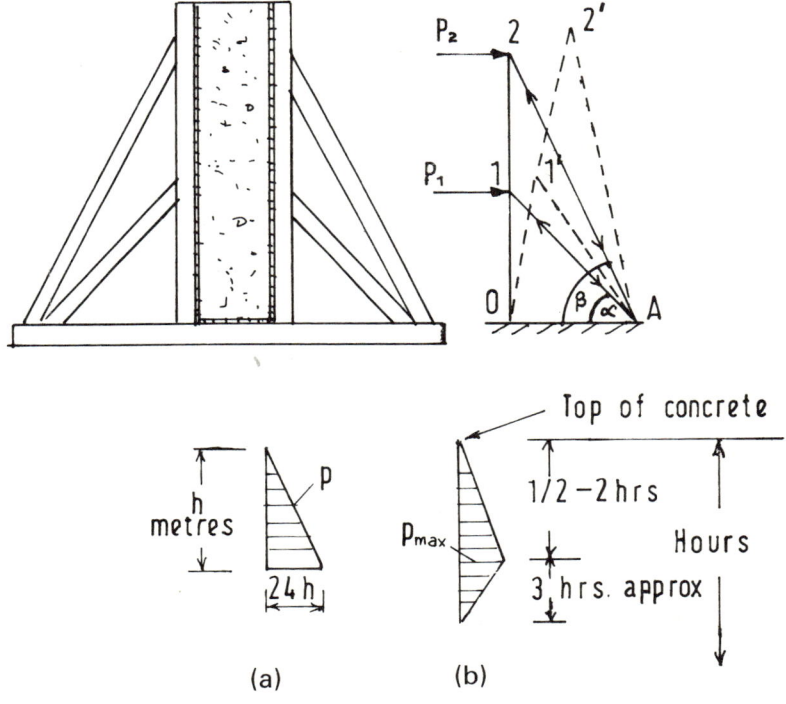

Fig. 14.1.

the pressure equals 24 kN/m² multiplied by the depth of the pour in metres. Below this level the concrete starts to solidify.

In the case of shallow members, therefore, where the pour is completed before initial set takes place, it is usual to consider the concrete as a fluid with a density of 24 kN/m³. The pressure distribution is triangular so that the pressure at any depth h = 24 h kN/m² (Fig. 14.1(a)).

Where the height of formwork exceeds the above, a different pressure diagram results, somewhat as shown in Fig. 14.1(b). This takes into account the fact that, because of the time taken to fill the forms, the concrete in the lower part will have attained its initial set. Thus the maximum pressure is less than that obtained by treating the concrete as a fluid. It should be noted that Fig. 14.1(b) represents the situation at a particular stage of construction. As the concrete is placed, the formwork at each level is subject to the maximum pressure in turn. The deflected formwork is then prevented from returning to its original position by the hardened concrete. Thus each portion of the formwork must be capable of catering for the maximum pressure.

Tables and charts have been published showing the maximum pressures that can be expected in different situations. Formwork pressure data sheets have been published by the CIRIA in Britain, for instance. A report in 1985 extended the scope of the data sheets to include concretes with admixtures and faster rates of placement (Clear and Harnism, 1985).

Once the formwork pressures have been decided, we can determine the forces acting at the strut positions. Forces and displacements at the joints may be ascertained by the release-deformation method in the usual way. Thus, for the simple frame shown in Fig. 14.1, consisting of raking struts and vertical members, formwork pressure is assumed to be applied at the nodal intersections. These forces are marked P_1 and P_2 in Fig. 14.1.

Equilibrium equations are formed by resolving the forces acting at the joints, and displacement equations by using the method and sign convention described in previous chapters. The shape of the frame with the joints released is shown as a dashed deformation diagram in Fig. 14.1. Note that, due to the forces in the struts, nodes 1 and 2 are displaced upwards. The arrowheads assist in determining the appropriate sign in the equations. Hence:

E.1X:	$F_{1A} \cos \alpha = P_1$
E.1Y:	$F_{1A} \sin \alpha - F_{10} + F_{12} = 0$
E.2X:	$F_{2A} \cos \beta = P_2$
E.2Y:	$F_{2A} \sin \beta - F_{12} = 0$
L.1A:	$r\Delta_{1X} \cos \alpha - r\Delta_{1Y} \sin \alpha - F_{1A} S_{1A} = 0$
L.10:	$r\Delta_{1Y} - F_{10} S_{10} = 0$
L.2A:	$r\Delta_{2X} \cos \beta - r \Delta_{2Y} \sin \beta - F_{2A} S_{2A} = 0$
L.21:	$-r\Delta_{1Y} + r\Delta_{2Y} - F_{21} S_{21} = 0$

The equations are summarised in Table 14.1.

FORMWORK AND FALSEWORK

Table 14.1 Strutted formwork

Equation	F				Δ				$\overline{W}$
	1A	10	2A	12	1X	1Y	2X	2Y	
E.1X	$\cos \alpha$								P
E.1Y	$\sin \alpha$	-1		1					0
E.2X			$\cos \beta$						P
E.2Y			$\sin \beta$	-1					0
L.1A	$-S$				$r \cos \alpha$	$-r \sin \alpha$			0
L.10		$-S$				r			0
L.2A			$-S$				$r \cos \beta$	$-r \sin \beta$	0
L.21				$-S$		$-r$		r	0

If we consider the example in Fig. 14.2 where the raked members have a cross-section area of 1000 mm² and the vertical posts 500 mm² and E = 10 kN/mm² for both, then using actual L/EA values so that r = 1 we have

$$S_{1A} = \frac{1414}{10 \times 10^3} = 0.1414, \quad S_{01} = S_{12} = \frac{500}{10 \times 10^3} = 0.05$$

and $S_{2A} = \dfrac{2236}{10 \times 10^3} = 0.2236$

The data statements and results obtained with MATIN are indicated below.

```
1000 DATA 0.7071,0,0,0,0,0,0,0
1005 DATA 0.7071,-1,0,1,0,0,0,0
1010 DATA 0,0,0.4473,0,0,0,0,0
1015 DATA 0,0,0.8945,-1,0,0,0,0
```

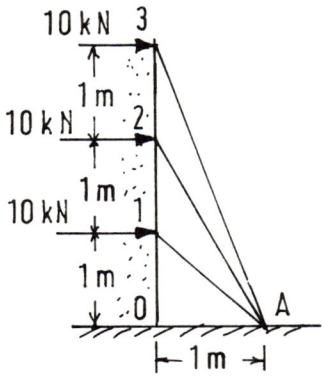

Fig. 14.2.

1020 DATA −0.1414,0,0,0,0.7071,−0.7071,0,0
1025 DATA 0,−0.05,0,0,0,1,0,0
1030 DATA 0,0,−0.2236,0,0,0,0.4473,−0.8945
1035 DATA 0,0,0,−0.05,0,−1,0,1
1040 DATA 10,0,10,0,0,0,0,0
Results: 14.1,   30,   22.36, 20 kN
         4.33, 1.5, 16.17,   2.5 mm

In other constructions, framework supporting the sheeting may consist of horizontal walings and continuous vertical studs held in position by tie rods and spacers, as shown in Fig. 14.3. In this case, the vertical studs may be analysed as a continuous beam system spanning between tie rods simply supported at each end and subject to the calculated concrete pressure averaged over each span. The procedure using the release-deformation method will be the same as described in Chapter 5 for continuous beams.

Falsework and centring

Figure 14.4 illustrates an example of raking struts used in falsework. Variations of this arrangement may be adopted; in each case the essential requirement to ensure a good quality finish in the concrete superstructure is that undue settlement does not occur during placing. This also ensures that the correct profile is imparted to the finished work. For this reason, the temporary foundations at A and B must be firm and the members selected so as to keep joint displacement below the specified value.

In the dashed deformation diagram in Fig. 14.4 loads are assumed to be applied at the nodal points so that bending is not considered. The arrows indicate the direction of the member forces for equilibrium. (An incorrect assumption will be shown up as a negative sign in the results, indicating that the assumed settlement pattern should be re-examined.) For the bridge falsework shown in the figure, the equations for one half trestle, assuming symmetrical loading and support conditions, are:

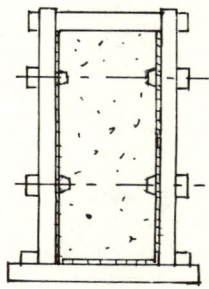

Fig. 14.3.

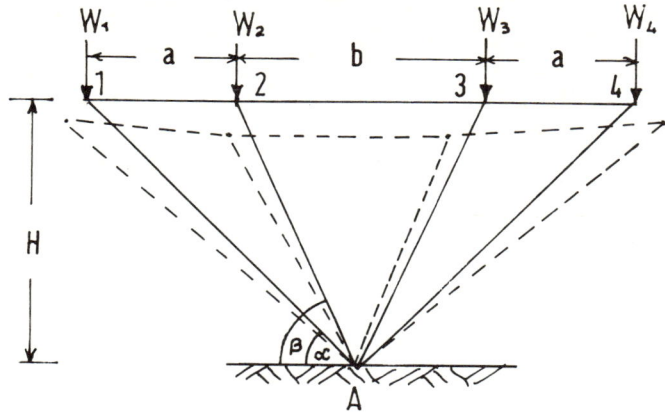

Fig. 14.4.

E.1Y: $\quad F_{1A} \sin \alpha = W_1$
E.1X: $\quad F_{1A} \cos \alpha - F_{12} = 0$
E.2Y: $\quad F_{2A} \sin \alpha = W_2$
E.2X: $\quad F_{2A} \cos \alpha + F_{12} - F_{23} = 0$
L.12: $\quad r\Delta_{1X} - F_{12}S_{12} = 0$
L.1A: $\quad r\Delta_{1X} \cos \alpha + r\Delta_{1Y} \sin \alpha - F_{1A}S_{1A} = 0$
L.23: $\quad r\Delta_{2X} - F_{23}S_{23}/2 = 0$
L.2A: $\quad -r\Delta_{2X} \cos \beta + r\Delta_{2Y} \sin \beta - F_{2A}S_{2A} = 0$

The equations are summarised in Table 14.2.

Taking the case where the load from the concrete formwork is 80 kN/m, $a = 3$ m, $b = 6$ m, $H = 8$ m and the falsework consists of tubular components with $E = 210$ kN/mm^2 and cross-sectional area $= 2500$ mm^2, then $W_1 = 120$ kN and $W_2 = 360$ kN. Selecting 12 as the reference member, then

Table 14.2 Trestle falsework

Equation	F				Δ				$\overline{W}$
	1A	12	2A	23	1X	1Y	2X	2Y	
E.1Y	sin α								W_1
E.1X	cos α	−1							0
E.2Y			sin α						W_2
E.2X		1	cos α	−1					0
L.12		−S			r				0
L.1A	−S				r cos α	r sin α			0
L.23				−$\frac{S}{2}$			r		0
L.2A			−S				−r cos β	r sin β	0

$S_{12} = 1$, $S_{1A} = 3.333$, $S_{23} = 2$, $S_{2A} = 2.85$ and $r = \dfrac{210 \times 2500}{3000} = 175$

From the given dimensions, $\alpha = 53°8'$ and $\beta = 69°27'$. The tabular matrix for solving the problem using MATIN and the results obtained are included in the printout below.

MATRIX SIZE?   8

DATA:−

```
  0.8      0     0      0    0    0      0       0
  0.6     −1     0      0    0    0      0       0
  0        0     0.8    0    0    0      0       0
  0        1     0.6   −1    0    0      0       0
  0       −1     0      0  175    0      0       0
 −3.333    0     0      0  105  140      0       0
  0        0     0     −1    0    0    175       0
  0        0    −2.85   0    0    0    −65.63  163.87
```

120
0
360
0
0
0
0
0

RESULTS:−

1   150
2   90
3   450
4   360
5   0.51
6   3.19
7   2.06
8   8.65

Displacements (in mm) are shown to two decimal places in items 5 to 8; member forces (in kN) are given in items 1 to 4.

Strutted beams

Several formwork systems incorporate strongbacks or telescopic floor centring that make use of the 'strutted beam' concept. In its original form, a heavy timber beam was braced against bending by stressed tie rods bearing against the ends of the beam and against one or more short cast iron struts. The single strut or 'king post' beam is shown in Fig.

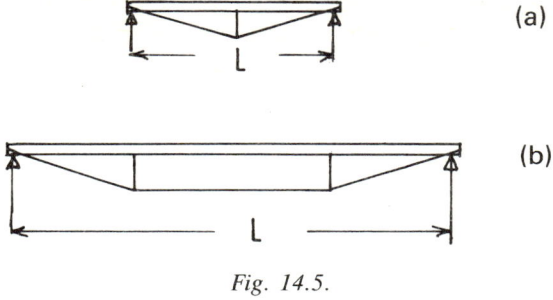

Fig. 14.5.

14.5(a) and the two strut arrangement in Fig. 14.5(b). Modern versions of this construction employ different materials, often using cables or triangulated frames for lightness.

Strutted beams are one example of statically indeterminate structures and their analysis illustrates the way in which that kind of structure can be tackled. In this section the 'two strut' beam is selected for analysis by the release-deformation method; the same approach may be adopted for solving most types of indeterminate structure using the appropriate combination of equations.

The deformation diagram for one half-span of a strutted beam carrying a uniformly distributed load is shown dashed in Fig. 14.6. As the struts are short and very stiff, it is assumed that $\Delta_{2Y} = \Delta_{3Y}$ and $\Delta_{4Y} = \Delta_{5Y}$ for symmetrical conditions. Supports 1 and 6 are taken to allow horizontal movements only (e.g. slender column supports or neoprene bearings). The direction of the moments required to restore angular compatibility on release of the joints at 3 and 5 are shown by arrows on the diagram, as well as the assumed direction of the member forces. The deformation diagram is based on the assumption that the top members are compressed and the lower members extended when the load is applied.

Equations for equilibrium, linear and angular displacements, using the same convention as before, are given below for one half of the strutted beam.

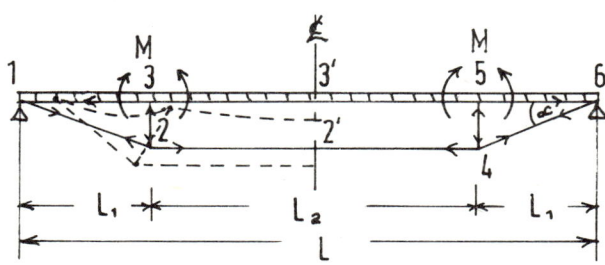

Fig. 14.6.

E.1X: $F_{12} \cos \alpha - F_{13'} = 0$

E.1Y: $F_{12} \sin \alpha - \dfrac{M_3}{L_1} = \dfrac{w}{2}(L - L_1)$

E.2X: $F_{22'} - F_{12} \cos \alpha = 0$

E.2Y: $F_{12} \sin \alpha - F_{23} = 0$

L.13': $r\Delta_{1X} - F_{13'}S_{13'} = 0$

L.12: $-r\Delta_{1X} \cos \alpha - r\Delta_{2X} \cos \alpha + r\Delta_{2Y} \sin \alpha - F_{12} S_{12} = 0$

L.22': $r \Delta_{2X} - F_{22'}S_{22'} = 0$

A.3: $\dfrac{L_1}{3EI} M_3 + \dfrac{L_2}{3EI} M_3 + \dfrac{L_2}{6EI} M_5 = \theta_{31} + \theta_{35} - \dfrac{\Delta_{3Y}}{L_1} - \dfrac{\Delta_{3Y}}{L_2} + \dfrac{\Delta_{3Y}}{L_2}$

$$2M_3 \dfrac{L_1}{L} + 2M_3 \dfrac{L_2}{L} + M_3 \dfrac{L_2}{L} = \left(\dfrac{wL_1^3}{24EI} + \dfrac{wL_2^3}{24EI} \right) \dfrac{6EI}{L} - \dfrac{6EI}{L} \dfrac{\Delta_{3Y}}{L_1} = 0$$

$$M_3 \dfrac{2L_1 + 3L_2}{L} + \dfrac{6EI}{L} \dfrac{1}{L_1} \Delta_{3Y} = \dfrac{w}{4L}(L_1^3 + L_2^3)$$

Let $P = \dfrac{6EI}{L}$ then, since $\Delta_{3Y} = \Delta_{2Y}$

$$M_3 \dfrac{2L_1 + 3L_2}{L} + \dfrac{P}{L_1} \Delta_{2Y} = \dfrac{w}{4L}(L_1^3 + L_2^3)$$

The equations are summarised in Table 14.3.

As a practical example, assume that the strutted beam is composed of a 102 × 51 RS channel with one 32 mm diameter tie rod supporting a concrete slab, including construction load, of 20 kN/m. The beam spans 6 m, has a depth of 0.5 m and $L_1 = L_2 = 2$ m. The beam has properties $A_s = 1328$ mm², and $I = 29.1$ cm⁴; cross-sectional area of the tie rod is 804 mm².

From the given dimensions, $\alpha = 14°02'$. Member 16 is chosen as the reference member, so

$$S_{16} = 1, \; S_{13'} = 0.5, \; S_{12} = \dfrac{2061}{6000} \times \dfrac{1328}{804} = 0.567$$

and $S_{22'} = \dfrac{1000}{6000} \times \dfrac{1328}{804} = 0.275$

$$r = \dfrac{200 \times 1328}{6000} = 44.267; \; P = \dfrac{6 \times 200 \times 29.1 \times 10^4}{6000} = 58\,200$$

The data table and results using MATIN are shown in Table 14.4. Since kN and mm units are used in the data figures, the moment in the printout has been divided by 1000 to show it in kNm units.

Table 14.3 Strutted beam

Equation	F				Δ			M	$\overline{W}$
	13'	12	22'	23	1X	2Y	2X	3	
E.1X	−1	$\cos \alpha$							0
E.1Y		$\sin \alpha$						$-\dfrac{1}{L_1}$	$\dfrac{w}{2}(L - L_1)$
E.2X		$-\cos \alpha$	1						0
E.2Y		$\sin \alpha$		−1					0
L.13'	−S				r				0
L.12		−S			$-r \cos \alpha$	$r \sin \alpha$	$-r \cos \alpha$		0
L.22'			−S				r		0
A.3						$\dfrac{P}{L_1}$		$\dfrac{2L_1 + 3L_2}{L}$	$\dfrac{w}{4L}(L_1^3 + L_2^3)$

Table 14.4 Data table for strutted beam example

Equation	F				Δ			M	
	13'	12	22'	23	1X	2Y	2X	3	
E.1X	−1	0.9702						—	
E.1Y		0.2425						−0.0005	
E.2X		−0.9702	1						
E.2Y		0.2425		−1					
L.13'	−0.5				44.267				
L.12		−0.567			−42.948	10.735	−42.948		
L.22'			−0.275				44.267	1.667	
A.3		40				29.1		13 333	
W									
MATIN N = 8	175.27	180.65	175.27	43.81	1.98	21.82	1.09	7.62	kN mm kNm

Statically indeterminate structures

The key feature in the analysis of statically indeterminate structures by the release-deformation method is the construction of the deformation diagram. It is important that the diagram represents the behaviour of the structure as accurately as can be predicted, especially in regard to direction of nodal displacements. It is up to the user to decide what angular or linear movements are to be incorporated in the analysis. In this regard, it is often feasible to carry out an initial approximate analysis using a small size matrix; then, when the member sizes have been selected, one can introduce additional effects to obtain a more accurate answer. Thus, a premium is placed on the user's skill in identifying the really relevant displacements in the structure, i.e. those that substantially alter the final results.

If a negative value appears in the printout, it is usually an indicator that an incorrect assumption has been made in preparing the data. A check should be made on the joint displacements and direction of the member forces. Another fundamental point to bear in mind is that all matrix terms must be calculated in the same units. The factor, r, in the displacement equations is introduced to permit use of comparative stiffness ratios; if actual stiffness values are used for the members, this puts $r = 1$.

Examples of some indeterminate forms are included in the Problems at the end of the book.

PART 3
LATTICES AND GRILLAGES

15 Lattice girders and trusses

N-girder

The analysis of simple pin-jointed frames has been described in previous sections of the book. The use of the release-deformation method to solve the forces and movements in flying shores was one example. In this chapter the method is used to determine the forces and joint displacements in some common statically determinate frameworks such as lattice girders and roof trusses.

The N-girder in Fig. 15.1 has loads applied at the lower panel points. The member forces may be obtained by simple resolution of the forces at each joint in the XX and YY directions. Assuming the physical properties, support conditions and loading of the girder are symmetrical, only half the frame need be considered. The equilibrium equations may be expressed in tabular form. This is shown as the force matrix in Table 15.1. By entering the terms as a series of data statements in a program such as MATIN, one can determine the member forces for the N-girder. (It is assumed the reader is familiar with this process so the derivation of the table is not described.)

To obtain the joint displacements, the directions of the member forces are marked on the frame diagram and the deformation diagram is drawn

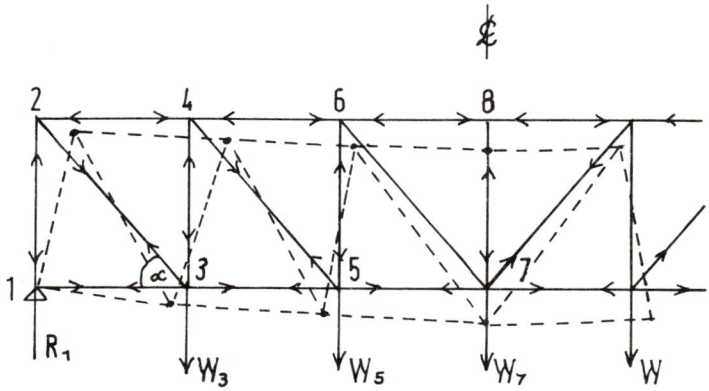

Fig. 15.1.

Table 15.1 Force matrix for N-girder

Equation E	12	13	23	24	34	35	45	46	56	57	67	68	78	W̄
1Y	1													R_1
1X		1												0
2Y	1		$-\sin\alpha$											0
2X			$\cos\alpha$	-1										0
3Y		-1	$\sin\alpha$		-1									W
3X			$-\cos\alpha$											0
4Y				1			$-\sin\alpha$							0
4X					1		$\cos\alpha$	-1						0
5Y						-1	$\sin\alpha$		-1					W
5X						1	$-\cos\alpha$							0
6Y								1		1	$-\sin\alpha$			0
6X									1		$\cos\alpha$	-1		0
7Y													-1	W/2

for the frame assuming releases at the nodal intersections. This diagram is shown dashed in Fig. 15.1. It should be noted that the supports are permitted lateral but not vertical movement. Thus $\Delta_{1Y} = 0$; also, from symmetry, $\Delta_{7X} = \Delta_{8X} = 0$. (If one support were fixed and the other 'free' it would be necessary to analyse the entire frame.) The equations for linear compatibility are listed below for the half frame using the conventions described in previous chapters.

L.12: $\quad r\Delta_{2Y} = F_{12} S_{12}$

L.13: $\quad r\Delta_{1X} - r\Delta_{3X} = F_{13} S_{13}$

L.23: $\quad -r\Delta_{2X}\cos \alpha - r\Delta_{2Y}\sin \alpha - r\Delta_{3X}\cos \alpha + r \Delta_{3Y}\sin \alpha = F_{23} S_{23}$

L.24: $\quad r\Delta_{2X} - r\Delta_{4X} = F_{24} S_{24}$

L.34: $\quad -r\Delta_{3Y} + r\Delta_{4Y} = F_{34} S_{34}$

L.35: $\quad r\Delta_{3X} - r\Delta_{5X} = F_{35} S_{35}$

L.45: $\quad -r\Delta_{4X}\cos \alpha - r\Delta_{4Y}\sin \alpha - r\Delta_{5X}\cos \alpha + r\Delta_{5Y}\sin \alpha = F_{45} S_{45}$

L.46: $\quad r\Delta_{4X} - r\Delta_{6X} = F_{46} S_{46}$

L.56: $\quad -r\Delta_{5Y} + r\Delta_{6Y} = F_{56} S_{56}$

L.67: $\quad -r\Delta_{6X}\cos \alpha - r\Delta_{6Y}\sin \alpha + r\Delta_{7Y}\sin \alpha = F_{67} S_{67}$

L.57: $\quad r\Delta_{5X} = F_{57} S_{57}$

L.68: $\quad r\Delta_{6X} = F_{68} S_{68}$

L.78: $\quad r\Delta_{8Y} - r\Delta_{7Y} = F_{78} S_{78}$

where F = member force

$r = \dfrac{E_n A_n}{L_n}$ for the reference member, n, and

S = comparative flexibility factor $= \dfrac{L}{EA} \cdot \dfrac{E_n A_n}{L_n}$

The equations are summarised in Table 15.2, where they are arranged with terms on the main diagonal to permit solution by MATIN or a similar program. Member forces must have been determined in advance, of course. The printout will supply the displacements in the XX and YY directions directly if actual FS values are entered in the vector column. If relative flexibility factors are used then the results must be divided by the value of r $\left(\text{i.e. } \dfrac{E_n A_n}{L_n}\right)$ as indicated by the tabular heading, $r\Delta$.

The force and displacement matrices may be combined (compare with Table 14.3) in which case all results are obtained in one run. To keep down the amount of data to be handled to manageable proportions, it is handier in many cases to feed in the matrices separately and use two runs to secure the results. Thus, if the N-girder has a span of 30 m, panel length of 5 m and height 6 m and carries a load on the top boom of 2 kN/m, then the member forces can be determined using the values sin α = 0.768 and cos α = 0.64 in the force matrix table. The results may then be inserted in the displacement matrix vector, as shown in Table 15.3. The cross-sectional area of the top and bottom booms is taken as 2500 mm^2

Table 15.2 Displacement matrix for N-girder

						$r\Delta$								
1X	2X	2Y	3X	3Y	4X	4Y	5X	5Y	6X	6Y	7Y	8Y		FS
1			−1											13
	1				−1									24
−cos α		−sin α												12
			1				−1							35
		1		−1										23
					1				−1					46
			−cos α	sin α		1								34
							1				−1			57
					−cos α	−sin α		1						45
									1			−1		68
							−cos α	sin α		1				56
									−cos α	−sin α	1			67
											−sin α	sin α		78

LATTICE GIRDERS AND TRUSSES

Table 15.3 N-girder problem data table

Equation	rΔ													F	$\dfrac{L_1}{L_n}$	$\dfrac{A_n}{A_1}$		$\overline{FS}$
L	1X	2X	2Y	3X	3Y	4X	4Y	5X	5Y	6X	6Y	7Y	8Y			A_1		
13	1													0	1	1	1	0
24		1												20.83	1	1	1	20.83
12			1	−1										15	1.2	2	1	36
35		−0.64	−0.768	−0.64	0.768									19.53	1	1	1	19.53
23				1	−1									16.29	1.562	2	1	50.89
46						−0.64	−0.768	−0.64	0.768					33.33	1	1	1	33.33
34						1		−1						12.50	1.2	2	1	30
57								1	−1					33.33	1	1	1	33.33
45								−0.64	−0.768	−0.64	0.768			9.77	1.562	2	1	30.52
68										1				37.5	1	1	1	37.5
56										−0.64	−0.768			7.5	1.2	2	1	18
67												0.768		3.26	1.562	2	2	10.18
78												−1	1	0	1.2	2	2	0
rΔ	52.9	91.7	36	52.9	222.7	70.8	252.7	33.3	379.2	37.5	397.2	441.7	441.7					
Δ	0.5	0.9	0.4	0.5	2.2	0.7	2.5	0.3	3.8	0.4	4.0	4.4	4.4		mm			

and internal members 1250 mm² (E_s = 200 kN/mm²). Member 13 is the reference element so the comparative values are multiplied by the forces to give the $\overline{FS}$ vector terms. This follows from the fact that, where E is constant,

$$S_1 = \frac{L_1}{A_1} \cdot \frac{A_n}{L_n} = \frac{L_1}{L_n} \cdot \frac{A_n}{A_1}$$

It will be seen that the use of relative values considerably simplifies computation of the $\overline{FS}$ terms.

The MATIN program with the coefficients and $\overline{FS}$ vector as data will print out the rΔ values; dividing these by r = (200 × 2500)/5000 = 100 gives the lateral and vertical joint displacements. These are shown in the bottom line of the table.

When preparing the deformation diagram it is useful to remember that the general tendency under vertical loads is for all nodes to move downwards and, where the end supports can move laterally, for the bottom nodes to be displaced outwards and top nodes displaced inwards progressively less as one goes from the supports to the centre of the span. With different support conditions the displacement pattern will be different and may be hard to predict. If a negative sign appears in the output, it is an indicator that the diagram has been incorrectly drawn; the equations should be re-examined with this in mind.

Common roof truss

One traditional form of roof truss is shown in Fig. 15.2. This is usually referred to as a 'common truss' and used for spans up to about 15 m. Loads from the roof cladding are assumed to be applied at the rafter nodes. The diagram indicates dead or vertical superimposed loads, but components of wind loading acting horizontally or vertically upwards must be considered also. Whatever the type of applied loading, however, the method of analysis is fundamentally the same. Equilibrium equations obtained by resolution of the forces acting at each node in turn are used to construct the force matrix. Due to the extra member at joint 4, some special device is needed to solve the forces there. One method is to consider a section through members 23, 24 and 14 and then take moments about the support. This gives rise to the equation for 2Y,

$$W_2 \frac{L}{6} = F_{24} \frac{L}{3} \sin \alpha, \text{ hence } 2 \sin \alpha \, F_{24} = W_2$$

The equations are summarised in the force matrix in Table 15.4.

Truss deflections are found, as before, by writing down the linear compatibility equations. Referring to the dashed deformation diagram in Fig. 15.2 for symmetrical conditions in which the end supports are permitted

LATTICE GIRDERS AND TRUSSES

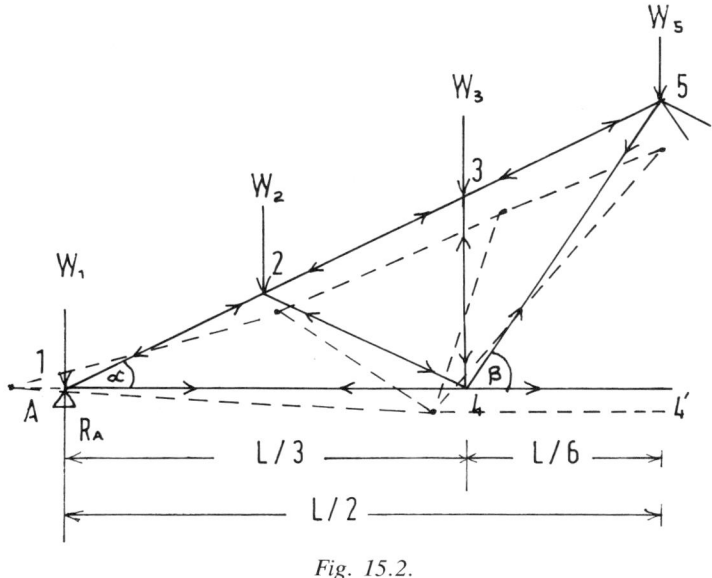

Fig. 15.2.

lateral but not vertical movement, the equations are set out below. The factor r permits use of relative terms in computing S values. The reference member should be one of the bottom members of the truss, preferably 14, since the angular inclinations α and β are measured from the horizontal.

L.12: $-r\Delta_{1X}\cos\alpha - r\Delta_{2X}\cos\alpha + r\Delta_{2Y}\sin\alpha = F_{12} S_{12}$
L.14: $r\Delta_{1X} - r\Delta_{4X} = F_{14} S_{14}$
L.23: $r\Delta_{2X}\cos\alpha - r\Delta_{3X}\cos\alpha - r\Delta_{2Y}\sin\alpha + r\Delta_{3Y}\sin\alpha = F_{23} S_{23}$
L.24: $r\Delta_{2X}\cos\alpha + r\Delta_{4X}\cos\alpha + r\Delta_{2Y}\sin\alpha - r\Delta_{4Y}\sin\alpha = F_{24} S_{24}$
L.34: $r\Delta_{3Y} - r\Delta_{4Y} = F_{34} S_{34}$
L.35: $r\Delta_{3X}\cos\alpha - r\Delta_{3Y}\sin\alpha + r\Delta_{5Y}\sin\alpha = F_{35} S_{35}$
L.45: $r\Delta_{4X}\cos\beta + r\Delta_{4Y}\sin\beta - r\Delta_{5Y}\sin\beta = F_{45} S_{45}$
L.44': $r\Delta_{4X} = F_{44'} S_{44'}$

Table 15.4 Force matrix for roof truss

Equation	F								$\overline{W}$
E	12	14	23	24	34	35	45	46	
1Y	$\sin\alpha$								$R_A - W_1$
1X	$-\cos\alpha$	1							0
2X	$\cos\alpha$		$-\cos\alpha$	$-\cos\alpha$					0
2Y				$2\sin\alpha$					W_2
3Y			$\sin\alpha$		1	$-\sin\alpha$			W_3
3X			$\cos\alpha$			$-\cos\alpha$			0
4Y				$-\sin\alpha$	-1		$\sin\beta$		0
4X		-1		$\cos\alpha$			$\cos\beta$	1	0

Table 15.5 Displacement matrix for roof truss

			$r\Delta$					$\overline{FS}$
1X	2X	2Y	3X	3Y	4X	4Y	5Y	
1					−1			14
−cos α	−cos α	sin α						12
	cos α	sin α			cos α	−sin α		24
	cos α	−sin α	−cos α	sin α				23
			cos α	−sin α			sin α	35
					1			44′
				1		−1		34
					cos β	sin β	−sin β	45

The general displacement matrix is set out in Table 15.5.

For the particular case of a roof truss of 20 m span with a rise to span ratio of 1:4 (that is, $\alpha = 26°34'$ and $\beta = 56°18'$) and panel point loads W = 20 kN, the member forces may be determined by inserting the appropriate values in Table 15.4 and running PIVOT with these as data. If, in addition, we know that all members except the rafter have a cross-sectional area of 1500 mm^2 (that for the rafter 15 being 3000 mm^2) and E_s = 210 kN/mm^2, then we can perform the same exercise for finding joint displacements using Table 15.5. Adopting 14 as the reference member, relative values for L and A are entered with the member forces and the vector terms determined as shown in the final column of Table 15.6. Results obtained from the printout are divided manually by r = (210 × 1500)/6667 = 47.25 to give the displacements displayed in the bottom line of the table.

Indeterminate trusses

In the case of indeterminate trusses and frames, member forces cannot be established from equilibrium equations alone, since results are dependent on displacement or bending effects also. This increases the number of equations to be handled, as seen already in the analysis of the strutted beam in Chapter 14. Analysis of cross-braced or redundant frames similarly will demand both equilibrium and linear compatibility equations; results obtained with MATIN will then give both forces and joint displacements. These will enable reactions, shear forces and axial forces to be determined from first principles without great difficulty.

Although the method of analysis by release-deformation is basically the same whether the structure is statically determinate or highly redundant, certain difficulties can arise in the latter case. One is in ensuring that the

Table 15.6 Displacement data table for truss problem

Equation	$r\Delta$								F	$\dfrac{L_1}{L_n}$	$\dfrac{A_n}{A_1}$	$\overline{FS}$
L	1X	2X	2Y	3X	3Y	4X	4Y	5Y				
14	1								100	1	1	100
12	−0.895	−0.895	0.447						111.9	0.559	0.5	31.27
24		0.895	0.447	−0.895					22.4	0.559	1	12.82
23		0.895	−0.447	0.895	0.447				89.5	0.559	0.5	25
35					−0.447	0.895			89.5	0.559	0.5	25
44'						−1			60	0.5	1	30
34					1		−1	0.447	20	0.5	1	10
45						0.555	0.832	−0.832	36.1	0.901	1	32.53
$r\Delta$	130	24.9	380	42.5	471.2	30	481.2	442.1	mm			
Δ	2.8	0.5	8.0	0.9	10.0	0.6	9.8	9.4				

deformation diagram interprets the behaviour of the structure correctly. Absence of a negative sign in the output is usually an indication that this aim has been achieved. Another problem is the size of the matrix to be handled, since the equations increase in number with the complexity of the structure. On this account, the programs listed in the book are really only applicable to indeterminate structures of modest size or simple configuration.

16 Beam frameworks

Beam framework types

In the following chapters we consider systems of interconnected beams carrying vertical loads applied at or between the intersection points. Three arrangements are shown in Fig. 16.1. The upper diagram shows two interconnecting beams with a column load on the centre node; in diagram (b) the interconnecting beams carry vertical loads both at the nodes and between them so that bending occurs in some members. (Torsion-free joints are assumed at this initial stage.) In diagram (c) a 'gridiron' or grillage arrangement is used to support vertical loads applied at the intersection points. In respect of foundations, the diagrams might represent ground beams transmitting column loads to firmer strata at the supports or an 'eggcrate' construction to form a stiff foundation for a

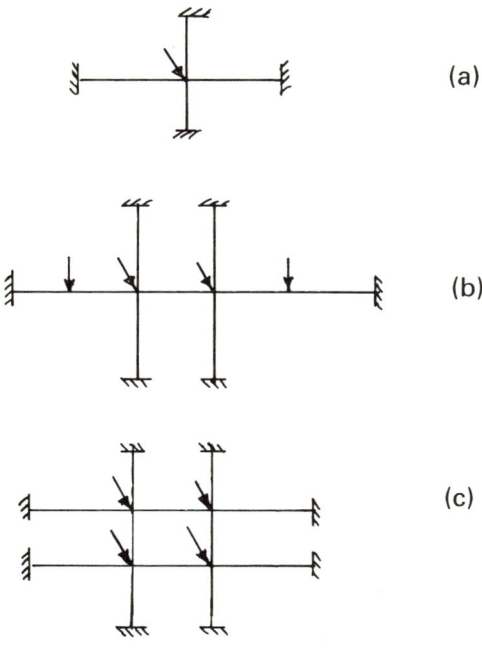

Fig. 16.1.

Manual analysis

It is proposed to commence by examining the simple arrangement illustrated in Fig. 16.2. In this case, the applied load, W, produces a vertical deflection, Δ_1, at node 1. If we introduce a release or 'hinge' at the node to permit angular changes to occur there, the frame will assume the shape shown by two straight lines in the deformation diagrams, Fig. 16.2. This is on the proviso that the deflection is maintained by a temporary propping force, R'_1, at the displaced node on release. Angular discontinuities may be identified from the deformation diagrams. We may then apply moments at the node so as to eliminate these, as shown by dashed arrows in the figure. In this initial case, the load is assumed to be applied at the nodal intersection, so that no bending occurs in the members.

In Fig. 16.2, supports A, B, C and D are assumed simply supported; node 1 supports the vertical load W_1. A general case is assumed with beams of different panel lengths L_1, L_2, L_3 and L_4 and different moments of inertia I_1, I_2, I_3 and I_4. The directions of moments M_{1A}, M_{1B}, M_{1C} and M_{1D} applied to the released structure with reaction R'_1 at node 1 are shown by dashed arrows. Their direction depends on the rule embodied

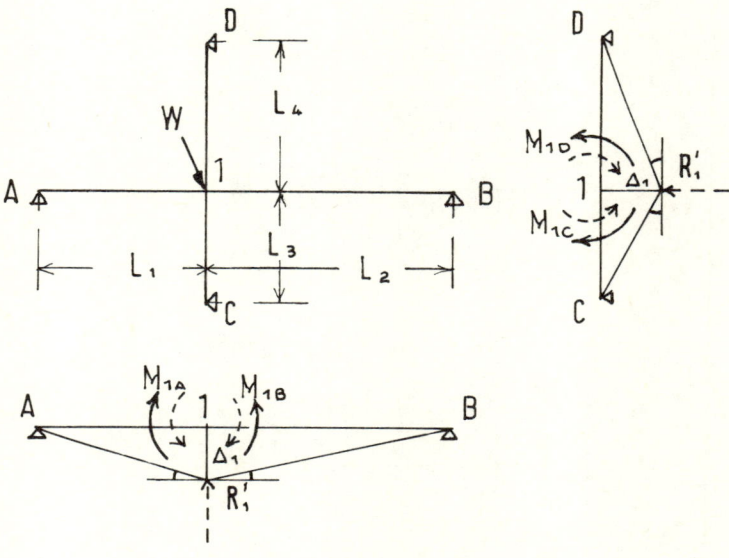

Fig. 16.2.

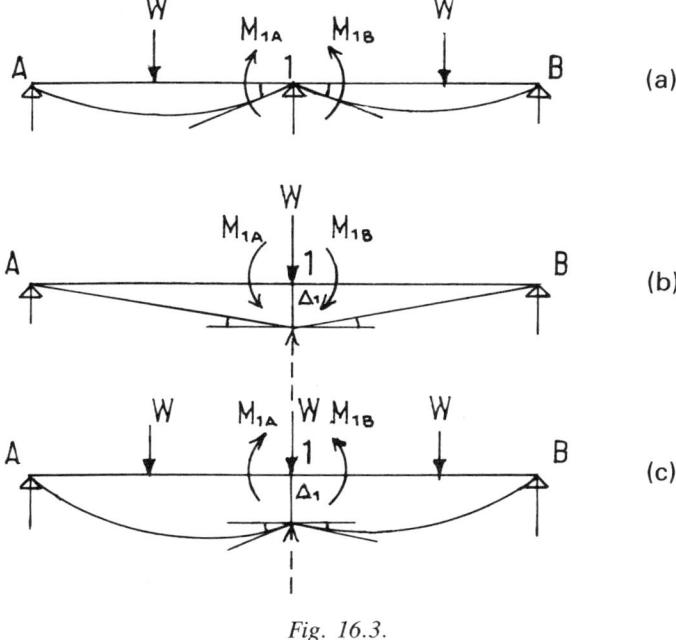

Fig. 16.3.

in the release-deformation method that moments or forces are applied to eliminate the discontinuities created by the introduction of releases.

If the loads are applied away from the nodes so that bending occurs, then the release moments are applied to counteract sag in the span, as shown in Fig. 16.3(a). As we have seen, where vertical displacement only occurs at the node, the release moments act in the opposite direction, as in Fig. 16.3(b). Where both effects occur together one has a choice of convention.

For beam frameworks subject to downward loading, the convention in Fig. 16.3(a) has been adopted, which is in line with that used previously for continuous beams. This is shown with bending and settlement effects combined in Fig. 16.3(c). For uniformity, the same convention is adopted for grillages without bending. As usual, where the release moment *increases* the angular change due to vertical displacement of a node, as may be seen from the deformation diagram, then a *negative* sign is to be attached to the angular displacement term.

Figure 16.2 will now be considered with the release moments applied as shown by the full arrows. The equilibrium equation for joint 1 is:

E.1: $$\frac{M_{1A}}{L_1} + \frac{M_{1B}}{L_2} + \frac{M_{1C}}{L_3} + \frac{M_{1D}}{L_4} = -W_1$$

But $M_{1B} = M_{1A}$ and $M_{1D} = M_{1C}$, so that

$$-M_{1A}\left(\frac{1}{L_1}+\frac{1}{L_2}\right) - M_{1C}\left(\frac{1}{L_3}+\frac{1}{L_4}\right) = W_1$$

The angular displacement along beam 1A due to Δ_1 is Δ_1/L_1 and along beam 1B is Δ_1/L_2,

$$\text{so the total angular displacement} = \frac{\Delta_1}{L_1}+\frac{\Delta_1}{L_2}$$

similarly, the total angular displacement along beam CD $= \dfrac{\Delta_1}{L_3}+\dfrac{\Delta_1}{L_4}$

Incorporating these results in the equations for angular displacement gives, first of all for beam A1B:

A.A1B: $\quad M_{1A}\left(\dfrac{L_1}{3EI_1}+\dfrac{L_2}{3EI_2}\right) = -\dfrac{\Delta_1}{L_1}-\dfrac{\Delta_1}{L_2}$

Multiplying by $\dfrac{6EI_1}{L_1}$ and putting $K_2 = \dfrac{I_1}{L_1}\cdot\dfrac{L_2}{I_2}$, $K_3 = \dfrac{I_1}{L_1}\cdot\dfrac{L_3}{I_3}$ and $K_4 = \dfrac{I_1}{L_1}\cdot\dfrac{L_4}{I_4}$ and $P = \dfrac{6EI_1}{L_1}$ gives:

$$2M_{1A} + 2K_2M_{1A} = -P\Delta_1\left(\frac{1}{L_1}+\frac{1}{L_2}\right)$$

$$2(1+K_2)M_{1A} = -P\Delta_1\left(\frac{1}{L_1}+\frac{1}{L_2}\right)$$

Similarly for beam C1D

A.C1D: $\quad M_{1C}\left(\dfrac{L_3}{3EI_3}+\dfrac{L_4}{3EI_4}\right) = -\dfrac{\Delta_1}{L_3}-\dfrac{\Delta_1}{L_4}$

$$2K_3M_{1C} + 2K_4M_{1C} = -P\Delta_1\left(\frac{1}{L_3}+\frac{1}{L_4}\right)$$

$$2(K_3+K_4)M_{1C} = -P\Delta_1\left(\frac{1}{L_3}+\frac{1}{L_4}\right)$$

Expressing these equations in tabular form produces Table 16.1.

One method of solving the equations is to assume unit deflection for Δ_1 and solve for M_{1A} and M_{1C}. Next use the equilibrium equation E.1 to determine the factor by which the unit moments must be multiplied to suit the actual loading. The unit moments multiplied by this factor give the required values. Instead of 'unit moments' and 'unit deflections' we can use any relative values, in fact, and select them so as to eliminate the term P from the analysis.

Consider the example shown in Fig. 16.4 where $W_1 = 100$ kN and E is constant. If $I_1 = I_2$ and $I_3 = I_4 = 0.5I_1$, then

BEAM FRAMEWORKS

Table 16.1 Simply supported frame with nodal load

Equation	Moment		$\overline{\Delta}$
	1A	1C	
A.A1B	$2(1 + K_2)$		$-\dfrac{1}{L_1} - \dfrac{1}{L_2}$
A.C1D		$2(K_3 + K_4)$	$-\dfrac{1}{L_3} - \dfrac{1}{L_4}$
E.1	$-\dfrac{1}{L_1} - \dfrac{1}{L_2}$	$-\dfrac{1}{L_3} - \dfrac{1}{L_4}$	W_1

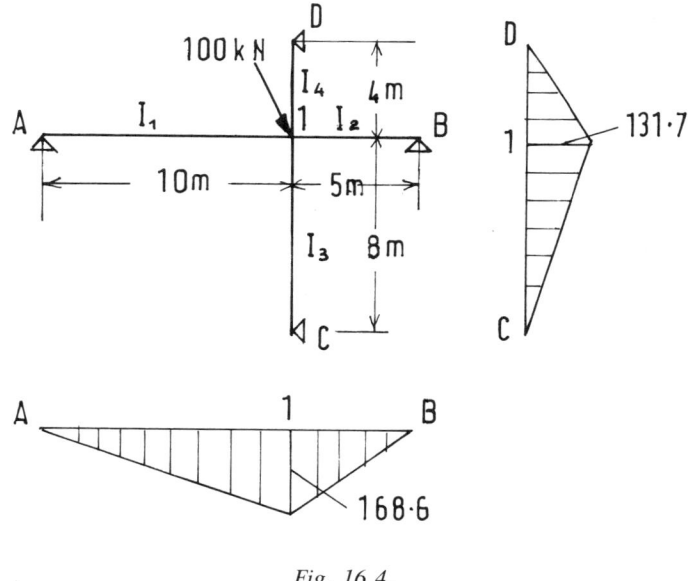

Fig. 16.4.

Table 16.2 Data for the problem in Fig. 16.4

Equation	Moment		$\overline{\Delta}$
	1A	1C	
A.A1B	3	0	−0.3
A.C1D	0	4.8	−0.375
E.1	−0.3	−0.375	100

$$K_2 = \frac{I_1}{10} \frac{5}{I_1} = 0.5, \quad K_3 = \frac{I_1}{10} \frac{8}{0.5I_1} = 1.6 \quad \text{and} \quad K_4 = \frac{I_1}{10} \frac{4}{0.5I_1} = 0.8$$

The tabular matrix is shown in Table 16.2. The first two equations may be solved manually to give:

$$M_{1A} = -0.1; \quad M_{1C} = -0.0781$$

From the equilibrium equation E.1, if F is the multiplication factor,

$$(-0.3 \times F \times -0.1) + (-0.375 \times F \times -0.0781) = 100$$

$$F = \frac{100}{0.0593} = 1686$$

so that $M_{1A} = -168.6$ kNm and $M_{1C} = -131.7$ kNm.

The BM diagram is drawn on the tension flanges in Fig. 16.4, but it must be borne in mind that positive moments in the output indicate tension in the top flange. To determine the nodal deflections, we merely substitute values for the moments in one of the equations for angular movement using the relevant E and I values. Thus, if $E = 200$ kN/mm^2 and $I = 50$ cm$^4 = 50 \times 10^4$ mm^4, then

$$-0.3P\Delta_1 = 3M_{1A} = 3 \times -168.6 \times 10^3 \text{ kNmm}$$

$$\text{But } P = \frac{6 \times 200 \times 50 \times 10^4}{10\,000}$$

$$\text{so } \Delta_1 = \frac{3 \times 168.6 \times 10^3 \times 10^4}{0.3 \times 6 \times 200 \times 50 \times 10^4} = 28 \text{ mm}$$

Beam frameworks with bending

The more general case in which the loads are applied asymmetrically and between nodes is shown in Fig. 16.5. The frame is composed of three simply supported beams with a single longitudinal beam 123 simply supported at each end and continuous with beam CD at joint 2. (In this chapter the effects of torsion will be ignored.)

The action of load W causes nodal deflections Δ_1, Δ_2 and Δ_3 and bending in member 1B. Temporary reactions R_1, R_2 and R_3 are assumed to be applied at the nodes to hold them in their deflected state when releases are introduced at 1, 2 and 3. The deformation diagram in Fig. 16.5 describes the position at release for member A1B. This shows the angular discontinuities and end rotations in 1B and the release moments at node 1.

In considering the nodal reactions for equilibrium conditions, these may be divided into two components, assuming a prop is applied at each node on release:

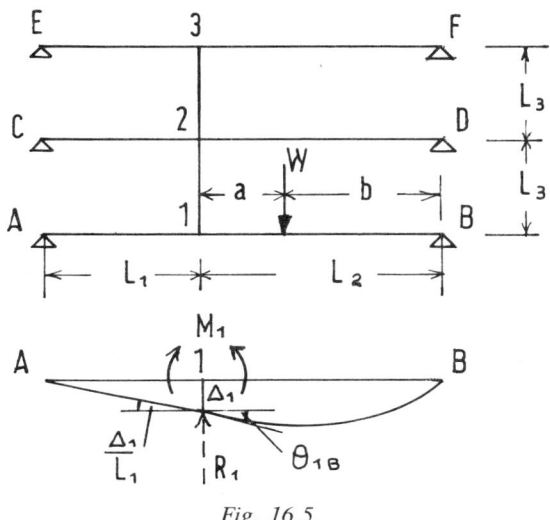

Fig. 16.5.

(1) reactions R'_1, R'_2, etc. due to the vertical loads (Fig. 16.6(a));
(2) reactions due to moments applied at the nodes to counteract bending under the external load system (Fig. 16.6(b)).

The former can be calculated at the outset, but the latter can be calculated only when the 'loads' moments have been determined.

The complete deformation diagram for the frame is reproduced in Fig. 16.7. Note that moments are applied at each node along the continuous members – non-continuous members are assumed to have 'pinned' ends – and in accordance with the convention in Fig. 16.3(c). This corrects the angular change in 1B due to bending but increases the angular change due to downward displacement of the node; hence the latter attracts a negative sign in the equations.

Nodal bending moments are denoted by suffixes. Where bending occurs

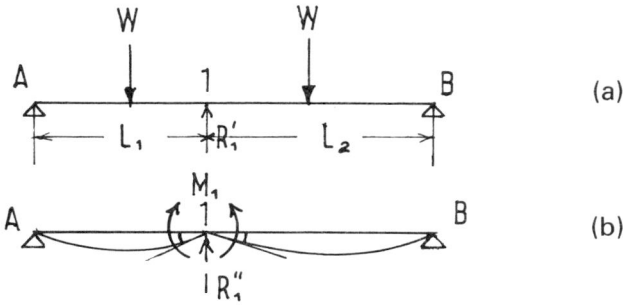

Fig. 16.6.

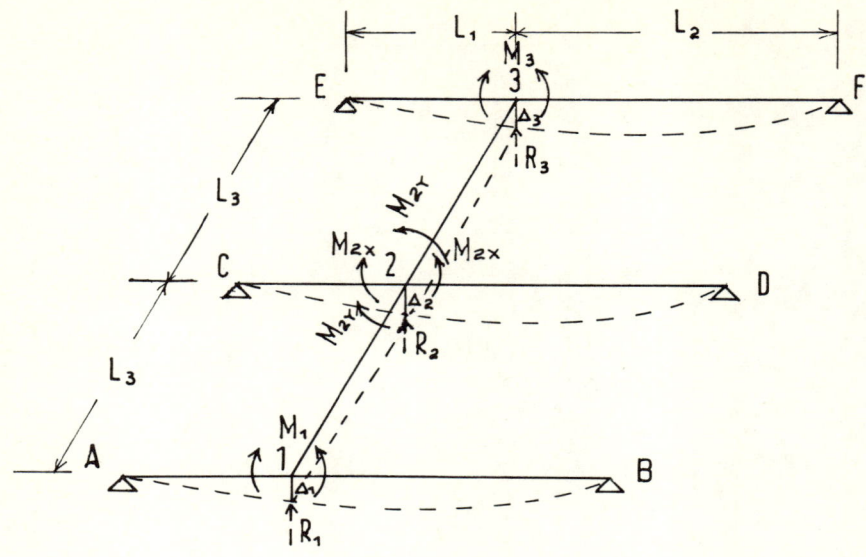

Fig. 16.7.

in one direction only on release, such as at nodes 1 and 3, moments are denoted by M_1, M_3, etc. Where moments occur in both XX and YY directions, as at node 2, they are distinguished by use of the terms M_{2X}, M_{2Y}, etc.

The relevant equations for this example are:

A.A1B:
$$M_1 \left(\frac{L_1}{3EI_1} + \frac{L_2}{3EI_2}\right) = \theta_{1B} - \frac{\Delta_1}{L_1} - \frac{\Delta_1}{L_2}$$

$$2M_1 (1 + K_2) = \frac{W_1 ab}{L_2^2}(L_2 + b) - P\Delta_1 \left(\frac{1}{L_1} + \frac{1}{L_2}\right)$$

A.C2D:
$$M_{2X} \left(\frac{L_1}{3EI_1} + \frac{L_2}{3EI_2}\right) = -\frac{\Delta_2}{L_1} - \frac{\Delta_2}{L_2}$$

$$2M_{2X} (1 + K_2) = -P\Delta_2 \left(\frac{1}{L_1} + \frac{1}{L_2}\right)$$

A.E3F:
$$M_3 \left(\frac{L_1}{3EI_1} + \frac{L_2}{3EI_2}\right) = -\frac{\Delta_3}{L_1} - \frac{\Delta_3}{L_2}$$

$$2M_3 (1 + K_2) = -P\Delta_3 \left(\frac{1}{L_1} + \frac{1}{L_2}\right)$$

A.123:
$$M_{2Y} \left(\frac{L_3}{3EI_3} + \frac{L_3}{3EI_3}\right) = -\left(\frac{\Delta_2 - \Delta_1}{L_3}\right) - \left(\frac{\Delta_2 - \Delta_3}{L_3}\right)$$

$$4K_3 M_{2Y} = P \left(\frac{\Delta_1}{L_3} - \frac{2\Delta_2}{L_3} + \frac{\Delta_3}{L_3}\right)$$

E.1: $\quad -\dfrac{M_1}{L_1} - \dfrac{M_1}{L_2} + \dfrac{M_{2Y}}{L_3} = R_1$

$\quad -M_1\left(\dfrac{1}{L_1} + \dfrac{1}{L_2}\right) + M_{2Y}\dfrac{1}{L_3} = R_1$

E.2: $\quad -\dfrac{M_{2X}}{L_1} - \dfrac{M_{2X}}{L_2} - \dfrac{2M_{2Y}}{L_3} = R_2 = 0$

$\quad -M_{2X}\left(\dfrac{1}{L_1} + \dfrac{1}{L_2}\right) - M_{2Y}\dfrac{2}{L_3} = 0$

E.3: $\quad -\dfrac{M_3}{L_1} - \dfrac{M_3}{L_2} + \dfrac{M_{2Y}}{L_3} = R_3 = 0$

$\quad -M_3\left(\dfrac{1}{L_1} + \dfrac{1}{L_2}\right) + M_{2Y}\dfrac{1}{L_3} = 0$

The equations in tabular form are shown in Table 16.3.

Analytical procedure

To solve these equations manually we could adopt the following procedure, based on the principle of superposition.

(1) Calculate the 'applied loads' reaction at each node ignoring bending (compare with Fig. 16.6(a)) and enter in the matrix table opposite the equilibrium equations.
(2) With temporary props holding the nodes in their deflected positions on release, determine the moments required to counteract the angular changes due to the external load system. (Use the coefficient matrix with the vector, $\overline{W}$.) This provides the 'load moments'.
(3) Evaluate the full nodal reactions, R_1, R_2, etc., by multiplying the 'loads moments' by the equilibrium coefficients and subtracting from the 'applied loads' reactions at each node.
(4) Next apply a force at each node in turn to produce unit deflection with the structure unloaded and the remaining nodes still held in place by temporary props. Determine the moments at the joints due to each unit deflection in turn. (Use the coefficient matrix with the vectors $\overline{\Delta}_1$, $\overline{\Delta}_2$, etc.)
(5) We have now to find the factors, say X_1, X_2, etc., by which the unit deflection moments must be multiplied so that when added to the undeflected 'loads' moments they will equal the moments due to the actual loading. These factors may be determined by considering the reactions at each node in turn in accordance with the equilibrium conditions. Thus, if E.1 denotes the equilibrium coefficients for joint 1, as shown in Table 16.3, and $U.\Delta_1$, $U.\Delta_2$, etc. denote the moments

Table 16.3 Tabular matrix for (2 × 2) beam framework

Equation	Moment				$\overline{W}$	$\overline{\Delta}_1$	$\overline{\Delta}_2$	$\overline{\Delta}_3$
	1	2X	3	2Y				
A.A1B	$2(1+K_2)$				$\dfrac{W_1 ab}{L_2^2}(L_2+b)$	$-\dfrac{1}{L_1}-\dfrac{1}{L_2}$		
A.C2D		$2(1+K_2)$					$-\dfrac{1}{L_1}-\dfrac{1}{L_2}$	
A.E3F			$2(1+K_2)$					$-\dfrac{1}{L_1}-\dfrac{1}{L_2}$
A.123				$4K_3$		$\dfrac{1}{L_3}$	$-\dfrac{2}{L_3}$	$\dfrac{1}{L_3}$
E.1	$-\dfrac{1}{L_1}-\dfrac{1}{L_2}$			$\dfrac{1}{L_3}$	R_1			
E.2		$-\dfrac{1}{L_1}-\dfrac{1}{L_2}$		$-\dfrac{2}{L_3}$	0			
E.3			$-\dfrac{1}{L_1}-\dfrac{1}{L_2}$	$\dfrac{1}{L_3}$	0			

obtained for $\Delta_1 = 1$, $\Delta_2 = 1$, etc., then we can write the equations for the reactions at node 1 in the form:

$$X_1 (\Sigma \text{ E.1} \times \text{U.}\Delta_1) + X_2 (\Sigma \text{ E.1} \times \text{U.}\Delta_2) \ldots = R_1$$

Similarly for node 2,

$$X_1 (\Sigma \text{ E.2} \times \text{U.}\Delta_1) + X_2 (\Sigma \text{ E.2} \times \text{U.}\Delta_2) \ldots = R_2, \text{ etc.}$$

By solving these simultaneous equations we obtain the factors X_1, X_2, etc. Multiplying each set of unit moments by X_1, X_2, etc. in sequence and summing the results gives the moments due to deflection of the frame alone. These added to the 'loads moments' already obtained in stage (1) give the final results.

Frame with nodal loads and bending

We may illustrate the procedure by the example shown in Fig. 16.8, which is a variant of the problem discussed in the previous section. The frame supports two concentrated loads, in this case of 100 kN each, one at node 1 and another 4 m away on member A1. The moment of inertia, I_2, of the lateral member is one-third that of the main longitudinal members. Member A1 is selected as the reference member. The following values are used in the data table, Table 16.4, derived from the previous table for generalised conditions.

$$K_2 = \frac{I_1}{10} \frac{5}{I_1} = 0.5; \quad K_3 = \frac{I_1}{10} \frac{4 \times 3}{I_1} = 1.2;$$

$$\overline{W}_{A1} = 1 \times \frac{100 \times 24 \times 16}{100} = 384$$

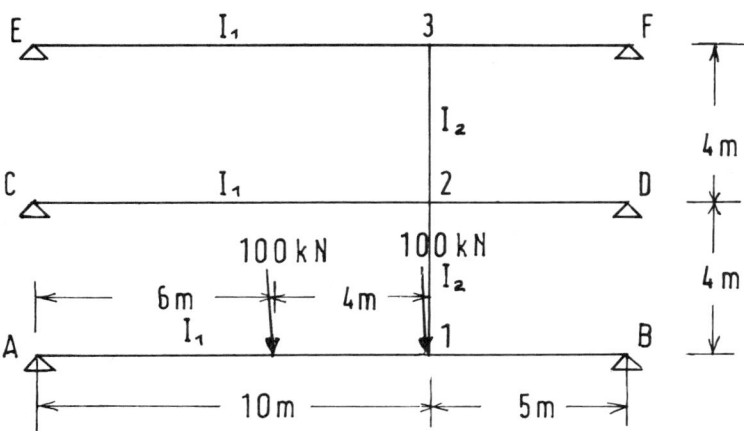

Fig. 16.8.

$$R_1' = W_1 + \frac{W_2 \times 6}{10} = 160; \quad R_2' = 0; \quad R_3' = 0; \quad \frac{1}{L_1} + \frac{1}{L_2} = 0.3$$

Table 16.4 Data for the problem in Fig. 16.8

Equation	\multicolumn{4}{c}{Moment}				
	1	2X	3	2Y	$\overline{W}$	$\overline{\Delta}_1$	$\overline{\Delta}_2$	$\overline{\Delta}_3$
A.A1B	3				384	−0.3		
A.C2D		3					−0.3	
A.E3F			3					−0.3
A.123				4.8		0.25	−0.5	0.25
E.1	−0.3			0.25	160			
E.2		−0.3		−0.5				
E.3			−0.3	0.25				

In solving this problem we shall see, first of all, how the MATIN program can assist at various stages in obtaining the results, even though manual methods would suffice in this simple example.

(1) Enter the coefficient matrix and $\overline{W}$ vector as data. Results from running the MATIN program are:

$$M_1 = 128; \quad M_{2X} = M_3 = M_{2Y} = 0$$

Substitute these values in the equilibrium equations and subtract from the right-hand side, i.e.

E.1: $160 - (-0.3 \times 128) = 160 + 38.4 = 198.4 = R_1$
E.2: $R_2 = 0$
E.3: $R_3 = 0$

(2) Enter the coefficient matrix and $\overline{\Delta}$ vectors in turn as data in the MATIN program and run it. The results are:

For $\Delta_1 = 1$: $M_1 = -0.1$; $M_{2X} = 0$; $M_3 = 0$; $M_{2Y} = 0.0521$
For $\Delta_2 = 1$: $M_1 = 0$; $M_{2X} = -0.1$; $M_3 = 0$; $M_{2Y} = -0.1042$
For $\Delta_3 = 1$: $M_1 = 0$; $M_{2X} = 0$; $M_3 = -0.1$; $M_{2Y} = 0.0521$

(3) The equations for the factors X_1, X_2 and X_3 are constructed from consideration of these results and the equilibrium equations E.1, E.2 and E.3. First, multiply each result for $\Delta_1 = 1$ by its coefficient from E.1 and sum the results; this will give the first term. Repeat using the results for $\Delta_2 = 1$ for the second term and $\Delta_3 = 1$ for the third. The other equations are constructed similarly using the coefficients from E.2 and E.3. The values to be inserted in the right-hand side of the equations are the previously determined R_1, R_2, R_3 values. This procedure gives the following set of equations:

From E.1: $-0.043X_1 + 0.026X_2 - 0.013X_3 = 198.4$
From E.2: $0.026X_1 - 0.0821X_2 + 0.026X_3 = 0$
From E.3: $-0.013X_1 + 0.026X_2 - 0.043X_3 = 0$

We can enter these values once more as data into MATIN (coefficients first followed by the constant terms) to obtain the factors $X_1 = -5816$, $X_2 = -1589$ and $X_3 = 797$.

(4) Finally, multiply the moments obtained for $\Delta_1 = 1$ by X_1, those for $\Delta_2 = 1$ by X_2 and for $\Delta_3 = 1$ by X_3; sum these and add to the 'loads' moments to get the final results as shown in Table 16.5. A check may be made on the final results by substituting in the equilibrium equations.

Table 16.5 Summary of nodal moments

Case	Moment			
	1	2X	3	2Y
Δ_1	−582			303
Δ_2		−159		−166
Δ_3			80	−41
Σ	−582	−159	80	96
W	128			
Total	−454	−159	80	96 kNm

The output contains both positive and negative bending moments. In constructing the BM diagram positive values are drawn above the baseline and, in accordance with the convention adopted in Fig. 16.3(c), indicate tension in the top surface. The BM diagrams in Fig. 16.9 are drawn below the baseline for members A1B and C2D and above for members E3F and 123, showing the effect of eccentric loading on the frame. The moment under the load at 4 m from node 1 may be ascertained from the 'free' BM value plus the relevant proportion of M_1, i.e.

$$-\frac{100 \times 4 \times 6}{10} = -240$$
$$\frac{6}{10} \times -454 = -272.4$$
$$\text{BM} = -512.4 \text{ kNm}$$

Flow chart for XBEAM program

Let us now see how the microcomputer can assist in solving this type of problem more efficiently. The first step in writing a program of any length is to set out the various stages in the analysis in as clear and concise a manner as possible. A flow chart, such as that in Fig. 16.10, can be of

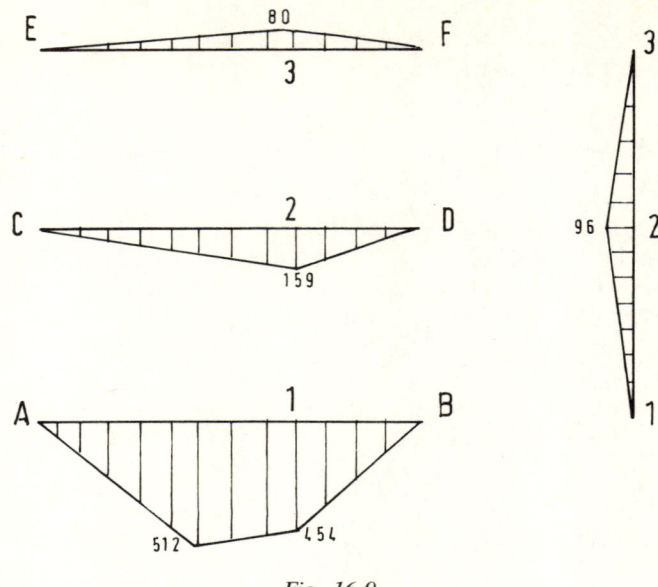

Fig. 16.9.

invaluable assistance in this regard. This shows the main sequence of events, the information that must be fed to the computer and the way in which we want the results displayed as the analysis progresses. (For the sake of clarity, many minor loops have been omitted from the diagram, such as those involving READ or PRINT statements.)

The chart shows MATSUB being used as a subroutine to solve the various sets of simultaneous equations met in the analysis. (We could with equal facility call up PIVOT as a subroutine instead, modifying it in the same way as in the case of MATIN.)

The chart shows the use of two 'flag' indicators as programming devices. The 'flag' F is used to store the output from MATSUB each time in the appropriate array and the 'flag' U to indicate whether or not another deflected node has to be considered.

The program based on the flow chart and entitled XBEAM is listed in Appendix B. Before use it must be merged with the MATSUB subroutine as described in Chapter 4. With the IBM PC load and use

SAVE "MATSUB", A

to save the subroutine in ASCII format on disk. Next, either load XBEAM into memory or key it in and use

MERGE "MATSUB"

Check that you have the combined program by LISTing it; then save it under the filename XBEAM. Should you wish to retain the listed program

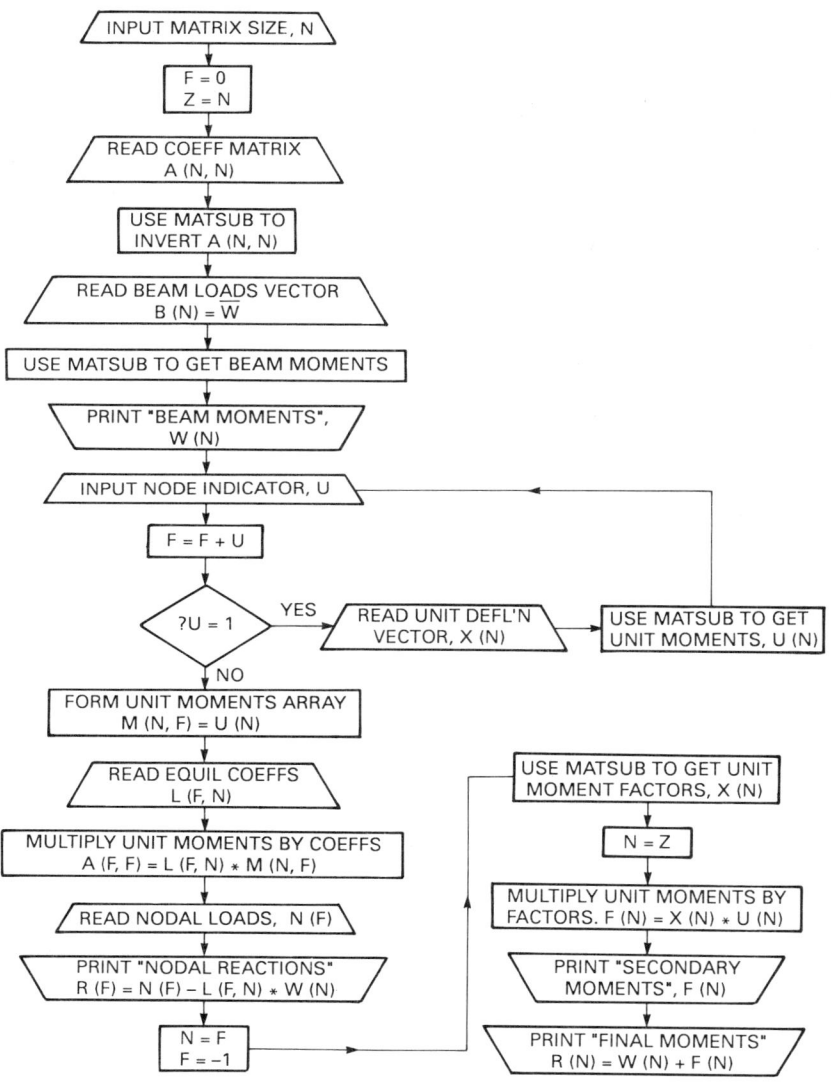

Fig. 16.10.

without subroutine on disk, then it must be given a different title before saving, of course.

A detailed explanation of the listing is not considered necessary, but it may be helpful to observe that statements 215–240 embody a standard routine for multiplication of matrices, the basis of which is described in many computer text books (Holloway, 1985). Lines 400–475 incorporate a subroutine for determining nodal reactions once the end moments in the beam-loaded members have been determined.

The usefulness of the program may be checked in a practical way by using it to solve the previous problem. Once XBEAM has been loaded into memory, data statements must be added in accordance with the program instructions (lines 30 and 35) before it is run. It is desirable to set out these values in a data table, as described previously, before keying them in. The table will indicate clearly the values to enter with the selected program.

When the program is RUN, key in 4 for the matrix size. This causes the initial data and 'loads' moments to be displayed, which should check with those obtained previously. When the prompt signal 'Deflected node?' appears, key in 1 on three occasions and next time key in zero (any number except 1 will do, in fact). The equilibrium coefficients, nodal loads and reactions will be seen next, followed by the secondary moments due to nodal deflection and finally the combined moments preceded by the numbers 1, 2, 3, etc. for identification and with the description 'Final moment' immediately after each value. A copy of the printout is shown below with values to two decimal places and a heading typed in after the program had been loaded. The printout records all the data as well as the input values, in this case $N = 4$ and $U = 1, 1, 1, 0$.

SOLUTION TO PROBLEM, FIG. 16.8

MATRIX SIZE? 4

DATA:–
3	0	0	0
0	3	0	0
0	0	3	0
0	0	0	4.8

384
0
0
0

BEAM MOMENTS:–
1	128
2	0
3	0
4	0

? DEFLECTED NODE. FOR YES PRESS 1
? 1

−0.3
  0
  0
  0.025

? ANOTHER DEFLECTED NODE. FOR YES PRESS 1
? 1

  0
−0.3
  0
−0.5

? ANOTHER DEFLECTED NODE. FOR YES PRESS 1
? 1

  0
  0
−0.3
 0.25

? ANOTHER DEFLECTED NODE. FOR YES PRESS 1
? 0

EQUILIBRIUM COEFFS:−
−0.3	0	0	0.25
0	−0.3	0	−0.5
0	0	−0.3	0.25

NODAL LOADS:−
1	160
2	0
3	0

NODAL REACTIONS:−
1	198.4
2	0
3	0

SECONDARY MOMENTS:−
1	−581.69	0
2	0	−159.28
3	0	0
4	302.97	−165.92

RESULTS:−
1	−453.69	FINAL MOMENT
2	−159.28	FINAL MOMENT
3	79.64	FINAL MOMENT
4	95.57	FINAL MOMENT

It will be seen that use of the program results in considerable saving in time and effort as well as increasing the accuracy of the results. The fact that all data are displayed and printed out allows the values to be verified by the program user or checker. Further examples of XBEAM in use are given in the next chapter.

17 Solving beam frameworks

Using the XBEAM program

The XBEAM program listed in Appendix B has been written to cater for frames with beam moments and up to four deflected nodes. Obviously we could extend the scope of the program, but XBEAM is considered sufficient as it stands to show its practical use in dealing with simple frameworks. This is illustrated in this Chapter by means of a few examples.

Figure 17.1 represents a beam framework carrying a series of column loads ranged along the transverse member 14. The four longitudinal members are simply supported at each end, the lateral member has a pin connection at both nodes 1 and 4 and a rigid, torsionless connection at each of nodes 2 and 3. Column loads W_1, W_2 and W_3 are carried on bay 23 as shown; W_2 is at a distance a from node 2 and b from node 3. All members have the same moment of inertia.

The reference element is taken to be member A1. Hence, for members 1B, etc. the relative flexibility factor is $K_2 = L_2/L_1$ and for members 12, etc. $K_3 = L_3/L_1$. Using the same notation as before we can write the following equations:

A.A1B: $\quad 2M_1 (1 + K_2) = -P\Delta_1 \left(\dfrac{1}{L_1} + \dfrac{1}{L_2}\right)$

A.C2D: $\quad 2M_{2Y} (1 + K_2) = -P\Delta_2 \left(\dfrac{1}{L_1} + \dfrac{1}{L_2}\right)$

A.E3F: $\quad 2M_{3Y} (1 + K_2) = -P\Delta_3 \left(\dfrac{1}{L_1} + \dfrac{1}{L_2}\right)$

A.G4H: $\quad 2M_4 (1 + K_2) = -P\Delta_4 \left(\dfrac{1}{L_1} + \dfrac{1}{L_2}\right)$

A.123: $\quad 4K_3 M_{2X} + K_3 M_{3X} = P \left(\dfrac{\Delta_1}{L_3} - \dfrac{2\Delta_2}{L_3} + \dfrac{\Delta_3}{L_3}\right) + \dfrac{Wab}{L_3^2}(L_3 + b)$

A.234: $\quad 4K_3 M_{3X} + K_3 M_{2X} = P \left(\dfrac{\Delta_2}{L_3} - \dfrac{2\Delta_3}{L_3} + \dfrac{\Delta_4}{L_3}\right) + \dfrac{Wab}{L_3^2}(L_3 + a)$

E.1: $\quad -M_1 \left(\dfrac{1}{L_1} + \dfrac{1}{L_2}\right) + M_{2X} \dfrac{1}{L_3} = R_1$

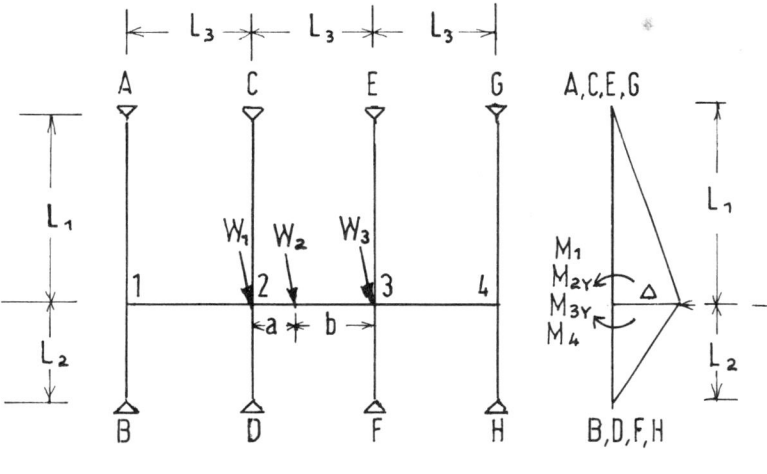

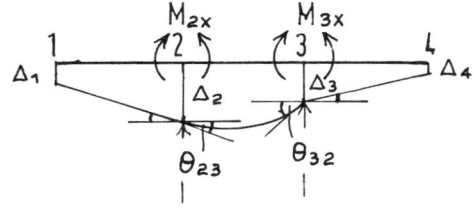

Fig. 17.1.

E.2: $-M_{2Y}\left(\dfrac{1}{L_1}+\dfrac{1}{L_2}\right)-M_{2X}\dfrac{2}{L_3}+M_{3X}\dfrac{1}{L_3}=R_2$

E.3: $-M_{3Y}\left(\dfrac{1}{L_1}+\dfrac{1}{L_2}\right)+M_{2X}\dfrac{1}{L_3}-M_{3X}\dfrac{2}{L_3}=R_3$

E.4: $-M_4\left(\dfrac{1}{L_1}+\dfrac{1}{L_2}\right)+M_{3X}\dfrac{1}{L_3}=R_4$

These equations take the same form as before, but note the carry-over from M_{3X} to node 2 and from M_{2X} to node 3 in equations A.123 and A.234 respectively. From the equations we can construct Table 17.1.

Use of the table may be illustrated by taking $L_1 = 8$, $L_2 = 4$, $L_3 = 5$ and $a = 2$, in metres, with $W_1 = 100$, $W_2 = 200$ and $W_3 = 100$ (kN).

Then, $K_1 = 1$, $K_2 = \dfrac{4}{8} = 0.5$, $K_3 = \dfrac{5}{8} = 0.625$

$\dfrac{Wab}{L^2}(L+b) = \dfrac{200 \times 2 \times 3 \times 8}{5 \times 5} = 384$ and $\dfrac{Wab}{L^2}(L+a) = 336$

$R'_1 = R'_4 = 0$; $R'_2 = 100 + 200 \times \dfrac{3}{5} = 220$; $R'_3 = 100 + 200 \times \dfrac{2}{5} = 180$

Table 17.1 Matrix table for the problem in Fig. 17.1

Equation	Moment						$\overline{W}$	$\overline{\Delta}_1$	$\overline{\Delta}_2$	$\overline{\Delta}_3$	$\overline{\Delta}_4$
	1	2Y	3Y	4	2X	3X					
A.A1B	$2(1+K_2)$							$-\dfrac{1}{L_1}-\dfrac{1}{L_2}$			
A.C2D		$2(1+K_2)$							$-\dfrac{1}{L_1}-\dfrac{1}{L_2}$		
A.E3F			$2(1+K_2)$							$-\dfrac{1}{L_1}-\dfrac{1}{L_2}$	
A.G4H				$2(1+K_2)$							$-\dfrac{1}{L_1}-\dfrac{1}{L_2}$
A.123					$4K_3$	K_3	$\dfrac{Wab}{L_3^2}(L_3+b)$	$\dfrac{1}{L_3}$	$-\dfrac{2}{L_3}$	$\dfrac{1}{L_3}$	
A.234					K_3	$4K_3$	$\dfrac{Wab}{L_3^2}(L_3+a)$		$\dfrac{1}{L_3}$	$-\dfrac{2}{L_3}$	$\dfrac{1}{L_3}$
E.1	$-\dfrac{1}{L_1}-\dfrac{1}{L_2}$				$\dfrac{1}{L_3}$	$\dfrac{1}{L_3}$	R_1				
E.2		$-\dfrac{1}{L_1}-\dfrac{1}{L_2}$			$-\dfrac{2}{L_3}$	$\dfrac{1}{L_3}$	R_2				
E.3			$-\dfrac{1}{L_1}-\dfrac{1}{L_2}$		$\dfrac{1}{L_3}$	$-\dfrac{2}{L_3}$	R_3				
E.4				$-\dfrac{1}{L_1}-\dfrac{1}{L_2}$		$\dfrac{1}{L_3}$	R_4				

These values can be used to construct a data table for use with the XBEAM program. It will be seen from the general matrix, Table 17.1, that the deflection vectors read column-by-column coincide with the equilibrium coefficients read row-by-row. This fact is used for construction of the compact data table, Table 17.2. The $\overline{R}$ values are those due to the applied loads ignoring bending, as calculated above. Actual node reactions R_1, R_2, etc. are printed out during the course of the program.

In using the XBEAM program, the tabulated quantities are entered line by line in a series of data statements. A zero must be keyed in for each blank space in the table. The terms opposite nodes 1 to 4 are keyed in twice, as directed in the data instructions (lines 30 and 35 of the listed program) and the four $\overline{R}$ values entered last. When the program is RUN, values to input for this particular example are N = 6 and U = 1, 1, 1, 1 and zero. A printout on similar lines to that reproduced in the previous chapter should be obtained, the 'final moments' being

$$M_1 = -66, M_{2Y} = -497, M_{3Y} = -461, M_4 = -43,$$
$$M_{2X} = -124, M_{3X} = -80 \text{ (kNm)}$$

These moments will appear below the baseline in the BM diagram, indicating tension in the lower surface of the members at the intersection points (Fig. 17.2).

Table 17.2 Compact data table

Ref.		Moment				
	1	2Y	3Y	4	2X	3X
A.A1B	3					
A.C2D		3				
A.E3F			3			
A.G4H				3		
A.123					2.5	0.625
A.234					0.625	2.5
$\overline{W}$					384	336
Node		Deflection and equilibrium coefficients				
1	−0.375				0.2	
2		−0.375			−0.4	0.2
3			−0.375		0.2	−0.4
4				−0.375		0.2
$\overline{R}$	0	220	180	0		

Symmetrically loaded frame

If the point load of 200 kN on beam 23 (Fig. 17.1) had been positioned at the centre of the span, we could have solved the previous problem using

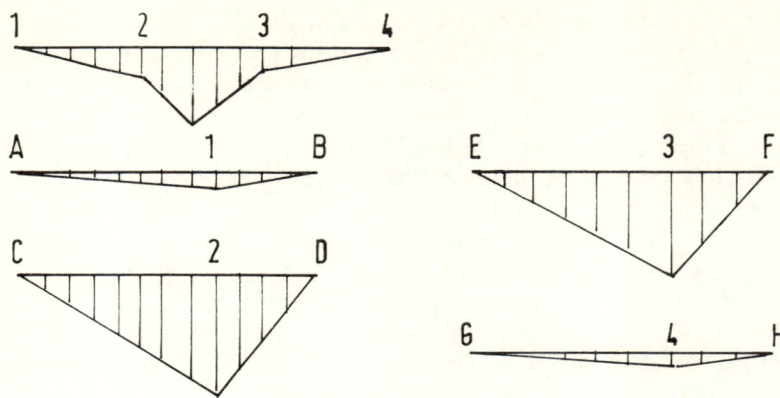

Fig. 17.2.

half the frame only and so reduced the size of the matrix considerably. To see the effect of this, assume the same frame construction but with loads of 200 kN applied at the centre of members 12, 23 and 34, as well as 100 kN loads at nodes 1, 2, 3 and 4 (Fig. 17.3).

The data table for the half frame is shown in Table 17.3. Equation A.122′ is derived from equation A.123 as follows:

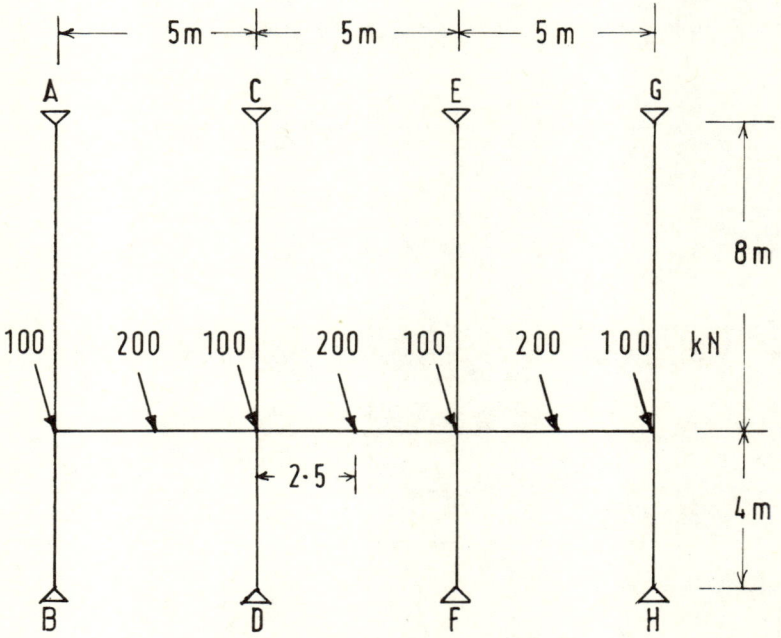

Fig. 17.3.

Table 17.3 Data table for Fig. 17.3

Ref.	Moment		
	1	2Y	2X
A.A1B	3		
A.C2D		3	
A.122'			3.125
$\overline{W}$			750
$\overline{\Delta_1}$	−0.375		0.2
Δ_2		−0.375	−0.2
E.1	−0.375		0.2
E.2		−0.375	−0.2
$\overline{R}$	200	300	

A.123: $\quad 4K_3M_{2X} + K_3M_{3X} = P\left(\dfrac{\Delta_1}{L_3} - \dfrac{2\Delta_2}{L_3} + \dfrac{\Delta_3}{L_3}\right) + 2 \times \dfrac{3}{8} W_3L_3$

But $M_{3X} = M_{2X}$ and $\Delta_3 = \Delta_2$, giving

A.122': $\quad 5K_3M_{2X} = \dfrac{3}{4} W_3L_3 + P\left(\dfrac{\Delta_1}{L_3} - \dfrac{\Delta_2}{L_3}\right)$

The flexibility factors have the same values as before; in this case $R'_1 = 200$ and $R'_2 = 300$. These figures are incorporated in the data table, Table 17.3. The printout for this problem is reproduced below. This shows all the data with headings and main stages of the analysis, as well as the results.

BEAM FRAMEWORK, FIG. 17.3

MATRIX SIZE?   3

DATA:−
3   0   0
0   3   0
0   0   3.125

0
0
750

BEAM MOMENTS:−
1   0
2   0
3   240

?DEFLECTED NODE. FOR YES PRESS 1
? 1

−0.375
0
0.2

?ANOTHER DEFLECTED NODE. FOR YES PRESS 1
? 1

 0
−0.375
−0.2

?ANOTHER DEFLECTED NODE. FOR YES PRESS 1
? 0

EQUILIBRIUM COEFFS:−
−0.375   0   0.2
   0   −0.375   −0.2

NODAL LOADS:−
1   200
2   300

NODAL REACTIONS:−
1   152
2   348

SECONDARY MOMENTS:−
1   −497.64   0
2   0   −835.69
3   254.79   −427.87

RESULTS:−
1   −497.64   FINAL MOMENT
2   −835.69   FINAL MOMENT
3   66.92   FINAL MOMENT

Deflections

Once the bending moments at the nodes are known, it is a fairly simple process to determine the deflections by back-substituting in the tabulated equations using actual E and I values. This was done manually in Chapter 16, but where a number of deflected nodes and moments are involved one can rearrange the matrix to obtain a new coefficient matrix of Δ terms and then use MATIN or a similar routine to find the required deflections. A program segment may be added to XBEAM to perform this operation where deflections are required regularly, but the following example shows how simply these may be obtained once the nodal moments are known.

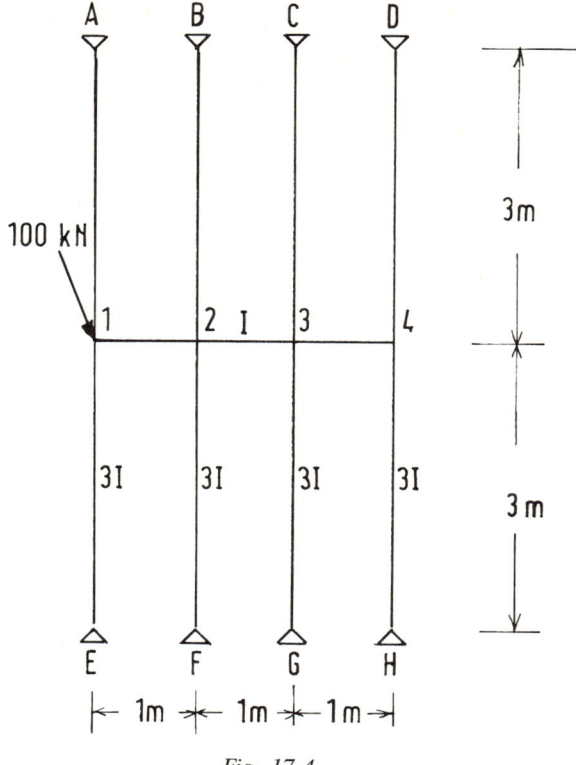

Fig. 17.4.

The frame in Fig. 17.4 is simply supported at A, B, C, D, E, F, G and H and carries a concentrated load of 100 kN at node 1. The moment of inertia of the main beams AE, etc. is three times that of the transverse beams 12, 23, etc. Given that $L_1 = L_2 = 3$ m and $L_3 = 1$ m, then

$$K_1 = K_2 = K_3 = 1 \quad \text{and} \quad \frac{1}{L_1} + \frac{1}{L_2} = 0.667$$

The tabular matrix for the frame is shown in Table 17.4 in compact form. The answers are shown in the bottom row of the table. Note that as there is no bending in the members, the $\overline{W}$ vector is composed of six zeroes and the nodal loads and reactions are equal. The program could be simplified on this account, as we shall see.

We can now construct a new matrix table based on the first four angular compatibility equations, as shown in Table 17.5. By substituting for the known moments, we can determined the nodal deflections directly. The MATIN program may be used for this purpose but in this simple case one can see clearly that for any node the following equation holds good:

$$\Delta = \frac{3}{2} \times 4 \frac{M}{P} = 6 \frac{M}{P}$$

Table 17.4 Compact data table for Fig. 17.4

Ref.	Moment					
	1Y	2Y	3Y	4Y	2X	3X
A.E1A	4					
A.F2B		4				
A.G3C			4			
A.H4D				4		
A.123					4	1
A.234					1	4
$\overline{W}$						
Node		Deflection and equilibrium coefficients				
1	−0.667				1	
2		−0.667			−2	1
3			−0.667		1	−2
4				−0.667		1
$\overline{R}$	100					
Results	−113.3	−51.1	−7.6	22.1	24.4	14.7

Table 17.5 New matrix table

Equation A	Δ_1	Δ_2	Δ_3	Δ_4	$\overline{M}$
E1A	−0.667				$4\dfrac{M_{1Y}}{P}$
F2B		−0.667			$4\dfrac{M_{2Y}}{P}$
G3C			−0.667		$4\dfrac{M_{3Y}}{P}$
H4D				−0.667	$4\dfrac{M_{4Y}}{P}$

If $E = 200$ kN/mm^2 and $I = 10$ cm^4, so that

$$P = \frac{6 \times 200 \times 10 \times 10^4}{3 \times 10^3} = 40 \times 10^3$$

then $\Delta_1 = \dfrac{6 \times 113.3 \times 10^3}{40 \times 10^3} = 17$ mm; $\Delta_2 = 8$ mm;

$\Delta_3 = 1$ mm and $\Delta_4 = -3$ mm

In the next chapter, a more suitable program for dealing with grids in which loads are applied at the nodes only is described.

18 Rectangular grillages

Uniform grillages

In this chapter we examine uniform grillages in which the members in each direction are similar and evenly spaced and in which loads are applied at the nodal points only. Grillage construction is fairly commonly used in foundations and temporary structures as well as permanent works in the shape of interconnected steel girders or continuous concrete beams, longitudinal beams with transverse diaphragms or cellular forms of construction. Practical examples are concrete 'eggcrate' foundations or cellular bases to buildings, concrete waffle floors, platforms or floors supported on a grid of steel or concrete beams and bridge decks composed of girders and transverse diaphragms.

Support conditions for the grillage framework can vary. Supports may be fixed or pinned, they may extend around the perimeter of the grillage or be located on two sides only. In this chapter it is assumed that torsional effects are absent at the joints. (These are considered later in Chapter 22.) Although the grillage itself is symmetrical, loads of any value may be applied at the joints. These are denoted by W_1, W_2, etc. The bending moment notation is the same as that described in Chapter 16, except that the 'X' suffix is dropped so that, for example,

M_2 represents bending at node 2 in the XX direction
M_{2Y} represents bending at node 2 in YY direction, etc.

Members are assumed to be spaced at intervals a in the XX direction with moment of inertia, I_1, and at intervals b in the YY direction, with moment of inertia, I_2. Therefore, the flexibility factor $k_1 = a/I_1$ and $k_2 = b/I_2$. Let $r = a/b$ so that the relative flexibility factor,

$$K_2 = \frac{k_2}{k_1} = \frac{I_1}{a}\frac{b}{I_2} = \frac{I_1}{r\,I_2}$$

Consider Fig. 18.1 in which the supports on all four sides are fixed. The load system causes vertical deflection of the joints to take place. A typical deflection pattern is shown for member BB' and moments applied on release of the joints, using the same convention as before, are indicated by arrows in Fig. 18.1(b); similarly for member AD, see Fig. 18.1 (c).

Equations for angular compatibility and equilibrium may be written as usual, thus:

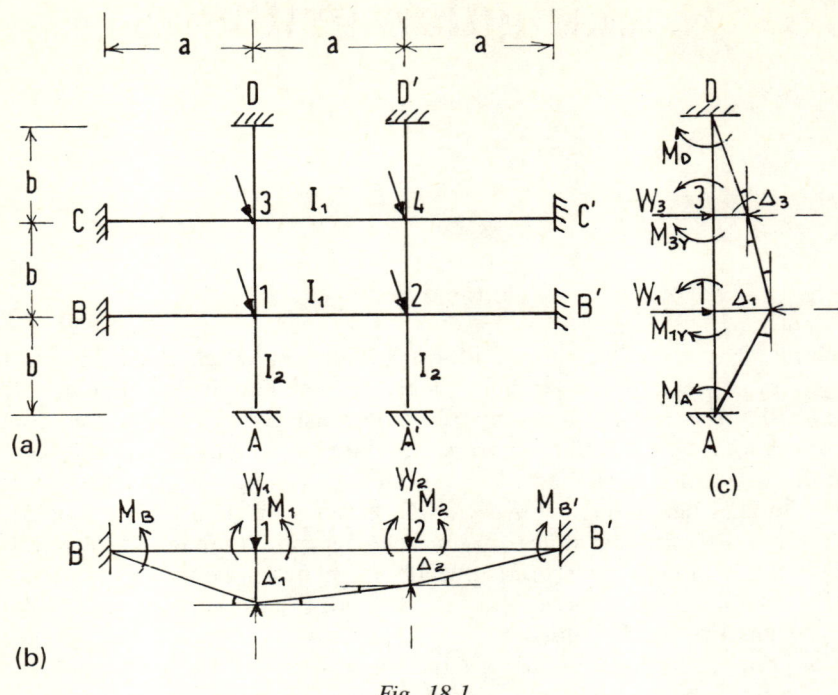

Fig. 18.1.

A.A: $\quad M_A \dfrac{b}{3EI_2} + M_{1Y} \dfrac{b}{6EI_2} = \dfrac{\Delta_1}{b}$

$2KM_A + KM_{1Y} = \dfrac{P}{b}\Delta_1 = \dfrac{P}{a} r \Delta_1$

A.B: $\quad M_B \dfrac{a}{3EI_1} + M_1 \dfrac{a}{6EI_1} = \dfrac{\Delta_1}{a}$

$2M_B + M_1 = \dfrac{P}{a} \Delta_1$

A.C: $\quad M_C \dfrac{a}{3EI} + M_3 \dfrac{a}{6EI} = \dfrac{\Delta_3}{a}$

$2M_C + M_3 = \dfrac{P}{a} \Delta_3$

A.D: $\quad M_D \dfrac{b}{3EI_2} + M_{3Y} \dfrac{b}{6EI_2} = \dfrac{\Delta_3}{b}$

$2KM_D + KM_{3Y} = \dfrac{P}{a} r \Delta_3$

A.B12: $\quad M_B \dfrac{a}{6EI_1} + M_1 \dfrac{2a}{3EI_1} + M_2 \dfrac{a}{6EI_1} = -\dfrac{\Delta_1}{a} + \dfrac{\Delta_2 - \Delta_1}{a}$

RECTANGULAR GRILLAGES 211

$$M_B + 4M_1 + M_2 = \frac{P}{a}(\Delta_2 - 2\Delta_1)$$

A.A13: $\quad M_A \dfrac{b}{6EI_2} + M_{1Y} \dfrac{2b}{3EI_2} + M_{3Y} \dfrac{b}{6EI_2} = -\dfrac{\Delta_1}{b} + \dfrac{\Delta_3 - \Delta_1}{b}$

$$KM_A + 4KM_{1Y} + KM_{3Y} = \frac{P}{a}(r\Delta_3 - 2r\Delta_1)$$

A.C34: $\quad M_C \dfrac{a}{6EI_1} + M_3 \dfrac{2a}{3EI_1} + M_4 \dfrac{a}{3EI_1} = \dfrac{\Delta_4 - \Delta_3}{a} - \dfrac{\Delta_3}{a}$

$$M_C + 4M_3 + M_4 = \frac{P}{a}(\Delta_4 - 2\Delta_3)$$

A.13D: $\quad M_{1Y} \dfrac{b}{6EI_2} + M_{3Y} \dfrac{2b}{3EI_2} + M_D \dfrac{b}{6EI_2} = \dfrac{\Delta_1 - \Delta_3}{b} - \dfrac{\Delta_3}{b}$

$$KM_{1Y} + 4KM_{3Y} + KM_D = \frac{P}{a}(r\Delta_1 - 2r\Delta_3)$$

E.1: $\quad \dfrac{M_B - M_1}{a} + \dfrac{M_2 - M_1}{a} + \dfrac{M_A - M_{1Y}}{b} + \dfrac{M_{3Y} - M_{1Y}}{b} = W_1$

$$M_B - 2M_1 + M_2 - 2rM_{1Y} + rM_A + rM_{3Y} = W_1 a$$

E.3: $\quad \dfrac{M_C - M_3}{a} + \dfrac{M_4 - M_3}{a} + \dfrac{M_{1Y} - M_{3Y}}{b} + \dfrac{M_D - M_{3Y}}{b} = W_3$

$$M_C - 2M_3 + M_4 + rM_{1Y} - 2rM_{3Y} + rM_D = W_3 a$$

Since no bending is assumed to occur in the members, the nodal reactions are equal to the applied load in each case. This set of equations is sufficient to enable the full-frame matrix table to be prepared, as shown in Table 18.1. From this one may also construct half-frame and quarter-frame matrices to be used where conditions of symmetry permit, as described in the next section.

Matrix size

The size of matrix required to obtain results will depend on the degree of symmetry of the frame and load pattern. For this problem, if the frame or loading is asymmetrical, a full-frame (16 × 16) size matrix with four deflection vectors will be needed. If the frame and loading is symmetrical about either the XX or YY axis, a half-frame (8 × 8) size matrix with two deflection vectors will suffice. If frame and loading are symmetrical about both axes, a quarter-frame (4 × 4) size matrix with one deflection vector is adequate. Equations for 'part' frames will need modification for members that are truncated. Thus, for a frame symmetrical about the YY axis, equations A.B12 and E.1 would be rewritten as:

212 FOUNDATION AND STRUCTURAL PROBLEMS

Table 18.1 Tabular matrix for the grillage in Fig. 18.1

Equation A	Moment														$\overline{\Delta}$			
	A	B	C	D	1	1Y	2	2Y	3	3Y	4	4Y	A'	B'	C'	D'	1	2	3	4
A	2K					K											r			
B	2	2			1												1			
C		2		2K			1												1	
D			2K						K											r
B12	1				4	4K	1										−2	−2		
A13	K				1	4K	4		K								−2r	−2r		
12B'								4K		K							1	1		
A'24								4K	1		4	K	K	1						
C34		1								4K	1								−2	−2r
13D			K	K					1		4	4K	K		1		r		−2r	1
34C'												2K						r	1	−2
24D'													K		2	2K		r	1	−2r
A'														1						1
B'															1			r	r	r
C'																1				
D'																				

Equation E																	$\overline{W}$			
1	r	1			−2	−2r	1										W_1a			
2		1			1		−2	−2r	1				r	1			W_2a			
3			1	r		r			−2	−2r	1			r	1		W_3a			
4				r				r	1		−2	−2r			1	r	W_4a			

RECTANGULAR GRILLAGES

A.B12: $M_B + 4M_1 + M_1 = \dfrac{P}{a}(\Delta_1 - 2\Delta_1)$ since $M_2 = M_1$ and $\Delta_2 = \Delta_1$

$M_B + 5M_1 = -\dfrac{P}{a}\Delta_1$

E.1: $M_B - 2M_1 + M_1 - 2rM_{1Y} + rM_A + rM_{3Y} = W_1 a$

$M_B - M_1 - 2rM_{1Y} + rM_A + rM_{3Y} = W_1 a$

Similarly, for a frame symmetrical about the XX axis, equations A.A13 and E.1 become:

A.A13: $KM_A + 4KM_{1Y} + KM_{1Y} = \dfrac{P}{a}(r\Delta_1 - 2r\Delta_1)$

$KM_A + 5KM_{1Y} = -\dfrac{P}{a}r\Delta_1$

E.1: $M_B - 2M_1 + M_2 - 2rM_{1Y} + rM_A + rM_{1Y} = W_1 a$

$M_B - 2M_1 + M_2 - rM_{1Y} + rM_A = W_1 a$

For the quarter frame, the angular equations will take the same form as above; the equilibrium equation for joint 1 becomes:

E.1: $M_B - 2M_1 + M_1 - 2rM_{1Y} + rM_A + rM_{1Y} = W_1 a$

$M_B - M_1 - rM_{1Y} + rM_A = W_1 a$

These modifications are incorporated in the half-frame and quarter-frame matrices in Tables 18.2 and 18.3.

If the supports are pinned or simply supported, the size of matrix to be

Table 18.2 Half-frame matrix for Fig. 18.1 (assumed symmetrical about the YY axis)

Equation A	Moment								$\overline{\Delta}$	
	A	B	C	D	1	1Y	3	3Y	1	3
A	2K					K			r	
B		2			1				1	
C			2				1			1
D				2K				K		r
B12		1			5				−1	
A13	K					4K		K	−2r	r
C34			1				5			−1
13D				K		K		4K	r	−2r

Equation E									$\overline{W}$	
1		r	1		−1	−2r		r	$W_1 a$	
3				1	r		−1	−2r	$W_3 a$	

Table 18.3 Quarter-frame matrix for Fig. 18.1

Equation A	Moment				$\overline{\Delta}$
	A	B	1	1Y	1
A	2K			K	r
B		2	1		1
B12		1	5		−1
A13	K			5K	−r
Equation E					$\overline{W}$
1	r	1	−1	−r	$W_1 a$

handled is reduced also, since moments M_A, M_B, etc. are not required and no angular change occurs on release at the supports. For this particular problem, the tabular matrix and deflection vectors required where the supports are not fixed are enclosed in Table 18.1 by two sets of dashed lines.

GRID program

As a first step in preparing a suitable computer program to solve the unknown moments and deflections in grillages, a flow chart should be drawn to indicate the main stages in the analysis. This will be similar to that used in connection with the XBEAM program in Chapter 16, but omitting references to the 'beam loads' and 'beam moments' since we are considering loads applied at the nodes only. A simplified version of Fig. 16.10 may be prepared on these lines; this will assist in the writing of a program specifically for solving grillages such as that in Fig. 18.1. Practical considerations will limit the size of grid that can be tackled with simple programs of the XBEAM type, however. A program that includes arrays for a total of nine deflected nodes is listed in Appendix B. This has been given the synoptic title GRID. One can, of course, insert additional statements to cater for extra nodes, if necessary.

The program does not warrant any special explanation as it is similar in many respects to XBEAM. One might note that, as the matrix inversion process is used a number of times, statements have to be included to zero the arrays each time the subroutine is called up in the main program. Line 405 incorporates a device to have results printed out to two decimal places. This is not only useful in practice, but also eliminates the false sense of precision given by having results displayed to several decimal places. If more accurate figures are needed for some reason, this statement

RECTANGULAR GRILLAGES 215

can be amended to provide extra decimal places or omitted altogether. Also, to conserve space in the printout, it is suggested that line 850 in MATSUB be amended to read

$$850 \text{ PRINT R (I);}$$

so that the vectors are printed in rows instead of columns. This has been done for the printout reproduced in Appendix B.

The program will be used initially to solve three problems involving the (3 × 3) grid depicted in Fig. 18.1.

Problem (1)   A load of 100 kN at joint 1.
Problem (2)   Loads of 100 kN each at joints 1 and 2.
Problem (3)   Loads of 100 kN each at joints 1,2,3 and 4.

In each case, a = 2 m, b = 1 m and $I_1 = 2I_2$.

Then, $$K = \frac{I_1}{2} \frac{2}{I_1} = 1 \text{ and } r = 2.$$

For Problem (1), the entire frame has to be considered so that values for a (16 × 16) coefficient matrix and four deflected nodes must be entered as data statements before the program is run. In accordance with the data instructions in the listing, the moment coefficients are keyed in first row-by-row, then the successive deflection vectors and equilibrium coefficients and finally the $\overline{W}$ vector terms. A half-frame matrix will suffice for Problem (2), since the loading is symmetrical about the YY axis; likewise, a quarter-frame matrix is adequate for solving Problem (3).

Use of the coefficient r = a/b simplifies data preparation and the coefficient matrix will be valid for all grillages of similar shape. However, it is important to remember, when using GRID, to multiply the nodal loads by the grid spacing a in entering the $\overline{W}$ vector terms in data tables incorporating r values. Where actual L_1, L_2, etc. values are used, as in the torsional problems in Chapter 20, nodal loads only are entered for $\overline{W}$. The heading in line 300 of GRID should be amended to suit, i.e. from NODAL LOAD × BAY LENGTH to NODAL LOADS.

It will be seen from the tables that in each case the deflection vectors are identical with the equilibrium coefficients and so can be entered by simply overtyping the line numbers. To illustrate the procedure for using the program, the data statements for Problem (2) are set out below. The values are based on Table 18.2.

```
1000 DATA 2,0,0,0,0,1,0,0
1005 DATA 0,2,0,0,1,0,0,0
1010 DATA 0,0,2,0,0,0,1,0
1015 DATA 0,0,0,2,0,0,0,1
1020 DATA 0,1,0,0,5,0,0,0
1025 DATA 1,0,0,0,0,4,0,1
1030 DATA 0,0,1,0,0,0,5,0
```

```
1035 DATA 0,0,0,1,0,1,0,4
1040 DATA 2,1,0,0,-1,-4,0,2
1045 DATA 0,0,1,2,0,2,-1,-4
1050 DATA 2,1,0,0,-1,-4,0,2
1055 DATA 0,0,1,2,0,2,-1,-4
1060 DATA 200,0
```

Values to input when the program is run are N = 8 and U = 1,0. The results printed out accompanied by the description 'Final moment' and identifying number are, to two decimal places:

$M_A = 37.28$; $M_B = 16.28$; $M_c = 10.38$; $M_D = 16.05$; $M_1 = -8.14$; $M_{1Y} = -25.72$; $M_3 = -5.19$ and $M_{3Y} = -0.94$

Fewer data are required for solving Problem (3) using a quarter-frame matrix. The following statements are sufficient, in fact

```
1000 DATA 2,0,0,1,0,2,1,0
1005 DATA 0,1,5,0,1,0,0,5
1010 DATA 2,1,-1,-2,2,1,-1,-2,200
```

INPUT N = 4 and U = 0 to obtain the results:

$M_A = 53.33$; $M_B = 26.67$; $M_1 = -13.33$; $M_{1Y} = -26.67$ kNm

Data statements for the full frame based on Table 18.1 may be used with GRID to obtain the results for Problem (1) in the same way. For comparison, these are given below in the same order as the column headings to the table:

30.18, 16.75, 9.35, 10.67, −14.1, −21.59, 6.0, −4.12, −7.45, 1.17, 2.26, −2.11, 7.11, 4.6, 1.03, 5.38.

See section on 'Errors in ouput'.

One-way grids

The previous section contained tabular matrices for two-way spanning grids with either fixed or simple supports. The same approach may be used for dealing with one-way grids, such as the (3 × 3) grid in Fig. 18.2. In this case supports A, B, C, etc. are assumed pinned.
Typical equations are:

A.A12: $\quad M_1 \dfrac{2a}{3EI_1} + M_2 \dfrac{a}{6EI_1} = -\dfrac{\Delta_1}{a} + \dfrac{\Delta_2 - \Delta_1}{a}$

$\quad 4M_1 + M_2 = \dfrac{P}{a}(\Delta_2 - 2\Delta_1)$

A.12A': $\quad M_1 \dfrac{a}{6EI_1} + M_2 \dfrac{2a}{3EI_1} = \dfrac{\Delta_1 - \Delta_2}{a} - \dfrac{\Delta_2}{a}$

RECTANGULAR GRILLAGES

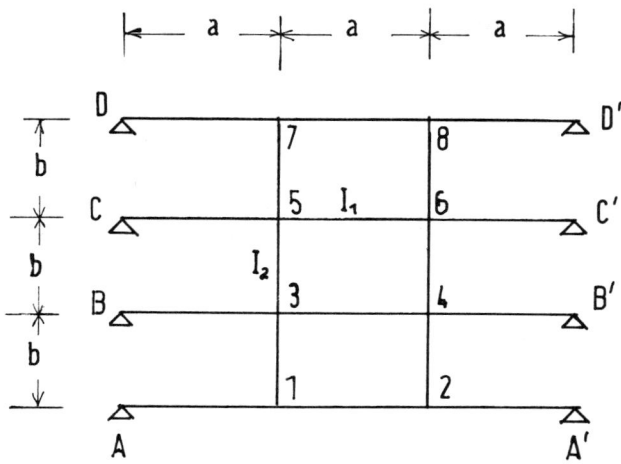

Fig. 18.2.

$$M_1 + 4M_2 = \frac{P}{a}(\Delta_1 - 2\Delta_2)$$

A.B34: $$4M_3 + M_4 = \frac{P}{a}(\Delta_4 - 2\Delta_3)$$

A.135: $$M_{3Y}\frac{2b}{3EI_2} + M_{5Y}\frac{b}{6EI_2} = \frac{\Delta_1 - \Delta_3}{b} + \frac{\Delta_5 - \Delta_3}{b}$$

$$4\frac{b}{a}\frac{I_1}{I_2}M_{3Y} + \frac{b}{a}\frac{I_1}{I_2}M_{5Y} = \frac{6EI}{ab}(\Delta_1 - 2\Delta_3 + \Delta_5)$$

$$4KM_{3Y} + KM_{5Y} = \frac{P}{a}(r\Delta_1 - 2r\Delta_3 + r\Delta_5)$$

A.357: $$M_{3Y}\frac{b}{6EI_2} + M_{5Y}\frac{2b}{3EI_2} = \frac{\Delta_3 - \Delta_5}{b} + \frac{\Delta_7 - \Delta_5}{b}$$

$$KM_{3Y} + 4KM_{5Y} = \frac{P}{a}(r\Delta_3 - 2r\Delta_5 + r\Delta_7)$$

E.1: $$-\frac{M_1}{a} + \frac{M_2 - M_1}{a} + \frac{M_{3Y}}{b} = W_1$$

$$-2M_1 + M_2 + rM_{3Y} = W_1 a$$

E.2: $$\frac{M_1 - M_2}{a} - \frac{M_2}{a} + \frac{M_{4Y}}{b} = W_2$$

$$-2M_2 + M_1 + rM_{4Y} = W_2 a$$

E.3: $$-\frac{M_3}{a} + \frac{M_4 - M_3}{a} - \frac{M_{3Y}}{b} + \frac{M_{5Y} - M_{3Y}}{b} = W_3$$

$$-2M_3 + M_4 - 2rM_{3Y} + rM_{5Y} = W_3 a$$

E.4: $$-2M_4 + M_3 - 2rM_{4Y} + rM_{6Y} = W_4 a$$

The equations are expressed in tabular matrix form in Table 18.4.

Taking the simple case where $I_1 = I_2$, $a = b = 2$ m and a concentrated load of 50 kN is applied at node 1, we can prepare a compact data table as in Table 18.5 to show the values to be entered in statements before the GRID program is run. The data statements are keyed in the prescribed order, remembering that one set of the figures must be entered twice, as indicated by the heading. Zeroes must be keyed in for all blanks in the table. Results to two decimal places are shown in the bottom line of the table.

Partitioned grillages

We have seen that, by use of a half- or quarter-frame matrix in the analysis of a symmetrical grillage, the number of data to be handled is reduced considerably. Tables 18.1, 18.2 and 18.3 indicate how equations for truncated members are modified where the partitioning axis of symmetry passes between nodes. It will be seen that *reflected* nodes result in the value 5 or 5K appearing in the angular equations and a single -1 or $-r$ value instead of two terms in the deflection vectors and corresponding equilibrium coefficients. These alterations hold good only for grillages with an odd number of bays.

Where a symmetrical grillage has an even number of bays the equations take on a different aspect. For instance, the (4×3) one-way grid in Fig. 18.3 may be analysed as a half frame about member 2468 under symmetrical conditions. The full-frame matrix can be written down in the usual manner. For the half-frame matrix the equations derived from these are shown below:

A.121': $\quad M_1 \dfrac{a}{6EI_1} + M_2 \dfrac{2a}{3EI_1} + M_{1'} \dfrac{a}{6EI_1} = \dfrac{\Delta_1 - \Delta_2}{a} + \dfrac{\Delta_{1'} - \Delta_2}{a}$

$\qquad 2M_1 + 4M_2 = \dfrac{P}{a}(2\Delta_1 - 2\Delta_2)$

$\qquad M_1 + 2M_2 = \dfrac{P}{a}(\Delta_1 - \Delta_2)$

A.343': $\quad M_3 \dfrac{a}{6EI_1} + M_4 \dfrac{2a}{3EI_1} + M_{3'} \dfrac{a}{6EI_1} = \dfrac{\Delta_3 - \Delta_4}{a} + \dfrac{\Delta_{3'} - \Delta_4}{a}$

$\qquad 2M_3 + 4M_4 = \dfrac{P}{a}(2\Delta_3 - 2\Delta_4)$

$\qquad M_3 + 2M_4 = \dfrac{P}{a}(\Delta_3 - \Delta_4)$

A.246: $\quad M_{4Y} \dfrac{2b}{3EI_2} + M_{6Y} \dfrac{b}{6EI_2} = \dfrac{\Delta_2 - \Delta_4}{b} + \dfrac{\Delta_6 - \Delta_4}{b}$

Table 18.4 Tabular matrix for the grid in Fig. 18.2

Equation A	Moment												$\overline{\Delta}$							
	1	2	3	3Y	4	4Y	5	5Y	6	6Y	7	8	1	2	3	4	5	6	7	8
A12	4	1											−2	1						
122'	1	4											1	−2						
B34			4		1										−2	1				
135			1	4K				K					r		−2r		r			
344'			1		4										1	−2				
246						4K	4		1	K				r		−2r		r		
C56							4		1								−2	1		
357				K				4K	1		4				r		−2r		r	
566'							1		4								1	−2		
468						K			4	4K	1					r		−2r		r
D78											4	1							−2	1
788'											1	4							1	−2

Equation E	1	2	3	4	5	6	7	8	$\overline{W}$
1	−2	1		r					$W_1 a$
2	1	−2							$W_2 a$
3			−2	1	−2r				$W_3 a$
4			1	−2		r			$W_4 a$
5			r		−2	1	−2r		$W_5 a$
6					1	−2		r	$W_6 a$
7					r		−2	1	$W_7 a$
8							1	−2	$W_8 a$

Table 18.5 Data table for the (3 × 3) grid problem

					Moment							
Member	1	2	3	3Y	4	4Y	5	5Y	6	6Y	7	8
A12	4	1										
122'	1	4										
B34			4		1							
135			4									
344'			1		4							
246						4			1			
C56							4		1			
357				1				4				
566'							1	4				
468						1			4	4		
D78											4	1
788'											1	4

Deflection and equilibrium coefficients

Node	1	2	3	3Y	4	4Y	5	5Y	6	6Y	7	8
1	−2	1										
2	1	−2										
3			−2	1								
4			1	−2	1							
5			1		−2	1	−2	1				
6						−2	1	−2	1			
7						1	−2	1	−2	1		
8										1		
W	100										−2	1
											1	−2

| Results: | −55.05 | −23.57 | −17.21 | 13.48 | −13.60 | 7.91 | −0.43 | 6.13 | −2.08 | 5.82 | 6.03 | 5.92 |

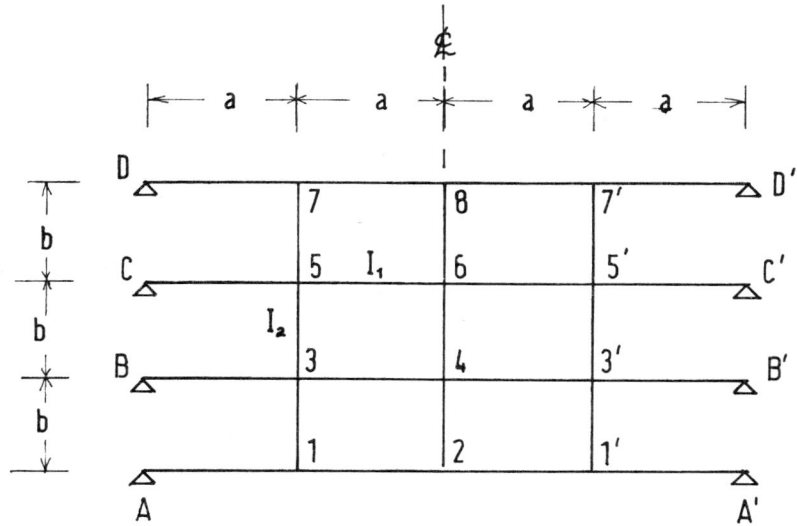

Fig. 18.3.

$$4KM_{4Y} + KM_{6Y} = \frac{P}{a}(r\Delta_2 - 2r\Delta_4 + r\Delta_6)$$

E.2: $\quad \dfrac{M_1 - M_2}{a} + \dfrac{M_{1'} - M_2}{a} + \dfrac{M_{4Y}}{b} = W_2$

$-2M_2 + 2M_1 + rM_{4Y} = W_2 a$

$M_1 - M_2 + \dfrac{r}{2} M_{4Y} = \dfrac{W_2 a}{2}$

E.4: $\quad \dfrac{M_3 - M_4}{a} + \dfrac{M_{3'} - M_4}{a} - \dfrac{M_{4Y}}{b} + \dfrac{M_{6Y} - M_{4Y}}{b} = W_4$

$-2M_4 + 2M_3 - 2rM_{4Y} + rM_{6Y} = W_4 a$

The equations are summarised in the tabular matrix in Table 18.6, arranged so that the deflection vectors and equilibrium coefficients coincide. It will be seen that angular compatibility equations for members along the axis of symmetry as well as those with 'reflected' nodes and equilibrium equations for nodes on the axis are all halved. Other equations are unaffected.

Similarly, we can produce a quarter-frame matrix, for a square grid with an even number of bays, such as the (4 × 4) grid in Fig. 18.4, where both axes of symmetry lie along the grid lines. Relevant equations for this case are:

A.E34: $\quad M_3 \dfrac{2a}{3EI_1} + M_4 \dfrac{a}{6EI_1} = -\dfrac{\Delta_3}{a} + \dfrac{\Delta_4 - \Delta_3}{a}$

222 FOUNDATION AND STRUCTURAL PROBLEMS

Table 18.6 Tabular matrix for the half-frame in Fig. 18.3

Equation A	Moment													$\overline{\Delta}$							
	1	2	3	3Y	4	4Y	5	5Y	6	6Y	7	8	1	2	3	4	5	6	7	8
A12	4	1											-2	1						
121'	1	2											1	-1						
B34			4		1										-2	1				
135			4K				K						r		$-2r$		r			
343'			1		2										1	-1				
246						2K				$\frac{K}{2}$								$\frac{r}{2}$	$\frac{r}{2}$		
C56							4		1										1		
357				K			1	4K	2							$-r$		$-2r$	-1		
565'									2								r	1	$-r$	r	
468						$\frac{K}{2}$				2K						$\frac{r}{2}$			$\frac{r}{2}$		
D78											4	1								-2	1
787'											1	2								1	-1

Equation E	Moment													$\overline{W}$
	1	2	3	3Y	4	4Y	5	5Y	6	6Y	7	8		
1	-2	1		r										W_1a
2	1	-1				$\frac{r}{2}$								$W_2a/2$
3			-2	$-2r$	1			r						W_3a
4			1	-1	$-r$									$W_4a/2$
5				r			-2	$-2r$	1	$\frac{r}{2}$				W_5a
6						$\frac{r}{2}$	1		-1	$-r$				$W_6a/2$
7								r			-2	1		W_7a
8										$\frac{r}{2}$	1	-1		$W_8a/2$

RECTANGULAR GRILLAGES

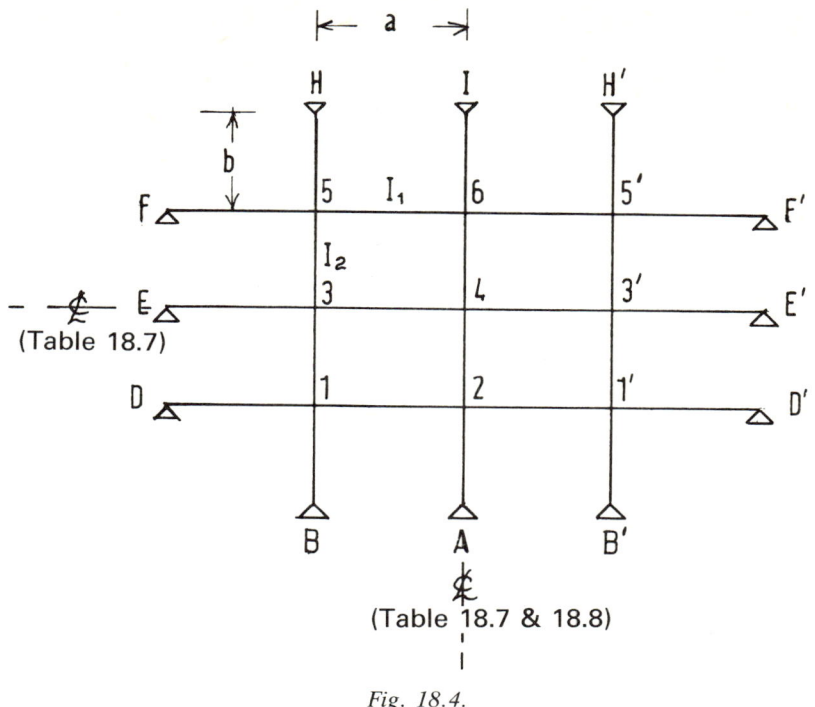

Fig. 18.4.

$$4M_3 + M_4 = \frac{P}{a}(\Delta_4 - 2\Delta_3)$$

A.343': $\quad M_3 \dfrac{a}{6EI_1} + M_4 \dfrac{2a}{3EI_1} + M_{3'} \dfrac{a}{6EI_1} = \dfrac{\Delta_3 - \Delta_4}{a} + \dfrac{\Delta_{3'} - \Delta_4}{a}$

$$2M_3 + 4M_4 = \frac{P}{a}(2\Delta_3 - 2\Delta_4)$$

$$M_3 + 2M_4 = \frac{P}{a}(\Delta_3 - \Delta_4)$$

A.A24: $\quad M_{2Y} \dfrac{2b}{3EI_2} + M_{4Y} \dfrac{b}{6EI_2} = -\dfrac{\Delta_2}{b} + \dfrac{\Delta_4 - \Delta_2}{b}$

$$4KM_{2Y} + KM_{4Y} = \frac{P}{a}(r\Delta_4 - 2r\Delta_2)$$

A.246: $\quad M_{4Y} \dfrac{2b}{3EI_2} + M_{2Y} \dfrac{b}{6EI_2} + M_{6Y} \dfrac{b}{6EI_2} = \dfrac{\Delta_2 - \Delta_4}{b} + \dfrac{\Delta_6 - \Delta_4}{b}$

$$4KM_{4Y} + 2KM_{2Y} = \frac{P}{a}(r\Delta_2 - 2r\Delta_4 + r\Delta_6)$$

$$2KM_{4Y} + KM_{2Y} = \frac{P}{a}(r\Delta_2 - r\Delta_4)$$

Table 18.7 Tabular matrix for the quarter-frame in Fig. 18.4

Equation A	Moment								$\overline{\Delta}$			
	1	1Y	2	2Y	3	3Y	4	4Y	1	2	3	4
D12	4		1						-2	1		
B13		4K				K			$-2r$		r	
121'	1		2						1	-1		
A24			2K					$\dfrac{K}{2}$		$-r$		$\dfrac{r}{2}$
E34					2	$\dfrac{1}{2}$					-1	$\dfrac{1}{2}$
135		K			2K				r	$-r$		
343'				$\dfrac{1}{2}$			1				$\dfrac{1}{2}$	$-\dfrac{1}{2}$
246				$\dfrac{K}{2}$				K		$\dfrac{r}{2}$		$-\dfrac{r}{2}$

Equation E									$\overline{W}$
1	-2	$-2r$	1			r			$W_1 a$
2	1		-1	$-r$				$\dfrac{r}{2}$	$W_2 a/2$
3		r			-1	$-r$	$\dfrac{1}{2}$		$W_3 a/2$
4				$\dfrac{r}{2}$	$\dfrac{1}{2}$		$-\dfrac{1}{2}$	$-\dfrac{r}{2}$	$W_4 a/4$

A.121': $M_1 \dfrac{a}{6EI_1} + M_2 \dfrac{2a}{3EI_1} + M_{1'} \dfrac{a}{6EI_1} = \dfrac{\Delta_1 - \Delta_2}{a} + \dfrac{\Delta_{1'} - \Delta_2}{a}$

$2M_1 + 4M_2 = \dfrac{P}{a}(2\Delta_1 - 2\Delta_2)$

$M_1 + 2M_2 = \dfrac{P}{a}(\Delta_1 - \Delta_2)$

A.135: $M_{3Y} \dfrac{2b}{3EI_2} + M_{1Y} \dfrac{b}{6EI_2} + M_{5Y} \dfrac{b}{6EI_2} = \dfrac{\Delta_1 - \Delta_3}{b} + \dfrac{\Delta_5 - \Delta_3}{b}$

$4KM_{3Y} + 2KM_{1Y} = \dfrac{P}{a}(2r\Delta_1 - 2r\Delta_3)$

$2KM_{3Y} + KM_{1Y} = \dfrac{P}{a}(r\Delta_1 - r\Delta_3)$

E.2: $\dfrac{M_1 - M_2}{a} + \dfrac{M_{1'} - M_2}{a} - \dfrac{M_{2Y}}{b} + \dfrac{M_{4Y} - M_{2Y}}{b} = W_2$

RECTANGULAR GRILLAGES 225

Table 18.8 Tabular matrix for the half-frame in Fig. 18.4

Equation A	Moment													$\overline{\Delta}$					
	1	1Y	2	2Y	3	3Y	4	4Y	5	5Y	6	6Y	1	2	3	4	5	6
D12	4	4K	1										-2	1				
B13	1		2			K							$-2r$		r			
121'													1	-1				
A24				2K	4		1	$\dfrac{K}{2}$								$\dfrac{r}{2}$		
E34		K				4K	2			K					-2	1		
135					1								r	$-r$	$-2r$	-1	r	
343'															1			
246				$\dfrac{K}{2}$		K	2	2K	4		1	$\dfrac{K}{2}$		$\dfrac{r}{2}$	r	$\dfrac{r}{2}$		
F56									4	4K	1						-2	1
35H									1		2						$-2r$	-1
565'																	1	-1
461								$\dfrac{K}{2}$				2K				$\dfrac{r}{2}$		$-r$

Equation E	1	2	2Y	3	3Y	4	4Y	5	5Y	6	6Y	$\overline{W}$
1	-2	1			r							$W_1 a$
2	1	-1	$-r$				$\dfrac{r}{2}$					$W_2 a/2$
3	r			-2	$-2r$	1			r			$W_3 a$
4			$\dfrac{r}{2}$	1		-1	$-r$				$\dfrac{r}{2}$	$W_4 a/2$
5					r			-2	$-2r$	1		$W_5 a$
6							$\dfrac{r}{2}$	1		-1	$-r$	$W_6 a/2$

E.4:
$$2M_1 - 2M_2 - 2rM_{2Y} + rM_{4Y} = W_2 a$$
$$\frac{M_3 - M_4}{a} + \frac{M_{3'} - M_4}{a} + \frac{M_{2Y} - M_{4Y}}{b} + \frac{M_{6Y} - M_{4Y}}{b} = W_4$$
$$2M_3 - 2M_4 + 2rM_{2Y} - 2rM_{4Y} = W_4 a$$
$$M_3 - M_4 + rM_{2Y} - rM_{4Y} = W_4 a/2$$

The equations are incorporated in Table 18.7, arranged so that the deflection vectors and the rows of equilibrium coefficients are identical. This permits the second data set to be entered merely by overtyping the line numbers; alternatively, one may use the RESTORE (line number) facility, but this requires care should the data be altered at any stage.

In a similar fashion, for a two-way spanning grid with an even number of bays, the half-frame matrix about the YY axis for symmetrical conditions will take the form shown in Table 18.8. This applies to the (4 × 4) grid with knife-edge supports shown in Fig. 18.4. Equations for axial or truncated members are given below:

A.121':
$$M_1 \frac{a}{6EI_1} + M_2 \frac{2a}{3EI_2} + M_{1'} \frac{a}{6EI_1} = \frac{\Delta_1 - \Delta_2}{a} + \frac{\Delta_{1'} - \Delta_2}{a}$$
$$2M_1 + 4M_2 = \frac{P}{a}(2\Delta_1 - 2\Delta_2)$$
$$M_1 + 2M_2 = \frac{P}{a}(\Delta_1 - \Delta_2)$$

A.A24:
$$M_{2Y} \frac{2b}{3EI_2} + M_{4Y} \frac{b}{6EI_2} = -\frac{\Delta_2}{b} + \frac{\Delta_4 - \Delta_2}{b}$$
$$4KM_{2Y} + KM_{4Y} = \frac{P}{a}(r\Delta_4 - 2r\Delta_2)$$

A.246:
$$M_{2Y} \frac{b}{6EI_2} + M_{4Y} \frac{2b}{3EI_2} + M_{6Y} \frac{b}{6EI_2} = \frac{\Delta_2 - \Delta_4}{b} + \frac{\Delta_6 - \Delta_4}{b}$$
$$KM_{2Y} + 4KM_{4Y} + KM_{6Y} = \frac{P}{a}(r\Delta_2 - 2r\Delta_4 + r\Delta_6)$$

A.461:
$$M_{4Y} \frac{b}{6EI_2} + M_{6Y} \frac{2b}{3EI_2} = \frac{\Delta_4 - \Delta_6}{b} - \frac{\Delta_6}{b}$$
$$KM_{4Y} + 4KM_{6Y} = \frac{P}{a}(r\Delta_4 - 2r\Delta_6)$$

These equations, together with the standard equations obtained previously, enable Table 18.8 to be constructed.

Using the principles outlined above, it is relatively simple to construct tabular matrices for any shape or size of rectangular grillage and for any support or loading conditions for solution by the GRID program. The main obstacle is the number of data that one has to handle, which places a practical limit on the size of structure that can be tackled in this way. Some examples of various grillages and their solution by the GRID program are given in the next chapter.

Errors in Output

Grillages with a number of deflected nodes may show different results on some computer makes than those given in the text — although usually well within practical limits — since results often depend on the manipulation of either very small or very large quantities. Mathematical means are available for improving this situation but this forms too specialised a subject for treatment here. If the equations are not well-conditioned, error messages such as 'Division by zero' or 'Overflow' can be thrown up on the screen. When this occurs, try rearranging the data or using a different sub-routine. One way of achieving this is to modify PIVOT so as to act as a sub-routine and then merge it with GRID. This is left as an exercise for the reader to perform. The combined program is included, however, in the software available as described in the Preface, with an example of its use taken from the text. This is useful also as a means of checking the output.

19 Grillage analysis examples

One-way grid example

In this chapter one or two examples of grillage construction have been selected to illustrate the general principles outlined in the previous chapter. The equations used for obtaining the tabular matrices or data tables are not set out since these follow the same pattern as described earlier and so can be determined by inspection in most cases. For example, assuming the (5 × 3) one-way grid in Fig. 19.1 is symmetrical about the YY axis, the half-frame matrix will be similar in most respects to that derived for the (3 × 3) grid in Fig. 18.2 and reproduced in Table 18.4. Truncated members such as 122′, 344′, etc. are dealt with as described in Chapter 18 under *Partitioned Grillages* by inserting 5 or 5K in the angular equations and a single -1 or $-r$ value instead of two terms in the deflection vectors and equilibrium coefficients. This follows naturally from the equations. The end result is shown in Table 19.1 in compact tabular format.

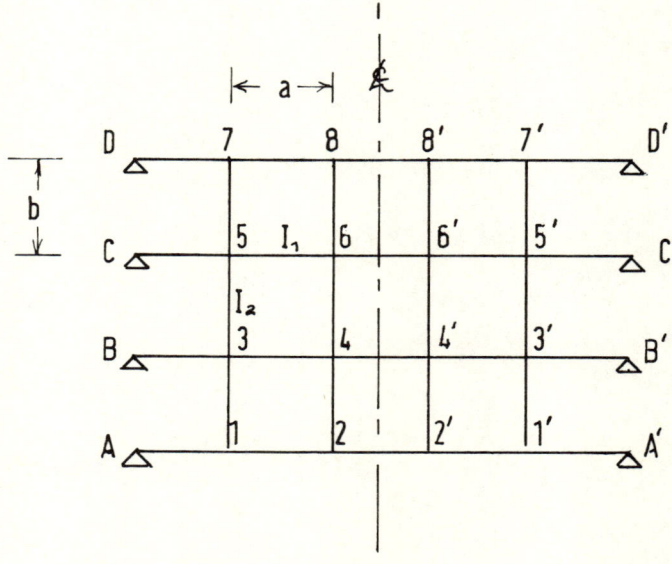

Fig. 19.1.

Table 19.1 Compact tabular matrix for Fig. 19.1

Equation A	Moment											
	1	2	3	3Y	4	4Y	5	5Y	6	6Y	7	8
A12	4	1										
122'	1	5										
B34			4		1							
135				4K			K					
344'			1		5							
246					4K			K				
C56						4			1			
357				K				4K				
566'						1			5			
468						K				4K		
D78											4	1
788'											1	5
Node	Deflection and equilibrium coefficients											
1	−2	1		2								
2	1	−1				2						
3			−2	−4	1			2				
4				1	−1	−4				2		
5					2		−2	−4	1			
6						2	1		−1	−4		
7								2			−2	1
8										2	1	−1
W	W₁a	W₂a	W₃a	W₄a	W₅a	W₆a	W₇a	W₈a				

To see how the data for a frame of this type are handled, consider the problem in which the grid supports a load of 100 kN at each of nodes 2, 4, 6 and 8 (as well, of course, as nodes 2', 4', 6' and 8') and has dimensions a = 2, b = 1 and moment of inertia $I_1 = 2I_2$. Therefore,

$$r = 2 \text{ and } K = \frac{I_1}{2}\frac{2}{I_1} = 1$$

The data statements to be keyed in with GRID are set out below, derived from Table 19.1. Note that lines 1100−1135 may be entered by overtyping 1060−1095 or by a RESTORE (line number) command in the program to permit the data to be re-read.

```
1000 DATA 4,1,0,0,0,0,0,0,0,0,0,0
1005 DATA 1,5,0,0,0,0,0,0,0,0,0,0
1010 DATA 0,0,4,0,1,0,0,0,0,0,0,0
1015 DATA 0,0,0,4,0,0,0,1,0,0,0,0
1020 DATA 0,0,1,0,5,0,0,0,0,0,0,0
1025 DATA 0,0,0,0,0,4,0,0,0,1,0,0
1030 DATA 0,0,0,0,0,0,4,0,1,0,0,0
1035 DATA 0,0,0,1,0,0,0,4,0,0,0,0
1040 DATA 0,0,0,0,0,0,1,0,5,0,0,0
```

```
1045 DATA 0,0,0,0,0,1,0,0,0,4,0,0
1050 DATA 0,0,0,0,0,0,0,0,0,0,4,1
1055 DATA 0,0,0,0,0,0,0,0,0,0,1,5
1060 DATA -2,1,0,2,0,0,0,0,0,0,0,0
1065 DATA -1,0,0,0,2,0,0,0,0,0,0,0
1070 DATA 0,0,-2,-4,1,0,0,2,0,0,0,0
1075 DATA 0,0,1,0,-1,-4,0,0,0,2,0,0
1080 DATA 0,0,0,2,0,0,-2,-4,1,0,0,0
1085 DATA 0,0,0,0,0,2,1,0,-1,-4,0,0
1090 DATA 0,0,0,0,0,0,0,2,0,0,-2,1
1095 DATA 0,0,0,0,0,0,0,0,0,2,1,-1
1100 DATA -2,1,0,2,0,0,0,0,0,0,0,0
1105 DATA -1,0,0,0,2,0,0,0,0,0,0,0
1110 DATA 0,0,-2,-4,1,0,0,2,0,0,0,0
1115 DATA 0,0,1,0,-1,-4,0,0,0,2,0,0
1120 DATA 0,0,0,2,0,0,-2,-4,1,0,0,0
1125 DATA 0,0,0,0,0,2,1,0,-1,-4,0,0
1130 DATA 0,0,0,0,0,0,0,2,0,0,-2,1
1135 DATA 0,0,0,0,0,0,0,0,0,2,1,-1
1140 DATA 0,200,0,200,0,200,0,200
```

When running GRID, input $N = 12$ and $U = 1,1,1,1,1,1,1,0$. The results for this problem to two decimal places are:

$M_1 = -200$, $M_2 = -400$, $M_3 = -200$, $M_{3Y} = 0$, $M_4 = -400$, $M_{4Y} = 0$, $M_5 = -200$, $M_{5Y} = 0$, $M_6 = -400$, $M_{6Y} = 0$, $M_7 = -200$, $M_8 = -400$ kNm.

When the moments are known, it is a simple matter to determine the end shears by dividing the difference of the end moments by the member length in each case, in accordance with equilibrium conditions. Deflections at the nodes can be obtained, first, by transforming the angular compatibility equations to permit back-substitution of the moments, as follows:

A.A12: $-2\Delta_1 + \Delta_2 = 4M_1 + M_2 = -1200$
A.122': $\Delta_1 - \Delta_2 = M_1 + 5M_2 = -2200$
A.B34: $-2\Delta_3 + \Delta_4 = 4M_3 + M_4 = -1200$
A.344': $\Delta_3 - \Delta_4 = M_3 + 5M_4 = -2200$
A.C56: $-2\Delta_5 + \Delta_6 = 4M_5 + M_6 = -1200$
A.566': $\Delta_5 - \Delta_6 = M_5 + 5M_6 = -2200$
A.D78 $-2\Delta_7 + \Delta_8 = 4M_7 + M_8 = -1200$
A.788': $\Delta_7 - \Delta_8 = M_7 + 5M_8 = -2200$

These equations assume a unit value for $P \ (= 6EI/L)$ and will give relative values of the nodal deflections. Thus, by solving the equations as they stand one can see that $\Delta_2 = 2\Delta_1$, $\Delta_4 = 2\Delta_3$, etc. To determine actual deflections we need to know the properties of at least the reference member. If, in this problem, $E = 200$ kN/mm^2 and $I = 100 \times 10^3$ mm^4 for member A12, then

$$P = \frac{6 \times 200 \times 100 \times 10^3}{1000} = 12 \times 10^4$$

By dividing the right-hand side terms in kNmm units by P, a data table may be prepared for solution by MATIN. This is shown in Table 19.2 with the results in the bottom line of the table. (Two equations would have sufficed for this problem, but more complex loading may demand that all the equations be examined as described.)

Most commercial computer programs output values for shear, deflection and torsion as well as moments at all nodes in the grillage. These are seldom of vital importance in the practical design of conventional structures. When required, critical values can be ascertained fairly easily once the nodal moments are known. For this reason, extension of the GRID program to print out these quantities has not been considered worthwhile.

Two-way grid printout

The two-way spanning (4 × 4) grid in Fig. 19.2 forms the next example. The tabular matrix, where conditions are symmetrical about the YY-axis, will be similar to that reproduced previously in Table 18.8. In this case, a load of 100 kN is applied at nodes 1, 2, 5 and 6 (and consequently nodes 1' and 5' also) and a = b = 1 m and $I_1 = I_2$. The printout for this problem is reproduced below. As usual, this gives all the data fed in by way of DATA and INPUT statements. The heading was typed in before the program was RUN.

Table 19.2 Data table for the deflections in Fig. 19.1

Equation A	Δ							
	1	2	3	4	5	6	7	8
A12	−2	1						
122'	1	−1						
B34			−2	1				
344'			1	−1				
C56					−2	1		
566'					1	−1		
D78							−2	1
788'							1	−1
M	−10	−18.33	−10	−18.33	−10	−18.33	−10	−18.33
Results (mm) MATIN N = 8	28.3	46.7	28.3	46.7	28.3	46.7	28.3	46.7

232   FOUNDATION AND STRUCTURAL PROBLEMS

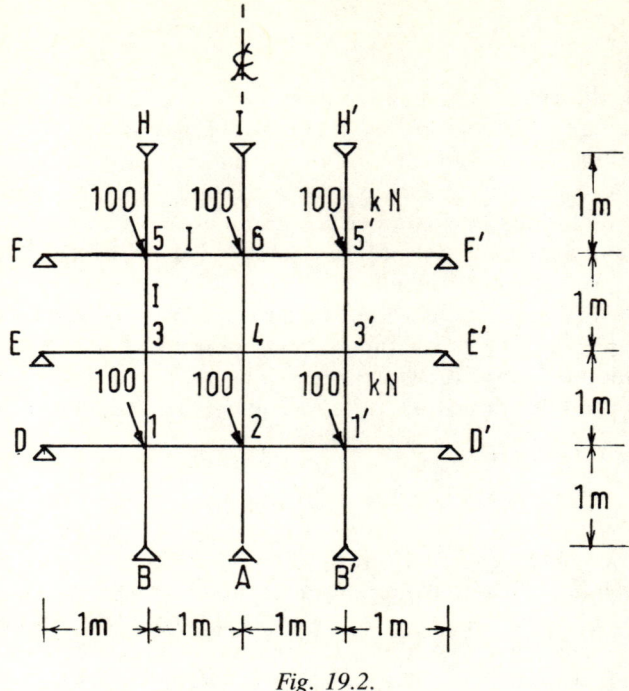

Fig. 19.2.

(4 × 4) GRID PROBLEM: HALF-FRAME MATRIX

MATRIX SIZE? 12

DATA:-
```
 4    0   1   0     0   0   0   0     0   0   0   0
 0    4   0   0     0   1   0   0     0   0   0   0
 1    0   2   0     0   0   0   0     0   0   0   0
 0    0   0   2     0   0   0  0.5    0   0   0   0
 0    0   0   0     4   0   1   0     0   0   0   0
 0    1   0   0     0   4   0   0     0   1   0   0
 0    0   0   0     1   0   2   0     0   0   0   0
 0    0   0  0.5    0   0   0   2     0   0   0  0.5
 0    0   0   0     0   0   0   0     4   0   1   0
 0    0   0   0     0   1   0   0     0   4   0   0
 0    0   0   0     0   0   0   0     1   0   2   0
 0    0   0   0     0   0   0  0.5    0   0   0   2

-2   -2   1   0     0   1   0   0     0   0   0   0
```

? ANOTHER DEFLECTED NODE. FOR YES PRESS 1
? 1

```
 1    0  -1  -1     0   0   0  0.5    0   0   0   0
```

? ANOTHER DEFLECTED NODE. FOR YES PRESS 1
? 1

0 1 0 0 −2 −2 1 0 0 1 0 0

? ANOTHER DEFLECTED NODE. FOR YES PRESS 1
? 1

0 0 0 0.5 1 0 −1 −1 0 0 0 0.5

? ANOTHER DEFLECTED NODE. FOR YES PRESS 1
? 1

0 0 0 0 0 1 0 0 −2 −2 1 0

? ANOTHER DEFLECTED NODE. FOR YES PRESS 1
? 1

0 0 0 0 0 0 0 0.5 1 0 −1 −1

? ANOTHER DEFLECTED NODE. FOR YES PRESS 1
? 0

EQUILIBRIUM COEFFS:-

```
−2  −2   1   0    0   1   0   0    0   0   0   0
 1   0  −1  −1    0   0   0   0.5  0   0   0   0
 0   1   0   0   −2  −2   1   0    0   1   0   0
 0   0   0   0.5  1   0  −1  −1    0   0   0   0.5
 0   0   0   0    0   1   0   0   −2  −2   1   0
 0   0   0   0    0   0   0   0.5  1   0  −1  −1
```

NODAL LOAD X BAY LENGTH:-

1 100
2   50
3   0
4   0
5 100
6   50

RESULTS:-

1   −40.62   FINAL MOMENT
2   −50   FINAL MOMENT
3   −48.44   FINAL MOMENT
4   −67.19   FINAL MOMENT
5   −51.56   FINAL MOMENT
6   −32.81   FINAL MOMENT
7   −68.75   FINAL MOMENT
8   −50   FINAL MOMENT
9   −40.62   FINAL MOMENT
10 −50   FINAL MOMENT
11 −48.44   FINAL MOMENT
12 −67.19   FINAL MOMENT

Space frames

The principles outlined for the solution of two-dimensional frames can readily be applied to three-dimensional space frames also. The main difficulty in this regard is the increased volume of data to be handled since,

(a) Space frame structures are economical only for covering large spans. Latticed space frames will contain a large number of members as a rule, therefore.
(b) Additional equations are required to deal with the resolution of moments and forces, etc. in the extra dimension.

Keying in the data becomes a tedious exercise if the procedures outlined in this chapter are adopted. It is preferable in most cases, it is suggested, to use one of the many commercial software packages for solving the space frame and to check the output by analysing an equivalent plane frame structure treated as a grillage as described in Chapters 18 and 21. This will provide a reasonable check on the results from the commercial program.

20 Torsion: cranked beams and cantilevers

Torsion in circular sections

Examination of torsion in structures usually starts with the simplest case, namely that of a solid circular section subjected to uniform torsion by the application of equal and opposite twisting moments at its ends (Fig. 20.1). As no warping occurs, plane sections remain plane. It may be shown that the relative rotation, θ, between the ends of the shaft is related to the applied torque, T, by the formula:

$$\theta = \frac{TL}{GJ} \qquad (20.1)$$

where J = second polar moment of inertia of the circular section = torsion constant

G = elastic shear modulus of the material = $\dfrac{E}{2(1+\lambda)}$ (20.2)

L = length of shaft
E = Young's modulus of elasticity
λ = Poisson's ratio

Proof of these equations will be found in structural textbooks on the subject (for example, Terzaghi and Peck, 1967). The equations hold good for other shapes also, except that for non-circular sections the constant J will have a different value.

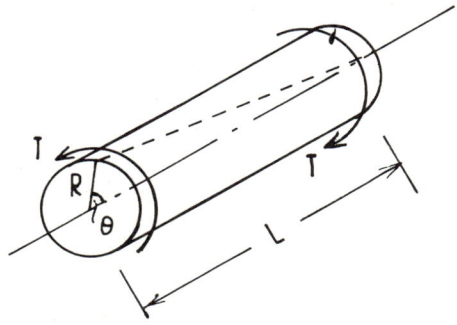

Fig. 20.1.

As the shaft suffers a uniform rate of twist along its length, we can denote the angular change per unit length under the torsion T by the term γ, so that

$$\gamma = \frac{\theta}{L} = \frac{T}{GJ} \quad (20.3)$$

Likewise, if the angular change per unit length under unit torsion is φ, then

$$\phi = \frac{\gamma}{T} = \frac{1}{GJ} \quad (20.4)$$

From equation (20.1) we obtain $\theta = TL\phi$ (20.5)

For a circular section only, $J = 2I$; therefore $\phi = \dfrac{1}{2GI}$ (20.6)

Later we will use the term $\Omega = 6EI\phi$

$$\therefore \Omega = \frac{6EI}{GJ} = \frac{12(1+\lambda)I}{J} \quad (20.7)$$

Note that Ω equals six times the ratio of the flexural rigidity (EI) to the torsional rigidity (GJ) of the section. For a circular section only,

$$\Omega = \frac{6EI}{2IG} = \frac{3E}{G} = \frac{6E(1+\lambda)}{E} = 6(1+\lambda) \quad (20.8)$$

Rectangular sections

For a rectangular section of depth d and width b, the torsion constant, J, may be described as:

$$J = \beta\, db^3 \quad (20.9)$$

where β is a factor varying with the ratio d/b (see Table 20.1). Substituting in equation (20.3),

$$\gamma = \frac{T}{G\beta db^3} = \frac{2(1+\lambda)T}{E\beta db^3}$$

Since $I = \dfrac{bd^3}{12}$, $\gamma = \dfrac{2T(1+\lambda)d^2}{E\beta I \cdot 12b^2} = \dfrac{T(1+\lambda)}{6EI\,\beta}\left(\dfrac{d}{b}\right)^2$

$$\phi = \frac{\gamma}{T} = \frac{1+\lambda}{6\beta EI}\left(\frac{d}{b}\right)^2$$

In this analysis, it is assumed that warping displacements are free to occur unrestrainedly under uniform torsion, as postulated in Saint-Venant's theory. Thus, for any rectangular beam of dimensions (b × d) and of a given material the term φ is constant.

Table 20.1 Factors for rectangular sections

$\dfrac{d}{b}$	1	1.2	1.5	2	2.5	3	4	5	10	∞
β	0.141	0.166	0.196	0.229	0.249	0.263	0.281	0.291	0.312	0.333

If we use $\Omega = 6EI\phi$ for the selected beam, then

$$\Omega = \frac{1+\lambda}{\beta}\left(\frac{d}{b}\right)^2 \qquad (20.10)$$

This equation is used in the release-deformation flexibility method for the 'reference element' (usually the first span or main bay member) in considering torsional effects on rectangular sections. This is normally the case with concrete structures, of course, but for other shapes reference should be made to Table 20.2, which gives values of I and J for both rectangular and other sections. Examples given in this chapter are for rectangular beams, but for beams of different section use the appropriate values in equation (20.7). The value of λ depends on the material used in the member; general values are, for steel $\lambda = 0.3$ and for reinforced concrete $\lambda = 0.2$. (The current Code of Practice, BS 8110, recommends a value of G equal to 0.4 times E_c and of J equal to one-half the Saint-Venant value for plain concrete sections. E_c varies with concrete grade but is usually taken as 30 kN/mm². For steel, BS 5400 gives $E_s = 205$ kN/mm², G = 80 kN/mm² and $\lambda = 0.3$.)

Note that the factor $\Omega_1 = 6EI_1\phi_1$ applies to the selected 'reference' member only or others identical to it. For any other member, using the suffix n as notation,

$$\Omega_n = \frac{6EI_1}{L_1} L_n \phi_n \qquad (20.11)$$

If the shear modulus, G, is constant for all frame members, then from equation (20.4),

$$\Omega_n = \frac{6EI_1}{L_1} L_n J_n = \Omega_1 \frac{L_n}{L_1} \frac{J_1}{J_n} \qquad (20.12)$$

For rectangular beams, referring to equation (20.10),

$$\Omega_n = \Omega_1 \frac{\beta_1}{\beta_n}\left(\frac{d_n}{b_n}\right)^2 \div \left(\frac{d_1}{b_1}\right)^2 \qquad (20.13)$$

Table 20.2 Section properties

Solid section	I_x	J
(circle, d)	$0.0491d^4$	$0.098d^4$
(square, a)	$0.083a^4$	$0.141a^4$
(rectangle, b×d)	$0.083bd^3$	βbd^3
(triangle, a)	$0.028a^4$	$0.0385a^4$

If the torsional rigidity (GJ) is the same for all members, then, since φ is constant,

$$\Omega_n = \frac{L_n}{L_1} \Omega_1 \qquad (20.14)$$

End cantilevers

The simplest structural example of torsion occurs where two cantilevers are interconnected at right angles, as shown in Fig. 20.5. Before examining this we will look at two common cases involving single cantilevers:

(1) end cantilever to a continuous beam;
(2) a cantilever carrying a uniformly distributed load.

In the first case, the cantilever moment is known but we want to determine its effect on the other support and span moments. Referring to Fig. 20.2, where the continuous beam is simply supported at each end A and D, it is clear that the cantilever moment, M_D, has no effect when a release is introduced in the normal way at support B; equation A.B will be unchanged, therefore. The compatibility equation for conditions at support C are affected, thus:

$$\text{A.C:} \quad M_B \frac{L_2}{6EI_2} + M_C \frac{L_2}{3EI_2} + M_C \frac{L_3}{3EI_3} + M_D \frac{L_3}{6EI_3} = \theta_{CB} + \theta_{CD}$$

$$K_2 M_B + 2(K_2 + K_3) M_C + K_3 M_D = \frac{1}{4} K_2 W_2 L_2 + \frac{1}{4} K_3 W_3 L_3$$

$$K_2 M_B + 2(K_2 + K_3) M_C = \frac{1}{4}(K_2 W_2 L_2 + K_3 W_3 L_3) - K_3 M_D$$

This introduces the additional quantity, $-K_3 M_D$ into the $\overline{W}$ vector. As the value of M_D will be known (or can easily be calculated), then in analysing the beam the $\overline{W}$ vector term at the penultimate support should be reduced by $K_3 M_D$ to take account of the end cantilever. The release moment is applied in the same direction as the cantilever, since this is treated as an externally applied moment. For example, if the three-span

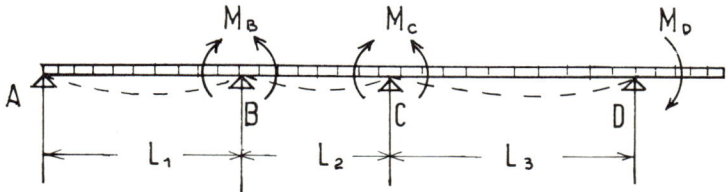

Fig. 20.2.

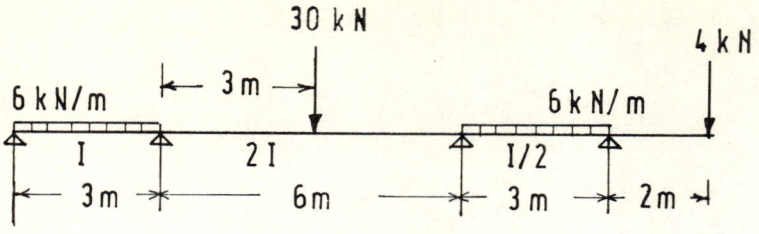

Fig. 20.3.

beam in Fig. 20.3 has a cantilever load of 4 kN applied at one end as shown then, taking the first span as reference,

$$K_1 = 1, \quad K_2 = \frac{I}{3}\frac{6}{2I} = 1 \text{ and } K_3 = \frac{I}{3}\frac{6}{I} = 2$$

$$\overline{W} \text{ at B} = \frac{1}{4} \times 18 \times 3 = \quad 13.5$$
$$+ \frac{3}{8} \times 30 \times 6 = \quad \underline{67.5}$$
$$\phantom{+ \frac{3}{8} \times 30 \times 6 = \quad} 81.0$$

$$\overline{W} \text{ at C} = \frac{3}{8} \times 30 \times 6 = \quad 67.5$$
$$+ \frac{1}{4} \times 18 \times 3 \times 2 = \quad 27.0$$
$$- 2 \times 8 = \quad \underline{-16.0}$$
$$ 78.5$$

These values are used to construct Table 20.3, which shows the results obtained by MATIN in the bottom row. Note that, as the cantilever is treated as an externally applied moment, the I value of the cantilever arm is irrelevant.

Cantilever deflection

The cantilever beam of length L in Fig. 20.4 carries a uniformly distributed load of intensity w per unit length, causing the end B to deflect by an amount Δ. If a release is introduced at A, the angular change there may be examined in two stages as shown in the deformation diagrams:

(a) angular change θ_{AB} due to the loading with a temporary prop acting at B;
(b) angular change due to the end deflection = Δ/L.

The equation for angular deformation is, therefore:

Table 20.3 Three-span beam problem in Fig. 20.3

Equation	Moment		$\overline{W}$
	B	C	
A.B	4	1	81
A.C	1	6	78.5
Result	7.72	10.13	MATIN N = 2

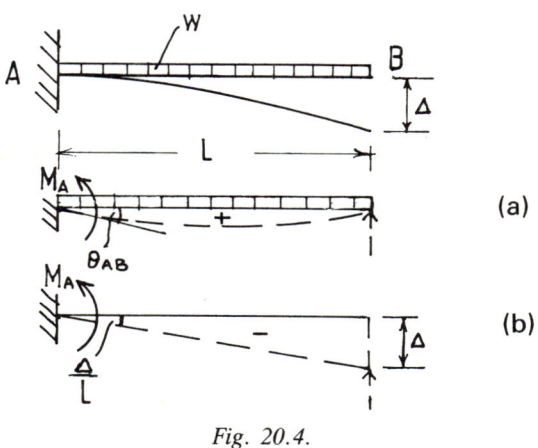

Fig. 20.4.

A.A:
$$M_A \frac{L}{3EI} = \theta_{AB} + \frac{\Delta}{L}$$

(A plus sign is used since the release moments act in the same direction; the diagrams, however, are of opposite sign, as shown.) From first principles,

$$M_A = \frac{WL}{2} \text{ and } \theta_{AB} = \frac{WL^3}{24EI} \text{ (due to UDL on AB)}$$

Thus
$$\Delta = \frac{WL^2}{2} \frac{L}{3EI} - \frac{WL^3}{24EI} = \frac{WL^3}{8EI}$$

Formulae for the deflection of other types of beams and loading systems may be obtained in like fashion.

Cranked beams

If two cantilevers are connected at right angles to each other, as in Fig. 20.5, by a continuous joint but one incapable of transmitting torsion, then

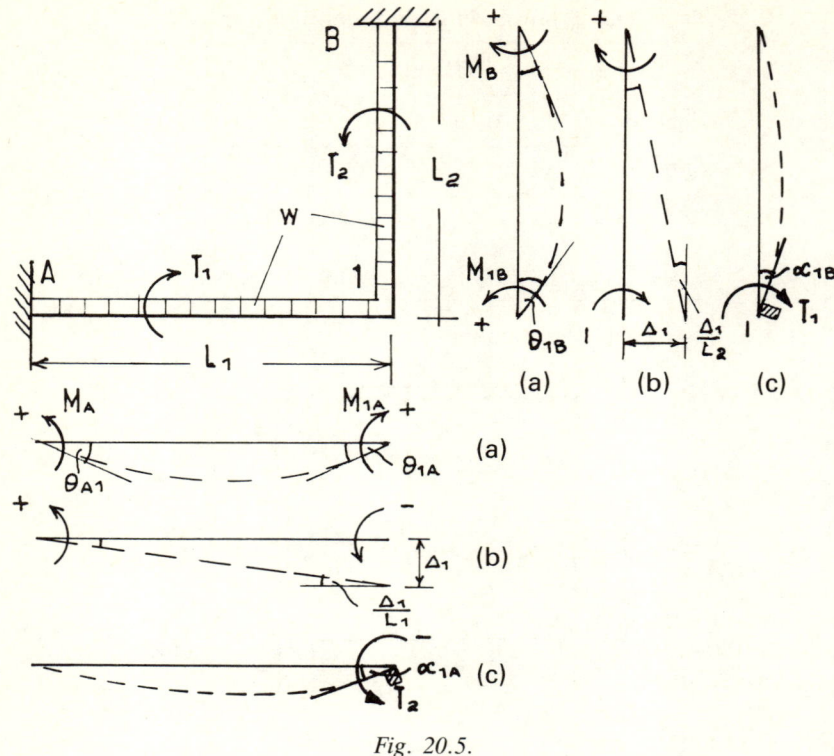

Fig. 20.5.

for moment equilibrium at node 1, $M_{1A} = 0$ and $M_{1B} = 0$. The equations for angular compatibility at A and B and equilibrium at joint 1 are:

A.A: $\quad M_A \dfrac{L_1}{3EI_1} = \theta_{A1} + \dfrac{\Delta_1}{L_1}$

$\quad\quad 2M_A = \dfrac{1}{4} W_1 L_1 + P\Delta_1 \dfrac{1}{L_1}$

A.B: $\quad M_B \dfrac{L_2}{3EI_2} = \theta_{B1} + \dfrac{\Delta_1}{L_2}$

$\quad\quad 2K_2 M_B = \dfrac{1}{4} K_2 W_2 L_2 + P\Delta_1 \dfrac{1}{L_2}$

E.1: $\quad \dfrac{M_A}{L_1} + \dfrac{M_B}{L_2} = \dfrac{1}{2}(W_1 + W_2)$

These are summarised in Table 20.4.

If the joint at 1 connecting the two cantilevers allows torsional movement to take place, then additional changes occur at 1 affecting the values of the bending moments in the cranked beam. As the beam is fixed at

Table 20.4 Cranked beam with no torsion

Equation	Moment A	Moment B	$\overline{W}$	$\overline{\Delta}$
A.A	2		$\frac{1}{4} W_1 L_1$	$\frac{1}{L_1}$
A.B		$2K_2$	$\frac{1}{4} K_2 W_2 L_2$	$\frac{1}{L_2}$
E.1	$\frac{1}{L_1}$	$\frac{1}{L_2}$	$\frac{W_1 + W_2}{2}$	

supports A and B, the torsion varies from a maximum at node 1 to zero at supports A and B. The additional deformation diagrams in Fig. 20.5(c) show that the torque in the adjoining member, indicated by an arrow, produces a rotation, α, at 1 that *increases* the bending angle, θ, in each case and so attracts a *negative* sign when the angular displacements are combined. (The diagram assumes that the beams are rectangular in shape, but torsion will occur in the same way in beams of any shape.) Although not shown in the figure, it is assumed that a temporary vertical prop is inserted at node 1 when released to maintain it in its deflected position. For clarity this has been omitted from this and subsequent figures.

If the properties of the two legs of the cranked beam conform with the Saint-Venant constant rate of twist theory, then we can write:

ϕ_1 = angular change per unit length in A1 under unit torsion
ϕ_2 = angular change per unit length in B1 under unit torsion

so that, $T_1 L_1 \phi_1$ = angular change at 1 in B1 due to torsion T_1 at 1 = α_{1B}

and $T_2 L_2 \phi_2$ = angular change at 1 in A1 due to torsion T_2 at 1 = α_{1A}

where α is the angular change along the length of the adjacent member at node 1.

If θ_{1A} and θ_{1B} indicate, as before, the angular changes at 1 in A1 and B1 under the applied loads with no torsional movement, then the total angular changes at 1 may be described as:

$$\text{total angular change in A1 at 1} = \theta_{1A} - \alpha_{1A} - \frac{\Delta_1}{L_1}$$

$$= \theta_{1A} - T_2 L_2 \phi_2 - \frac{\Delta_1}{L_1}$$

244 FOUNDATION AND STRUCTURAL PROBLEMS

$$\text{total angular change in B1 at } 1 = \theta_{1B} - \alpha_{1B} - \frac{\Delta_1}{L_2}$$

$$= \theta_{1B} - T_1 L_1 \phi_1 - \frac{\Delta_1}{L_2}$$

It follows that, due to the effect of torsion, M_{1A} is no longer equal to M_{1B}, unless, of course, $L_1 = L_2$. An extra equation is needed, therefore, to solve the cranked beam problem. This is provided by considering the angular changes on either side of node 1 separately; this is done in the equations designated T.A1/1B and T.B1/1A below.

At node 1, the angular change in A1 is a measure of both the bending moment in A1 and torsion in B1 at that point; similarly, the angular change in B1 is a measure of both the bending moment in B1 and torsion in A1, i.e. $M_{1A} = T_2$ and $M_{1B} = T_1$. This substitution is used in the equations below.

A.A:
$$M_A \frac{L_1}{3EI_1} + M_{1A} \frac{L_1}{6EI_1} = \theta_{A1} + \frac{\Delta_1}{L_1}$$

$$2M_A + M_{1A} = \frac{1}{4} W_1 L_1 + P \frac{\Delta_1}{L_1}$$

A.B:
$$M_B \frac{L_2}{3EI_2} + M_{1B} \frac{L_2}{6EI_2} = \theta_{B1} + \frac{\Delta_1}{L_2}$$

$$2K_2 M_B + K_2 M_{1B} = \frac{1}{4} K_2 W_2 L_2 + P \frac{\Delta_1}{L_2}$$

T.A1/1B: (i.e. angular change in A1 including torsion in 1B)

$$M_A \frac{L_1}{6EI_1} + M_{1A} \frac{L_1}{3EI_1} = \theta_{1A} - \frac{\Delta_1}{L_1} - T_2 L_2 \phi_2$$

$$M_A + 2M_{1A} = \frac{1}{4} W_1 L_1 - P \frac{\Delta_1}{L_1} - T_2 \frac{6EI_1}{L_1} L_2 \phi_2$$

But $\Omega_2 = \frac{6EI_1}{L_1} L_2 \phi_2$ and $T_2 = M_{1A}$

$$M_A + (2 + \Omega_2) M_{1A} = \frac{1}{4} W_1 L_1 - P \frac{\Delta_1}{L_1}$$

T.B1/1A: (i.e. angular change in B1 including torsion in 1A)

$$M_B \frac{L_2}{6EI_2} + M_{1B} \frac{L_2}{3EI_2} = \theta_{1B} - \frac{\Delta_1}{L_2} - T_1 L_1 \phi_1$$

$$K_2 M_B + 2K_2 M_{1B} = \frac{1}{4} K_2 W_2 L_2 - P \frac{\Delta_1}{L_2} - T_1 \, 6EI_1 \phi_1$$

But $\Omega_1 = 6EI_1 \phi_1$ and $T_1 = M_{1B}$

$$K_2 M_B + (2K_2 + \Omega_1) M_{1B} = \frac{1}{4} K_2 W_2 L_2 - P \frac{\Delta_1}{L_2}$$

TORSION: CRANKED BEAMS AND CANTILEVERS

E.1: $\quad \dfrac{M_A - M_{1A}}{L_1} + \dfrac{M_B - M_{1B}}{L_2} = W_1$

These are summarised in Table 20.5.

Table 20.5 Cranked beam with torsion

Equation	Moment				$\overline{W}$	$\overline{\Delta}$
	A	1A	1B	B		
A.A	2	1			$\frac{1}{4}W_1L_1$	$\dfrac{1}{L_1}$
T.A1/1B	1	$2 + \Omega_2$			$\frac{1}{4}W_1L_1$	$-\dfrac{1}{L_1}$
T.B1/1A			$2K_2 + \Omega_1$	K_2	$\frac{1}{4}K_2W_2L_2$	$-\dfrac{1}{L_2}$
A.B			K_2	$2K_2$	$\frac{1}{4}K_2W_2L_2$	$\dfrac{1}{L_2}$
E.1	$\dfrac{1}{L_1}$	$-\dfrac{1}{L_1}$	$-\dfrac{1}{L_2}$	$\dfrac{1}{L_2}$	W_1	

Equations T.A1/1B and T.B1/1A may be combined to express the angular changes both sides of node 1 as follows:

A.A1B: $\quad M_A + (2 + \Omega_2) M_{1A} + (2K_2 + \Omega_1)M_{1B} + K_2 M_B =$
$$\frac{1}{4}(W_1L_1 + K_2W_2L_2) - P\Delta_1 \left(\frac{1}{L_1} + \frac{1}{L_2}\right)$$

This may be substituted for either of the single member equations T.A1/1B or T.B1/1A as shown in Table 20.6. It is less efficient in use since it incorporates an equation used elsewhere in the table, but can provide a handy means of checking the results.

As an example, assume a rectangular concrete cranked beam of dimensions b = 300 mm, d = 600 mm, lengths L_1 = 1 m and L_2 = 1.5 m and with λ = 0.25, carrying a distributed load of 10 kN/m on each length. From Table 20.1 for d/b = 2, β = 0.229.

$$\Omega_1 = \frac{1.25}{0.229} \times 2^2 = 21.83; \quad \Omega_2 = \frac{1.5}{1} \times 21.83 = 32.75; \quad K_2 = 1.5$$

By entering the appropriate values in Tables 20.4, 20.5 and 20.6 the following data statements may be constructed for use with GRID. Results in the bottom row show the moments with and without torsion.

Table 20.6 Cranked beam with torsion (alternative tabular matrix)

Equation	Moment				$\overline{W}$	$\overline{\Delta}$
	A	1A	1B	B		
A.A	2	1			$\frac{1}{4}W_1L_1$	$\frac{1}{L_1}$
T.A1/1B	1	$2+\Omega_2$			$\frac{1}{4}W_1L_1$	$-\frac{1}{L_1}$
A.A1B	1	$2+\Omega_2$	$2K_2+\Omega_1$	K_2	$\frac{1}{4}(W_1L_1+K_2W_2L_2)$	$-\left(\frac{1}{L_1}+\frac{1}{L_2}\right)$
A.B			K_2	$2K_2$	$\frac{1}{4}K_2W_2L_2$	$\frac{1}{L_2}$
E.1	$\frac{1}{L_1}$	$-\frac{1}{L_1}$	$-\frac{1}{L_2}$	$\frac{1}{L_2}$	W_1	

(1) No torsion (Table 20.4)
    1000 DATA 2,0,0,3,2.5,8.44,1,0.667,1,0.667,12.5
    Results: 8.48, 6.03 kNm

(2) With torsion (Table 20.5)
    1000 DATA 2,1,0,0,1,34.75,0,0
    1005 DATA 0,0,24.83,1.5,0,0,1.5,3,2.5,2.5,8.44,8.44
    1010 DATA 1,−1,−0.667,0.667
    1015 DATA 1,−1,−0.667,0.667,12.5
    Results: 7.87,−0.52,−0.35, 5.82 kNm

(3) With torsion (Table 20.6)
    1000 DATA 2,1,0,0,1,34.75,0,0
    1005 DATA 1,34.75,24.83,1.5,0,0,1.5,3,2.5,2.5,10.94,8.44
    1010 DATA 1,−1,−1.667,0.667
    1015 DATA 1,−1,−0.667,0.667,12.5
    Results: 7.87,−0.52,−0.35,5.82 kNm

The overall effect, it will be seen, is a reduction in the peak moments at the supports where allowance for torsion is included.

Cranked beam with one free support

The cranked beam in Fig. 20.6 is fixed at support A and has a spherical bearing at B permitting free rotation in any direction. Bending in 1B

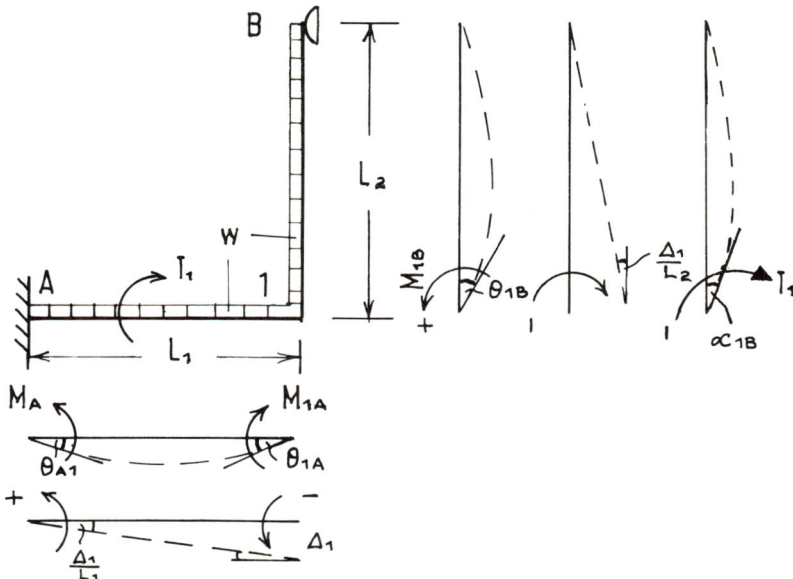

Fig. 20.6.

produces a torque T_1 in 1A, since it is restrained at A. Because 1B is free to rotate, there is no torsion in 1B. By referring to the deformation diagrams, the equations for angular displacement on release and for equilibrium at joint 1 may be written down as:

A.A: $\quad M_A \dfrac{L_1}{3EI_1} + M_{1A} \dfrac{L_1}{6EI_1} = \theta_{A1} + \dfrac{\Delta_1}{L_1}$

$\quad\quad\quad 2M_A + M_{1A} = \dfrac{1}{4} W_1 L_1 + P \dfrac{\Delta_1}{L_1}$

T.A1/1B: $\quad M_A \dfrac{L_1}{6EI_1} + M_{1A} \dfrac{L_1}{3EI_1} = \theta_{1A} - \dfrac{\Delta_1}{L_1}$

$\quad\quad\quad M_A + 2M_{1A} = \dfrac{1}{4} W_1 L_1 - P \dfrac{\Delta_1}{L_1}$

T.B1/1A: $\quad M_{1B} \dfrac{L_2}{3EI_2} = \theta_{1B} - \dfrac{\Delta_1}{L_2} - T_1 L_1 \phi_1$

$\quad\quad\quad 2K_2 M_{1B} + 6EI_1 \phi_1 T_1 = \dfrac{1}{4} K_2 W_2 L_2 - P \dfrac{\Delta_1}{L_2}$

$\quad\quad\quad (2K_2 + \Omega_1) M_{1B} = \dfrac{1}{4} K_2 W_2 L_2 - P \dfrac{\Delta_1}{L_2}$

E.1: $\quad \dfrac{M_A - M_{1A}}{L_1} - \dfrac{M_{1B}}{L_2} = \dfrac{W_1 + W_2}{2}$

The equations are summarised in Table 20.7. This table could have been derived directly from Table 20.5 by omitting the last column and row of the moment coefficient matrix in the normal way for a 'free' support as well as the torsional term, Ω_2, consequent on T_2 being equal to zero. The

Table 20.7 Cranked beam with one simple support

Equation	Moment			$\overline{W}$	$\overline{\Delta}$
	A	1A	1B		
A.A	2	1		$\dfrac{1}{4} W_1 L_1$	$\dfrac{1}{L_1}$
T.A1/1B	1	2		$\dfrac{1}{4} W_1 L_1$	$-\dfrac{1}{L_1}$
T.B1/1A			$2K_2 + \Omega_1$	$\dfrac{1}{4} K_2 W_2 L_2$	$-\dfrac{1}{L_2}$
E.1	$\dfrac{1}{L_1}$	$-\dfrac{1}{L_1}$	$-\dfrac{1}{L_2}$	$\dfrac{W_1 + W_2}{2}$	

same procedure can also be applied to grillages or frameworks with simply supported end bearings.

Cantilever balcony

Figure 20.7 depicts a building constructional problem in which a balcony projects from the main wall and is supported along its front edge by a U-shaped cranked cantilever with fixed supports at A and B. The balcony slab carries a uniformly distributed load and spans from the main wall to the front member, 12. As conditions are symmetrical, the problem may be solved by considering just two joints, as in the following equations for the cranked cantilever A12B.

A.A: $\quad M_A \dfrac{L_1}{3EI_1} = \dfrac{\Delta_1}{L_1} \quad$ (as there is no load on A1, $M_{1A} = 0$)

$$2M_A = P \dfrac{\Delta_1}{L_1}$$

T.21/1A: $\quad M_{12} \dfrac{L_2}{3EI_2} + M_{21} \dfrac{L_2}{6EI_2} = \theta_{12} + \dfrac{\Delta_2}{L_2} - \dfrac{\Delta_1}{L_2} - T_{1A}L_{1A}\phi_{1A}$

From symmetry, $M_{12} = M_{21}$, $\Delta_2 = \Delta_1$; also $T_{1A} = M_{12}$

$$M_{12} \dfrac{L_2}{2EI_2} = \theta_{12} - M_{12}L_{1A}\phi_{1A}$$

$$3K_2M_{12} + \Omega_1 M_{12} = \tfrac{1}{4} K_2 W_2 L_2$$

$$M_{12}(3K_2 + \Omega_1) = \tfrac{1}{4} K_2 W_2 L_2$$

E.1: $\quad \dfrac{M_A}{L_1} = \dfrac{W_2}{2}$

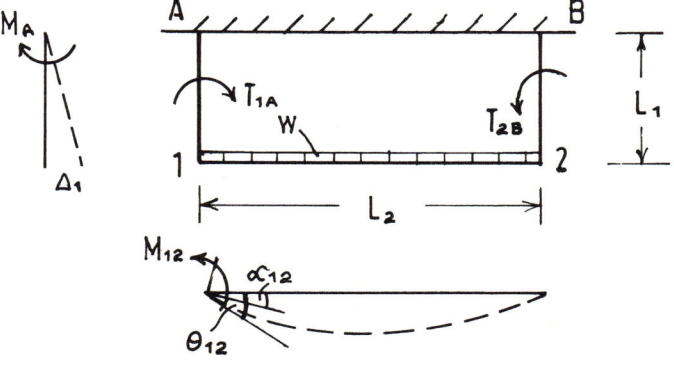

Fig. 20.7.

Table 20.8 Cantilever balcony

Equation	Moment		$\overline{W}$	$\overline{\Delta}$
	A	12		
A.A	2			$\dfrac{1}{L_1}$
T.21/1A		$3K_2 + \Omega_1$	$\dfrac{1}{4}K_2W_2L_2$	
E.1	$\dfrac{1}{L_1}$		$\dfrac{W_2}{2}$	

These equations give the half-matrix in Table 20.8.

If the balcony member, 12, supports a point load at its centre instead of a distributed load, the $\overline{W}$ vector term should be altered accordingly, of course. For instance, if $L_1 = 1$ m, $L_2 = 4$ m and the balcony beam dimensions are b = 200 mm and d = 600 mm and it carries a concentrated load of 20 kN at the centre of 12, then taking $\lambda = 0.2$ for reinforced concrete, from Table 20.1, $\beta = 0.263$ and from equation (20.10) $\Omega_1 = 41.06$. $K = 2$; $\tfrac{3}{8}K_2W_2L_2 = 60$.

The data statement to solve for the moments using XBEAM is:

    1000 DATA 2,0,0,47.06,0,60,1,0,1,0,10

This provides the results $M_A = 10$ and $M_{12} = 1.275$ kNm.

21 Torsion in beam frameworks and grillages

Three-member beam framework

In Chapter 17 the procedure for solving simple interconnected beam frameworks was described, ignoring the effect of torsion. Bending moments at the joints were obtained using the XBEAM program with data derived from the tabular matrix for that particular frame and loading conditions. The same program may be used with data statements that take account of torsional action.

Considering first the simple interconnected beam framework in Fig. 21.1, the initial step is to draw the deformation diagrams showing the situation with release moments applied at the joints and supports. The diagram in this instance is broken into three components (a) bending only, assuming the joint propped, (b) deflection at the joint and (c) angular movement due to torsion at the joint.

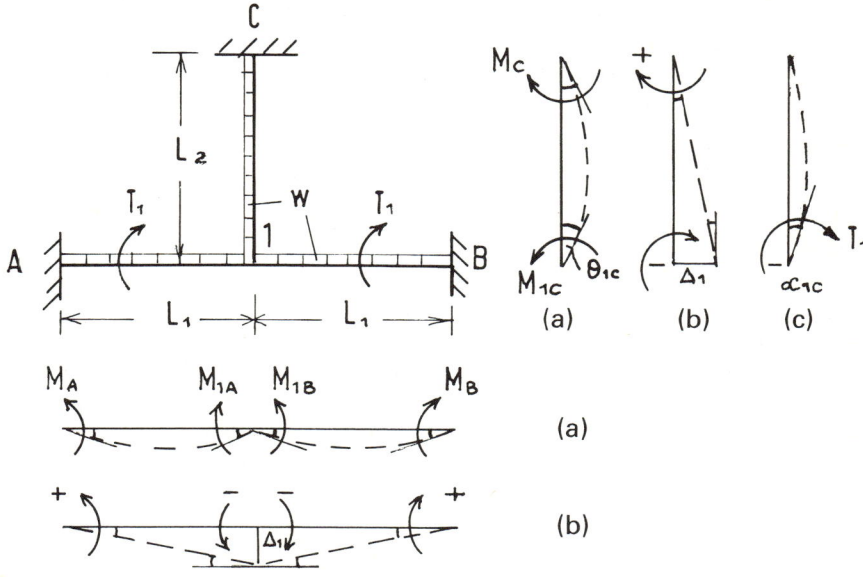

Fig. 21.1.

Equations for the frame ignoring torsion can be prepared as described in Chapter 17. In this case, from symmetry, $M_{1A} = M_{1B}$; also $M_{1C} = 0$ since $\Sigma M_{yy} = 0$ at joint 1. Thus, we have:

A.A: $\quad 2M_A + M_{1A} = \dfrac{1}{4}W_1L_1 + P\dfrac{\Delta_1}{L_1}$

A.B: $\quad 2M_B + M_{1A} = \dfrac{1}{4}W_1L_1 + P\dfrac{\Delta_1}{L_1}$

A.C: $\quad 2K_2M_c = \dfrac{1}{4}K_2W_2L_2 + P\dfrac{\Delta_1}{L_2}$

A.A1B: $\quad M_A + 4M_{1A} + M_B = -\dfrac{W_1L_1}{2} - 2P\dfrac{\Delta_1}{L_1}$

E.1: $\quad \dfrac{M_A - M_{1A} + M_B - M_{1A}}{L_1} + \dfrac{M_c}{L_2} = W_1 + \dfrac{W_2}{2}$

$\quad\quad \dfrac{M_A - 2M_{1A} + M_B}{L_1} + \dfrac{M_c}{L_2} = W_1 + \dfrac{W_2}{2}$

The equations are put in tabular format in Table 21.1.

If the joint at 1 is capable of suffering torsional as well as bending action, bending in member 1C will cause twisting to occur in A1B. As this is fixed at each end, the rotation at 1 relative to either end equals $T_1L_1\phi_1$, where

T_1 = torsion in A1B induced by the bending of 1C
L_1 = distance from the joint to the fixed support
ϕ_1 = rotation per unit length per unit torsion of A1B

Table 21.1 Three-member frame without torsion

Equation	Moment				$\overline{W}$	$\overline{\Delta}$
	A	B	C	1A		
A.A	2			1	$\dfrac{1}{4}W_1L_1$	$\dfrac{1}{L_1}$
A.B		2		1	$\dfrac{1}{4}W_1L_1$	$\dfrac{1}{L_1}$
A.C			$2K_2$		$\dfrac{1}{4}K_2W_2L_2$	$\dfrac{1}{L_2}$
A.A1B	1	1		4	$\dfrac{1}{2}W_1L_1$	$-\dfrac{2}{L_1}$
E.1	$\dfrac{1}{L_1}$	$\dfrac{1}{L_1}$	$\dfrac{1}{L_2}$	$-\dfrac{2}{L_1}$	W_1	

Torsional flexibility results in a moment M_{1C} occurring in 1C, which otherwise would be zero. From considerations of symmetry we can conclude that $M_{1A} = M_{1B}$ and $T_2 = 0$. (This would not be the case where 1A ≠ 1B.) The additional unknown, M_{1C}, is solved by considering the angular change at one side only of node 1; in our case the angular change in 1C at 1 due to the torsion in 1A (or 1B). The compatibility equation for this is designated T.C1/1A using the same convention as in the previous chapter. Note that, as the angle of twist, α, *increases* the angular change in 1C at node 1, the torsional factor is given a *negative* sign.

T.C1/1A: $\quad M_{1C} \dfrac{L_2}{3EI_2} + M_C \dfrac{L_2}{6EI_2} = \theta_{1C} - \dfrac{\Delta_1}{L_2} - T_1 L_1 \phi_1$

$\qquad 2K_2 M_{1C} + K_2 M_C = \dfrac{1}{4} K_2 W_2 L_2 - P \dfrac{\Delta_1}{L_2} - 6EI_1 T_1 \phi_1$

$\qquad \Omega_1 = 6EI_1 \phi_1$ and $T_1 = M_{1C}$

$\qquad K_2 M_C + (2K_2 + \Omega_1) M_{1C} = \dfrac{1}{4} K_2 W_2 L_2 - P \dfrac{\Delta_1}{L_2}$

A.C: $\quad 2K_2 M_C + 2K_2 M_{1C} = \dfrac{1}{4} K_2 W_2 L_2 + P \dfrac{\Delta_1}{L_2}$

E.1: $\quad \dfrac{M_A + M_B - 2M_{1A}}{L_1} + \dfrac{M_C - M_{1C}}{L_2} = W_1$

Other equations remain as before; the full set is summarised in Table 21.2.

Table 21.2 Three-member frame with torsion

Equation	Moment					$\overline{W}$	$\overline{\Delta}$
	A	B	C	1A	1C		
A.A	2			1		$\dfrac{1}{4} W_1 L_1$	$\dfrac{1}{L_1}$
A.B		2		1		$\dfrac{1}{4} W_1 L_1$	$\dfrac{1}{L_1}$
A.C			$2K_2$		K_2	$\dfrac{1}{4} K_2 W_2 L_2$	$\dfrac{1}{L_2}$
A.A1B	1	1		4		$\dfrac{1}{2} W_1 L_1$	$-\dfrac{2}{L_1}$
T.C1/1A			K_2		$2K_2 + \Omega_1$	$\dfrac{1}{4} K_2 W_2 L_2$	$-\dfrac{1}{L_2}$
E.1	$\dfrac{1}{L_1}$	$\dfrac{1}{L_1}$	$\dfrac{1}{L_2}$	$-\dfrac{2}{L_1}$	$-\dfrac{1}{L_2}$	W_1	

For an example to illustrate the effect of torsion on moments in the framework, assume $L_1 = 4$ m, $L_2 = 8$ m, $w = 10$ kN/m, $GJ = 0.8EI$ for all members, where GJ is the torsional rigidity and EI is the flexural rigidity of the members, then

$$K_2 = 2, \ W_1 = 80 \text{ kN and } \Omega_1 = \frac{6EI}{GJ} = \frac{6}{0.8} = 7.5$$

By inserting the appropriate values in Tables 21.1 and 21.2, the following data statements for XBEAM and results were obtained:

(1) No torsion
   1000 DATA 2,0,0,1,0,2,0,1,0,0,4,0,1,1,0,4
   1005 DATA 40,40,320,80
   1010 DATA 0.25,0.25,0.125,−0.5
   1015 DATA 0.25,0.25,0.125,−0.5,80
   Results: $M_A = 82.26$, $M_B = 82.26$, $M_C = 88.62$, $M_{1A} = -55.59$ kNm

(2) With torsion
   1000 DATA 2,0,0,1,0,0,2,0,1,0,0,0,4,0,2
   1005 DATA 1,1,0,4,0,0,0,20,11.5
   1010 DATA 40,40,320,80,320
   1015 DATA 0.25,0.25,0.125,−0.5,−0.125
   1020 DATA 0.25,0.25,0.125,−0.5,−0.125,80
   Results: $M_A = 84.14$, $M_B = 84.14$, $M_C = 83.76$, $M_{1A} = -57.47$, $M_{1C} = 10.18$ kNm

Four-member beam framework

The principles involved in the general analysis of beam frameworks with torsion can be illustrated by reference to the four-member interconnected beam system with fixed supports in Fig. 21.2. All members of the frame are of unequal length; as drawn, L_1 is less than L_3 and L_2 is less than L_4.

Considering conditions at joint 1, the net torsion in the YY direction is $T_1 - T_3$, so that

$$T_1 - T_3 = M_{1D} - M_{1C} \tag{21.1}$$

Similarly in the XX direction,

$$T_2 - T_4 = M_{1B} - M_{1A} \tag{21.2}$$

By making these substitutions one can solve for the moments and torsions at the nodes using just two torsional equations, as set out below. The other equations are derived in the usual way, so only a few typical ones are reproduced below.

From equations (21.1) and (21.2) it follows that, under uniform conditions, positive values are obtained when span 1D exceeds span 1C and span 1B exceeds 1A or, where the spans are equal, if C and A are fixed

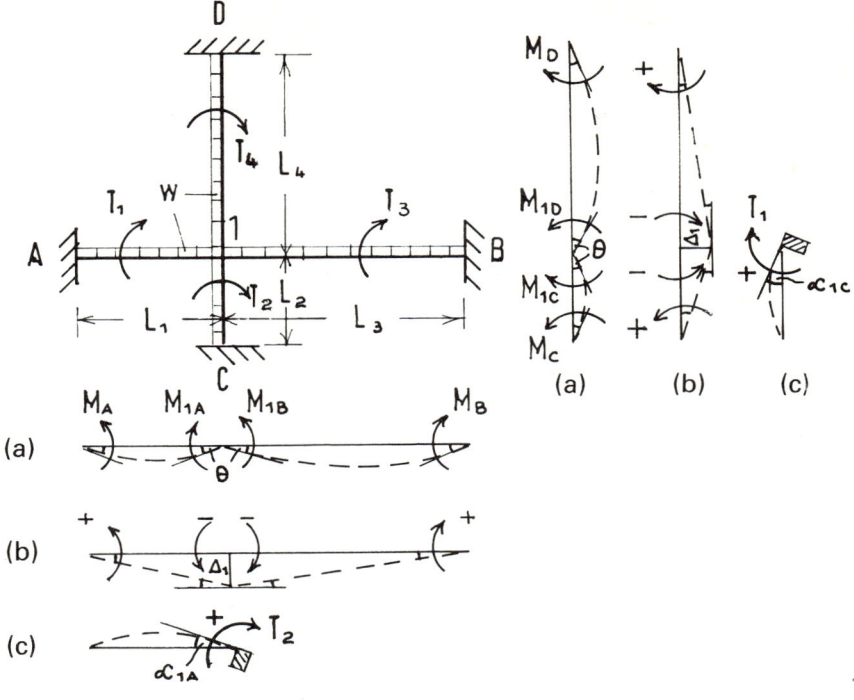

Fig. 21.2.

supports. In such cases it will be seen from the diagrams in Fig. 21.2 that the angle α is inclined above the baseline in the short or end spans and so attracts a positive sign, as it is of opposite sense to the angle θ in the same span.

A.A: $\quad 2M_A + M_{1A} = \dfrac{1}{4}W_1L_1 + P\dfrac{\Delta_1}{L_1}$

A.C: $\quad 2K_2M_C + K_2M_{1C} = \dfrac{1}{4}K_2W_2L_2 + P\dfrac{\Delta_1}{L_2}$

A.A1B: $\quad M_A + 2M_{1A} + 2K_3M_{1B} + K_3M_B = \dfrac{1}{4}(W_1L_1 + K_3W_3L_3)$
$\qquad - P\Delta_1\left(\dfrac{1}{L_1} + \dfrac{1}{L_3}\right)$

A.C1D: $\quad K_2M_C + 2K_2M_{1C} + 2K_4M_{1D} + K_4M_D$
$\qquad = \dfrac{1}{4}(K_2W_2L_2 + K_4W_4L_4) - P\Delta_1\left(\dfrac{1}{L_2} + \dfrac{1}{L_4}\right)$

T.A1/1C: $\quad M_A\dfrac{L_1}{6EI_1} + M_{1A}\dfrac{L_1}{3EI_1} = \theta_{1A} + T_2L_2\phi_2 - \dfrac{\Delta_1}{L_1}$

$$M_A + 2M_{1A} = \frac{1}{4}W_1L_1 + T_2\Omega_2 - P\frac{\Delta_1}{L_1}$$

But $T_2L_2 = T_4L_4$ and $T_2 - T_4 = M_{1B} - M_{1A}$

so that $T_2 = \dfrac{L_4}{L_4 - L_2}(M_{1B} - M_{1A})$

$$M_A + 2M_{1A} = \frac{1}{4}W_1L_1 + \frac{L_4}{L_4 - L_2}\Omega_2 M_{1B}$$

$$- \frac{L_4}{L_4 - L_2}\Omega_2 M_{1A} - P\frac{\Delta_1}{L_1}$$

$$M_A + M_{1A}\left(2 + \frac{L_4}{L_4 - L_2}\Omega_2\right) - \frac{L_4}{L_4 - L_2}\Omega_2 M_{1B}$$

$$= \frac{1}{4}W_1L_1 - P\frac{\Delta_1}{L_1}$$

T.C1/1A: $\quad M_C \dfrac{L_2}{6EI_2} + M_{1C}\dfrac{L_2}{3EI_2} = \theta_{1C} + T_1L_1\phi_1 - \dfrac{\Delta_1}{L_2}$

$$K_2 M_C + 2K_2 M_{1C} = \frac{1}{4}K_2 W_2 L_2 + T_1\Omega_1 - P\frac{\Delta_1}{L_2}$$

But $T_1L_1 = T_3L_3$ and $T_1 - T_3 = M_{1D} - M_{1C}$

so that $T_1 = \dfrac{L_3}{L_3 - L_1}(M_{1D} - M_{1C})$

$$K_2 M_C + 2K_2 M_{1C} = \frac{1}{4}K_2 W_2 L_2 + \frac{L_3}{L_3 - L_1}$$

$$(M_{1D} - M_{1C})\Omega_1 - P\frac{\Delta_1}{L_2}$$

$$K_2 M_C + M_{1C}\left(2K_2 + \frac{L_3}{L_3 - L_1}\Omega_1\right) - \frac{L_3}{L_3 - L_1}\Omega_1 M_{1D}$$

$$= \frac{1}{4}K_2 W_2 L_2 - P\frac{\Delta_1}{L_2}$$

E.1: $\quad \dfrac{M_A - M_{1A}}{L_1} + \dfrac{M_B - M_{1B}}{L_3} + \dfrac{M_C - M_{1C}}{L_2}$

$$+ \frac{M_D - M_{1D}}{L_4} = W_1$$

The equations are summarised in Table 21.3.

To illustrate use of the table, we can take the case of a framework composed of beams in which the ratio of torsional rigidity (GJ) to flexural rigidity (EI) equals 0.8 and carrying a distributed load of 10 kN/m on each member. If $L_1 = 4$ m, $L_2 = 2$ m, $L_3 = 6$ m and $L_4 = 8$ m (see

Table 21.3 Four-member frame with torsion

Equation	A	B	C	D	1A	1B	1C	1D	$\overline{W}$	$\overline{\Delta}$
A.A	2			1						$\frac{1}{L_1}$
A.B		$2K_3$				K_3			$\frac{1}{4}K_3W_3L_3$	$-\frac{1}{L_3}$
A.C			$2K_2$				K_2		$\frac{1}{4}K_2W_2L_2$	$-\frac{1}{L_2}$
A.D				$2K_4$				K_4	$\frac{1}{4}K_4W_4L_4$	$-\frac{1}{L_4}$
A.A1B	1	K_3			2	$2K_3$			$\frac{1}{4}(W_1L_1 + K_3W_3L_3)$	$-\left(\frac{1}{L_1}+\frac{1}{L_3}\right)$
T.A1/1C	1				$2+\dfrac{L_4}{L_4-L_2}\Omega_2$	$-\dfrac{L_4}{L_4-L_2}\Omega_2$			$\frac{1}{4}W_1L_1$	$-\frac{1}{L_1}$
T.C1/1A			K_2				$2K_2 + \dfrac{L_3}{L_3-L_1}\Omega_1$	$-\dfrac{L_3}{L_3-L_1}\Omega_1$	$\frac{1}{4}K_2W_2L_2$	$-\frac{1}{L_2}$
A.C1D			K_2	K_4			$2K_2$	$2K_4$	$\frac{1}{4}(K_2W_2L_2 + K_4W_4L_4)$	$-\left(\frac{1}{L_2}+\frac{1}{L_4}\right)$
E.1	$\frac{1}{L_1}$	$\frac{1}{L_3}$	$\frac{1}{L_2}$	$\frac{1}{L_4}$	$-\frac{1}{L_1}$	$\frac{1}{L_3}$	$\frac{1}{L_2}$	$\frac{1}{L_4}$	W_1	

Moment

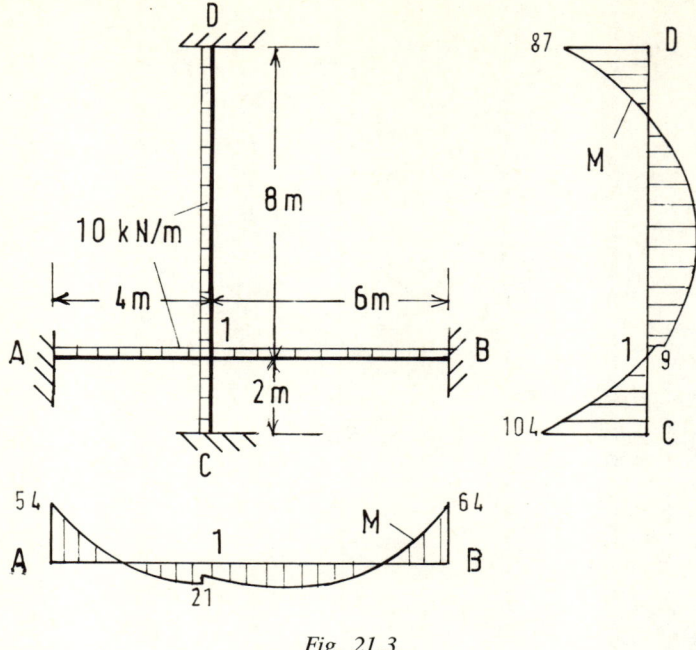

Fig. 21.3.

Fig 21.3), then the data table for use with XBEAM may be assembled using the values:

$$K_2 = 0.5, \quad K_3 = 1.5, \quad K_4 = 2, \quad \Omega_1 = \frac{6}{0.8} = 7.5 \quad \text{(from equation (20.7))}$$

$$\Omega_2 = \frac{2}{4} \times 7.5 = 3.75 \quad \text{(since GJ is constant)}$$

The data table and results for this problem are reproduced in Table 21.4.

Frame with equal-length members

If the four-member framework has two of its members in line of equal length, the tabulated terms in the torsional equation T.A1/1C or T.C1/1A will be indeterminate. These must be arranged in a different form, therefore.

(1) If $L_2 = L_4$ then, since $T_2L_2 = T_4L_4$, $T_2 = T_4 = M_{1B} - M_{1A}$. This enables the equation to be rewritten as:

$$\text{T.A1/1C:} \quad M_A + M_{1A}(2 + \Omega_2) - \Omega_2 M_{1B} = \frac{1}{4} W_1 L_1 - P \frac{\Delta_1}{L_1}$$

Table 21.4 Data table for Fig. 21.3

Equation	A	B	C	D	Moment 1A	1B	1C	1D	
A.A	2				1				
A.B		3				1.5			
A.C			1				0.5		
A.D				4				2	
A.A1B	1	1.5			2	3			
T.A1/1C	1				7	−5	23.5	−22.5	
T.C1/1A			0.5	2			1	4	
A.C1D			0.5				1	4	
W̄	40	135	5	320	175	40	5	325	
Δ	0.25	0.167	0.5	0.125	−0.417	−0.25	−0.5	−0.625	
E.1	0.25	0.167	0.5	0.125	−0.25	−0.167	−0.5	−0.125	100
Results	54.10	63.96	104.21	87.37	−20.80	−16.81	−8.80	−2.89	kNm

Table 21.5 Four-member frame with $L_2 = L_4$

Equation	A	B	C	D	Moment 1A	1B	1C	1D	$\overline{W}$	$\overline{\Delta}$
A.A	2				1				$\frac{1}{4}W_1L_1$	$\frac{1}{L_1}$
A.B		$2K_3$				K_3			$\frac{1}{4}K_3W_3L_3$	$-\frac{1}{L_3}$
A.C			$2K_2$				K_2		$\frac{1}{4}K_2W_2L_2$	$-\frac{1}{L_2}$
A.D				$2K_2$				K_2	$\frac{1}{4}K_2W_2L_2$	$-\frac{1}{L_2}$
A.A1B	1	K_3			2	$2K_3$			$\frac{1}{4}(W_1L_1 + K_3W_3L_3)$	$-\left(\frac{1}{L_1}+\frac{1}{L_3}\right)$
T.A1/1C	1				$2+\Omega_2$	$-\Omega_2$			$\frac{1}{4}W_1L_1$	$-\frac{1}{L_1}$
T.C1/1A			K_2				$2K_2 + \frac{L_3}{L_3-L_1}\Omega_1$	$-\frac{L_3}{L_3-L_1}\Omega_1$	$\frac{1}{4}K_2W_2L_2$	$-\frac{1}{L_2}$
A.C1D			K_2	K_2			$2K_2$	$2K_2$	$\frac{1}{2}K_2W_2L_2$	$-\frac{2}{L_2}$
E.1	$\frac{1}{L_1}$	$\frac{1}{L_3}$	$\frac{1}{L_2}$	$\frac{1}{L_2}$	$-\frac{1}{L_1}$	$-\frac{1}{L_3}$	$-\frac{1}{L_2}$	$-\frac{1}{L_2}$	W_1	

(2) If $L_1 = L_3$ then, since $T_1L_1 = T_3L_3$, $T_1 = T_3 = M_{1D} - M_{1C}$. Hence,

T.C1/1A: $K_2M_C + M_{1C}(2K_2 + \Omega_1) - \Omega_1 M_{1D} = \frac{1}{4}K_2W_2L_2 - P\frac{\Delta_1}{L_2}$

The other equations remain unchanged in each case.

Table 21.5 shows the general matrix for the four-member frame where $L_2 = L_4$. This has been used to solve the same frame as in the previous example, but with $L_2 = L_4 = 8$ m. Information for XBEAM took the form of the following data statements:

```
1000 DATA 2,0,0,0,1,0,0,0
1005 DATA 0,3,0,0,0,1.5,0,0
1010 DATA 0,0,4,0,0,0,2,0
1015 DATA 0,0,0,4,0,0,0,2
1020 DATA 1,1.5,0,0,2,3,0,0
1025 DATA 1,0,0,0,17,-15,0,0
1025 DATA 0,0,2,0,0,0,26.5,-22.5
1030 DATA 0,0,2,2,0,0,4,4
1035 DATA 40,135,320,320,175,40,320,640
1040 DATA 0.25,0.167,0.125,0.125,-0.417,-0.25,-0.125,-0.25
1045 DATA 0.25,0.167,0.125,0.125,-0.25,-0.167,-0.125,-0.125,130
```

Running the program with input $N = 8$ and $U = 1$ gave the results $M_A = 165.65$, $M_B = 135.59$, $M_C = 99.90$, $M_D = 99.90$, $M_{1A} = -105.02$, $M_{1B} = -98.23$, $M_{1C} = 6.77$ and $M_{1D} = 6.77$ kNm

It will be seen that $M_{1C} = M_{1D}$ and that $M_C = M_D$, as might be expected from the frame geometry and loading. (A half-frame matrix cannot be used, incidentally, unless L_1 also equals L_3, in which case no torsion is present in the frame.)

Bending moment diagrams for the various members may be prepared by combining the 'free' BM diagrams with the printed-out values. The program prints the reaction at node 1, in this case 131.96 kN. Other design criteria can be calculated manually. Thus, the torsion $T_2 = T_4 = M_{1B} - M_{1A} = 6.79$ kNm. Support reactions and shear forces may be calculated likewise from first principles. For example, at support A the shear force is 20 kN and the reaction is $20 + (165.65 - 105.02)/4 = 35.14$ kN. To evaluate the deflection, Δ_1, we need to know the flexural properties (i.e. E and I values) of the members in order to substitute for P in one of the equations.

It will be appreciated that, once the moments are determined, the other structural requirements can be ascertained without much difficulty by manual methods.

Asymmetrical three-member frame

Tables 21.3 and 21.5 will be found useful for the construction of tabular matrices for other sizes of frameworks. For example, if member 1D is omitted from the four-member frame, then equations A.D and A.C1D and columns headed D and 1D in Table 21.5 may be cancelled out. By so doing, one can obtain Table 21.6 for an asymmetrical three-member frame directly.

Simple supports result in a substantial reduction in the size of matrix needed to solve a given size of frame, of course, eliminating both bending and torsional angular changes on release at those points. Interconnected beam frameworks are only analysed as such where the frame is of limited extent as a rule. With complex or closely meshed frameworks it is more usual to assume that the loads are concentrated at the nodal points and to analyse them as rectangular grillages, as described in the next section.

Rectangular grillages

Rectangular grillages, in which the spacing of the members is uniform in each direction and which are subjected to point loading at the grid intersections only, are examined in this section. Grid spacings are designated 'a' in the XX direction and 'b' along the YY axis. The ratio a/b is denoted by 'r'.

The physical properties of the members are described in the same way as before. Thus,

$$K = \frac{k_x}{k_y} = \frac{I_x}{I_y}\frac{b}{a} = \frac{I_x}{rI_y}$$

(it is assumed that $E_x = E_y$)

Flexure factor $P = \dfrac{6EI_x}{a}$

Torsion factor $\Omega_x = 6EI_x \phi_x$

and $\Omega_y = \dfrac{6EI_x}{a} b\phi_y = \Omega_x \dfrac{b\,\phi_y}{a\,\phi_x} = \Omega_x \dfrac{\phi_y}{r\phi_x}$

From equation (20.4), $\qquad \Omega_y = \Omega_x \dfrac{G_x J_x}{rG_y J_y}$ \hfill (21.3)

If $G_x = G_y$, then $\qquad \Omega_y = \Omega_x \dfrac{J_x}{rJ_y}$ \hfill (21.4)

If, in addition, $J_x = J_y$, then $\quad \Omega_y = \dfrac{\Omega_x}{r}$ \hfill (21.5)

Table 21.6 Three-member asymmetrical frame

Equation	Moment						$\overline{W}$	$\overline{\Delta}$
	A	B	C	1A	1B	1C		
A.A	2			1			$\frac{1}{4}W_1L_1$	$\frac{1}{L_1}$
A.B		$2K_3$			K_3		$\frac{1}{4}K_3W_3L_3$	$\frac{1}{L_3}$
A.C			$2K_2$			K_2	$\frac{1}{4}K_2W_2L_2$	$\frac{1}{L_2}$
A.A1B	1	K_3		2	$2K_3$		$\frac{1}{4}(W_1L_1 + K_3W_3L_3)$	$-\left(\frac{1}{L_1}+\frac{1}{L_3}\right)$
T.A1/1C	1			$2+\Omega_2$	$-\Omega_2$		$\frac{1}{4}W_1L_1$	$-\frac{1}{L_1}$
T.C1/1A	1		K_2			$2K_2 + \frac{L_3}{L_3-L_1}\Omega_1$	$\frac{1}{4}K_2W_2L_2$	$-\frac{1}{L_2}$
E.1	$\frac{1}{L_1}$	$\frac{1}{L_3}$	$\frac{1}{L_2}$	$-\frac{1}{L_1}$	$-\frac{1}{L_3}$	$-\frac{1}{L_2}$	W_1	

Sample one-way and two-way spanning grillages are shown in Fig. 21.4. Release moments can be assumed to act as shown in Fig. 21.4(a). Moments acting at the supports tend to restore the end members to their original alignment. If the assumed direction of the release elsewhere is incorrect, this will be shown up by a negative sign in the results. As there is no question of combining bending and vertical displacement effects, the only signs to watch are those of the angular displacement terms. If the deformation diagram indicates that the angular displacements at the joint in question are *increased* by the release moments, then these terms should be given *negative* signs, as usual. This is illustrated in the examples that follow.

In the following examples it will be assumed that members in each direction have constant flexural and torsional rigidities, i.e. that EI and GJ are constant in each direction, which is normally the case. The basic features of the analysis may be illustrated by means of a simple example, as shown in Fig. 21.5. Assuming $W_1 = W_3$ and $W_2 = W_4$, equations for the left-hand half of the frame will suffice.

A.A: $\quad 2K.M_A + K.M_{1A} = P\dfrac{\Delta_1}{b} = \dfrac{P}{a}\Delta_1 r \qquad$ A.D similar

A.B: $\quad 2M_B + M_{1B} = \dfrac{P}{a}\Delta_1 \qquad$ A.C similar

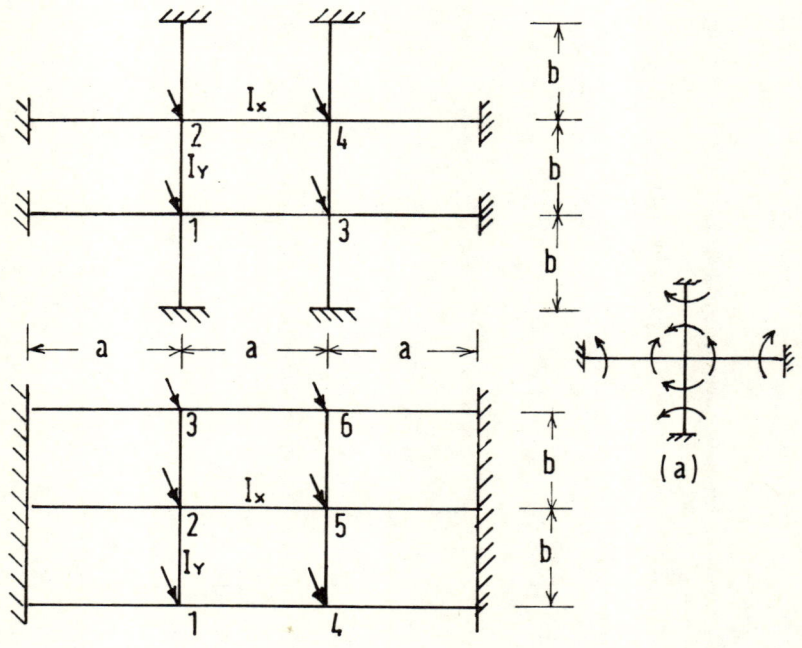

Fig. 21.4.

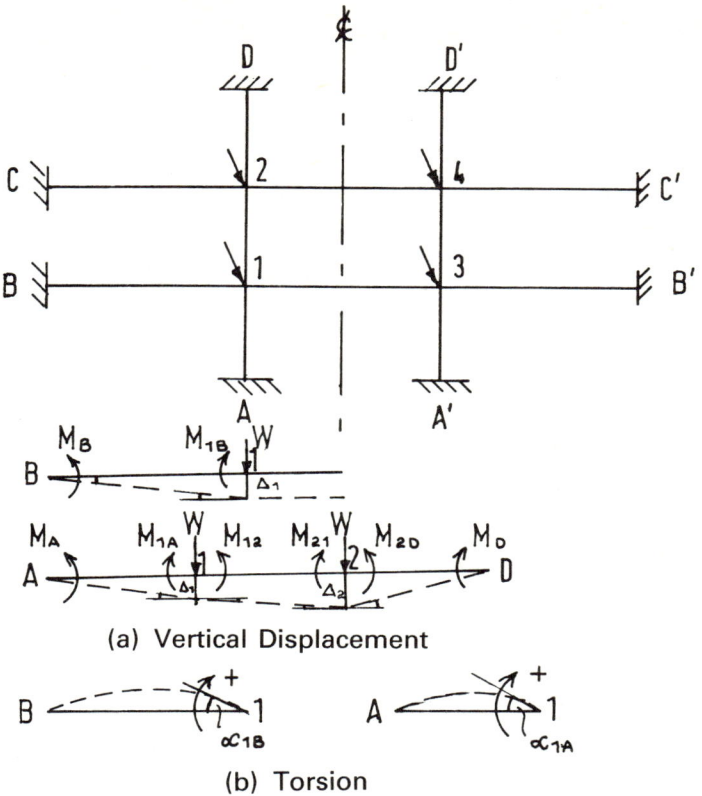

Fig. 21.5.

A.A12: $\quad K.M_A + 2K.M_{1A} + 2K.M_{12} + K.M_{21} = -\dfrac{P}{b}\Delta_1 +$

$\dfrac{P}{b}(\Delta_2 - \Delta_1) = \dfrac{P}{a}(-2r\Delta_1 + r\Delta_2)$ $\qquad$ A.D21 similar

A.B13: $\quad M_B + 2M_{1B} + 2M_{13} + M_{31} = -P\Delta_1\dfrac{2}{a} + P\Delta_3\dfrac{1}{a}$

$\qquad M_B + 2M_{1B} + 3M_{13} = -\dfrac{P}{a}\Delta_1$ $\qquad$ A.C24 similar

T.A1/1B: $\quad K.M_A + 2K.M_{1A} = -P\dfrac{\Delta_1}{b} + T_{1B}\,\Omega_{1B}$

$\qquad$ But $T_{1B} = M_{12} - M_{1A}$ and $\Omega_{1B} = \Omega_X$

$\qquad K.M_A + M_{1A}(2K + \Omega_X) - M_{12}\,\Omega_X = -\dfrac{P}{a}r\Delta_1$

$\qquad\qquad\qquad\qquad\qquad\qquad\qquad\qquad$ T.D2/2C similar

T.B1/1A: $M_B + 2M_{1B} = -\dfrac{P}{a}\Delta_1 + T_{1A}\Omega_{1A}$

But $T_{1A} = M_{13} - M_{1B}$ and $\Omega_{1A} = \Omega_Y$

$M_B + M_{1B}(2 + \Omega_Y) - M_{13}\Omega_Y = -\dfrac{P}{a}\Delta_1$ T.C2/2D similar

E.1: $\dfrac{M_A - M_{1A} + M_{21} - M_{12}}{b} + \dfrac{M_B - M_{1B} + M_{31} - M_{13}}{a} = W_1$

$r(M_A - M_{1A} + M_{21} - M_{12}) + M_B - M_{1B} = W_1 a$

E.2 similar

The equations are summarised in Table 21.7.

In order to compare results with and without torsion, the (3 × 3) rectangular grillage dealt with in Chapter 18, Problem (2), will be re-examined. The data for this problem were:

$a = 2$ m, $b = 1$ m, $I_X = 2I_Y$, $W_1 = 100$ kN, $W_3 = 100$ kN, no loads at joints 2 and 4 (note from Figs 18.1 and 21.5 that the positions of joints 2 and 3 have been interchanged)

Thus, $r = 2$, $K = \dfrac{I_X}{2}\dfrac{2}{I_X} = 1$, $W_1 a = 200$, $W_2 a = 0$

To allow for torsional effects, there must be sufficient information available to permit evaluation of Ω_X and Ω_Y. Taking the case where the grillage consists of two sets of concrete beams with Poisson's ratio $\lambda = 0.2$ and with dimensions of 100 × 200 (width × depth) in the XX direction and 50 × 200 at right angles to this, therefore

$$I_X = \dfrac{100 \times 200^3}{12} \text{ and } I_Y = \dfrac{50 \times 200^3}{12}$$

so that $I_X = 2I_Y$, as given. The torsional factors may be determined by reference to Table 20.1.

For $\dfrac{d}{b} = 2$, $\beta_X = 0.229$ and for $\dfrac{d}{b} = 4$, $\beta_Y = 0.281$

$\Omega_X = 6EI_X \phi_X = \dfrac{1 + \lambda}{\beta_X}\left(\dfrac{d}{b}\right)^2 = \dfrac{1.2}{0.229} \times 4 = 20.96$

$\Omega_Y = \dfrac{1 + \lambda}{\beta_Y}\left(\dfrac{d}{b}\right)^2 = \dfrac{1.2}{0.281} \times 16 = 68.33$

Data statements can now be prepared for use with the GRID program in order to determine the moments. The printout, which reproduces all the data for the program, is shown below.

Table 21.7 Equations for the (3 × 3) grillage in Fig. 21.5

Equation	Moment												$\bar{\Delta}$	
	A	B	C	D	13	1B	12	1A	24	2C	21	2D	1	2
A.A	2K							K					r	
A.B		2											1	
A.C			2			1				1				1
A.D				2K								K		r
A.B13	1				3	2								
T.B1/A	1				$-\Omega_Y$	$2 + \Omega_Y$								
A.A12	K						2K	2K					-1	
T.A1/1B	K						$-\Omega_X$	$2K + \Omega_X$					-1	
A.C24			1						3	2	K		$-2r$	
T.C2/2D			1						$-\Omega_Y$	$2 + \Omega_Y$				
A.D21				K							2K	2K	r	$-2r$
T.D2/2C				K			K				$-\Omega_X$	$2K + \Omega_X$		$-r$
E.1	r		1	r		-1	$-r$	$-r$			r		W_1a	
E.2		1	1	r			r			-1	$-r$	$-r$	W_2a	

268   FOUNDATION AND STRUCTURAL PROBLEMS

(3 × 3) GRID, FIG. 21.5.

MATRIX SIZE? 12

2	0	0	0	0	0	0	1	0	0	0	0
0	2	0	0	0	1	0	0	0	0	0	0
0	0	2	0	0	0	0	0	1	0	0	0
0	0	0	2	0	0	0	0	0	0	0	1
0	1	0	0	3	2	0	0	0	0	0	0
0	1	0	0	−68.33	70.33	0	0	0	0	0	0
1	0	0	0	0	0	2	2	0	0	1	0
1	0	0	0	0	0	−20.96	22.96	0	0	0	0
0	0	1	0	0	0	0	0	3	2	0	0
0	0	1	0	0	0	0	0	−68.33	70.33	0	0
0	0	0	1	0	0	1	0	0	0	2	2
0	0	0	1	0	0	0	0	0	0	−22.96	22.96

| 2 | 1 | 0 | 0 | −1 | −1 | −4 | −2 | 0 | 0 | 2 | 0 |

?ANOTHER DEFLECTED NODE. FOR YES PRESS 1
? 1

| 0 | 0 | 1 | 2 | 0 | 0 | 2 | 0 | −1 | −1 | −4 | −2 |

?ANOTHER DEFLECTED NODE. FOR YES PRESS 1
? 0

EQUILIBRIUM COEFFS:-

| 2 | 1 | 0 | 0 | 0 | −1 | −2 | −2 | 0 | 0 | 2 | 0 |
| 0 | 0 | 1 | 2 | 0 | 0 | 2 | 0 | 0 | −1 | −2 | −2 |

NODAL LOAD × BAY LENGTH:-

1   200
2   0

RESULTS:-

1	36.77	FINAL MOMENT
2	15.88	FINAL MOMENT
3	9.96	FINAL MOMENT
4	15.71	FINAL MOMENT
5	−7.77	FINAL MOMENT
6	−8.11	FINAL MOMENT
7	−24.74	FINAL MOMENT
8	−26.24	FINAL MOMENT
9	−4.87	FINAL MOMENT
10	−5.09	FINAL MOMENT
11	0.25	FINAL MOMENT
12	−1.75	FINAL MOMENT

From considerations of frame symmetry, we can make some checks on the results, as follows:

$$M_{13} + M_{1B} = -M_B = -15.88$$
$$M_{21} + M_{2D} = -M_C = -9.96$$
$$M_{1A} + M_{12} + M_{21} + M_{2D} = -M_A - M_D = -52.48$$

Comparing results with those obtained for the same grillage in Chapter 18 where torsional effects were ignored (Problem (2)), it can be seen that moments at the supports are reduced slightly and that nodal moments are reduced slightly in the XX direction and increased in the YY direction (see page 216).

Allowance for torsion

In Chapter 18, the equations for grillage of various sizes and shapes were derived without consideration to torsional effects. Table 18.2, for example, shows the half-frame matrix for a (3 × 3) grillage and is to be compared with Table 21.7 in which torsional effects are included. The latter shows the four torsional equations and the subdivision of the angular compatibility equations needed to solve the problem. Table 21.7 illustrates also the way in which the torsional constants Ω_X and Ω_Y are introduced. Using the same procedure, we can alter the tables in Chapter 18, or any similar tables, so as to take account of torsional effects, where required.

PART 4
RIGID FRAMES

22 Portal frames and box culverts

Symmetrical portal frames

Symmetrical rigid frames that are not subject to sway can be analysed in much the same way as continuous beams. Thus, the portal frame in Fig. 22.1 may be regarded as equivalent to a continuous beam ABCD with the end spans rotated into the vertical. Using the approach described for the solution of continuous beams in Chapters 5 and 6, we are able to compose equations for angular compatibility when releases are inserted at joints or supports. The only difference is that, for portals, the beam member BC is chosen to be the reference element. If its flexibility is denoted by L_B/I_B, then the two columns AB and CD, identified by the suffixes 1 and 2 respectively, have relative flexibilities

$$K_1 = \frac{H_1}{I_1} \frac{I_B}{L_B} \text{ and } K_2 = \frac{H_2}{I_2} \frac{I_B}{L_B}$$

also $K_B = 1$, by definition. (For a symmetrical portal $K_1 = K_2$, of course.)

The deformation diagram in Fig. 22.1 shows the angular changes due to bending only when the load system is applied to the frame. The arrows indicate the direction of the release moments in the usual way. Since the portal and the loading is symmetrical, we need consider only one-half of the frame when writing down the compatibility equations, thus:

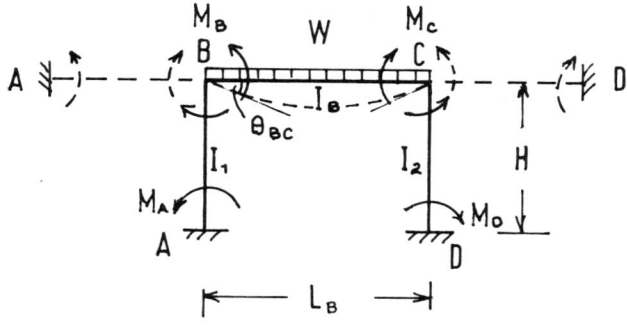

Fig. 22.1.

A.A: $\quad M_A \dfrac{H}{3EI_1} + M_B \dfrac{H}{6EI_1} = 0$

Multiplying across by $\dfrac{6EI_B}{L_B}$,

$$2K_1 M_A + K_1 M_B = 0$$

A.B: $\quad M_A \dfrac{H}{6EI_1} + M_B \dfrac{H}{3EI_1} + M_B \dfrac{L_B}{3EI_B} + M_B \dfrac{L_B}{6EI_B} = \theta_{BC}$

$$K_1 M_A + (3 + 2K_1) M_B = \tfrac{1}{4} W_B L_B$$

The two equations in tabular form are reproduced in Table 2.1. This table may be used with MATIN for the solution of any symmetrical frame with member BC carrying a uniformly distributed load. The requirements for symmetry apply to:

(1) the physical properties of the members, i.e. their E and I values;
(2) support conditions;
(3) the loading system.

For other types of symmetrical loading, the factors in Table 2.1 should be used to determine the appropriate $\overline{W}$ vector terms. For example, suppose $H = 6$ m, $L_B = 16$ m, $I_B = 3I_1$ and the portal carries a point load of 100 kN at midspan on BC, then

$$K_1 = \dfrac{6}{I_1} \dfrac{3I_1}{16} = 1.125, \quad \overline{W} \text{ for BC} = \dfrac{3}{8} \times 100 \times 16 = 600$$

From Table 22.1, the relevant data statement for obtaining the nodal moments using MATIN is:

1000 DATA 2.25,1.125,1.125,5.25,0,600

Input N = 2 to get the required answer,

$$M_A = -64 \text{ and } M_B = 128 \text{ kNm}$$

Table 22.1 Half-matrix for symmetrical portal

Equation	Moment		$\overline{W}$
	A	B	
A.A	$2K_1$	K_1	
A.B	K_1	$3 + 2K_1$	$\tfrac{1}{4} W_B L_B$

PORTAL FRAMES AND BOX CULVERTS 275

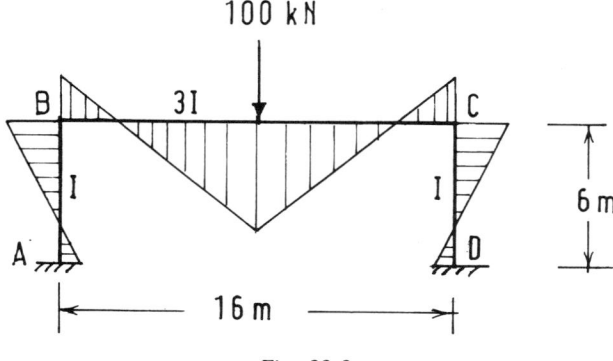

Fig. 22.2.

The BM diagram is reproduced in Fig. 22.2. Positive moments are drawn on the outside face of the columns and top surface of the portal beam, so that the diagram appears on the tension face in all cases.

Box culverts

Symmetrical box culverts can be tackled in the same fashion. If, for instance, the rectangular culvert in Fig. 22.3, of dimensions L × H, is assumed to carry a central point load of W kN on the roof and a total triangular earth pressure load of T kN on each side wall, the equations of angular compatibility may be written as before. It is assumed also that E is constant and that a uniform pressure is imposed on the soil. (This may need to be modified, especially for a clay soil as described in Chapter 7; however, non-uniform pressure will not seriously effect the corner moments.) Selecting BC as the reference element, we have:

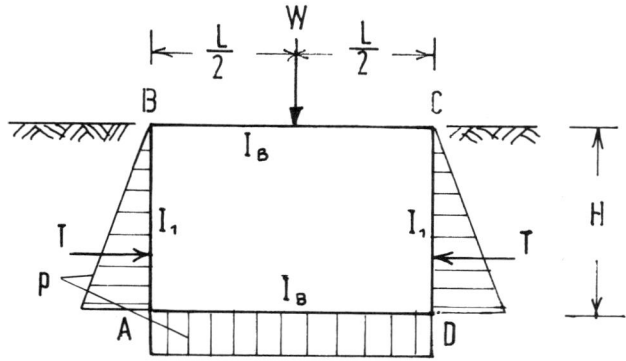

Fig. 22.3.

A.A: $M_D \dfrac{L}{6EI_B} + M_A \dfrac{L}{3EI_B} + M_A \dfrac{H}{3EI_1} + M_B \dfrac{H}{6EI_1} = \theta_{AD} + \theta_{AB}$

Since $M_D = M_A$ and $K_1 = \dfrac{H}{I_1} \dfrac{I_B}{L}$

$M_A + 2M_A + 2K_1 M_A + K_1 M_B = \dfrac{1}{4} WL + \dfrac{4}{15} K_1 TH$

$(3 + 2K_1) M_A + K_1 M_B = \dfrac{1}{4} WL + \dfrac{4}{15} K_1 TH$

A.B: $M_A \dfrac{H}{6EI_1} + M_B \dfrac{H}{3EI_1} + M_B \dfrac{L}{3EI_B} + M_C \dfrac{L}{6EI_B} = \theta_{BA} + \theta_{BC}$

$K_1 M_A + 2K_1 M_B + 3M_B = \dfrac{7}{30} K_1 TH + \dfrac{3}{8} WL$ (since $M_B = M_C$)

$K_1 M_A + (3 + 2K_1) M_B = \dfrac{7}{30} K_1 TH + \dfrac{3}{8} WL$

These are summarised in Table 22.2.

Table 22.2 Half-matrix for symmetrical box culvert

Equation	Moment		$\overline{W}$
	A	B	
A.A	$3 + 2K_1$	K_1	$\dfrac{1}{4} WL + \dfrac{4}{15} K_1 TH$
A.B	K_1	$3 + 2K_1$	$\dfrac{3}{8} WL + \dfrac{7}{30} K_1 TH$

For a slight variation on the above, consider the problem in Fig. 22.4 where the rectangular culvert has line loads applied on the sides and roof as shown. Values in the figure indicate the load per metre run of the culvert. A uniform bearing pressure on the soil is assumed, as before. The relevant terms are:

Flexibility factor, $K_1 = \dfrac{9}{2I} \dfrac{3I}{12} = 1.125$

$\overline{W}$ terms: load on BC, angular change at B $= \dfrac{3}{8} WL = 9$

load on AD, angular change at A $= \dfrac{1}{4} WL = 6$

load on AB, angular change at A $= K_1 \dfrac{Tab}{H^2} (H + a) = 3.75$

PORTAL FRAMES AND BOX CULVERTS

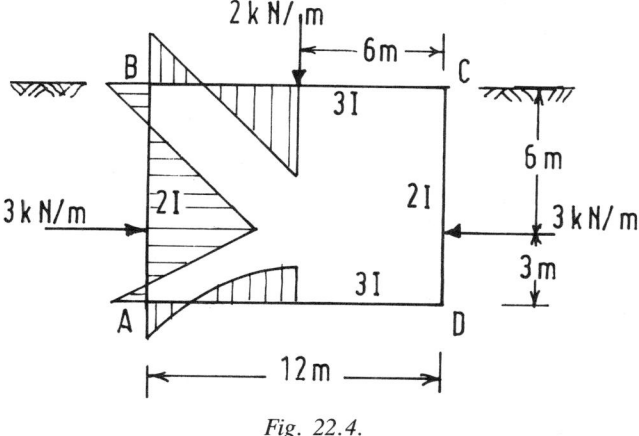

Fig. 22.4.

load on AB, angular change at B $= K \dfrac{Tab}{H^2} (H + b) = 3$

These values furnish the following data statement:

1000 DATA 5.25,1.125,1.125,5.25,9.75,12

Results obtained with MATIN are $M_A = 1.43$ and $M_B = 1.98$ kNm. The BM diagram for one-half of the culvert is shown in Fig. 22.4.

Multi-bay portals

For the two-bay symmetrical portal frame in Fig. 22.5, one may infer from conditions of symmetry that $M_{CB} = M_{CE} = M_C$ and that $M_{CD} = M_{DC} = 0$. Where the beam member BE carries a uniformly distributed load of W kN on each span, L, the appropriate equations are:

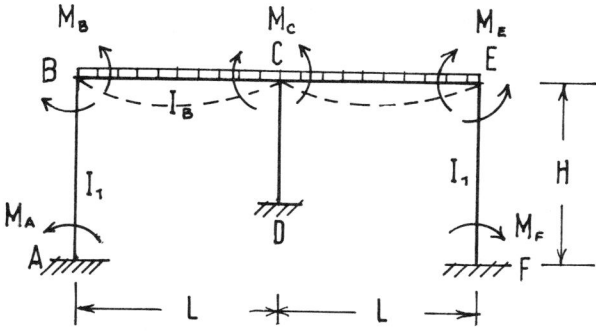

Fig. 22.5.

A.A:
$$M_A \frac{H}{3EI_1} + M_B \frac{H}{6EI_1} = 0$$
$$2K_1M_A + K_1M_B = 0$$

A.B:
$$M_A \frac{H}{6EI_1} + M_B \frac{H}{3EI_1} + M_B \frac{L}{3EI_B} + M_C \frac{L}{6EI_B} = \theta_{BC}$$
$$K_1M_A + 2K_1M_B + 2M_B + M_C = \frac{1}{4}WL$$
$$K_1M_A + 2(1+K_1)M_B + M_C = \frac{1}{4}WL$$

A.BCE:
$$M_B \frac{L}{6EI_B} + M_C \frac{2L}{3EI_B} + M_B \frac{L}{6EI_B} = \theta_{CB} + \theta_{CE}$$
$$2M_B + 4M_C = \frac{1}{2}WL$$

These provide the half-frame matrix in Table 22.3.

Table 22.3 Two-bay symmetrical portal

Equation	Moment			$\overline{W}$
	A	B	C	
A.A	$2K_1$	K_1		
A.B	K_1	$2(1+K_1)$	1	$\frac{1}{4}WL$
A.C		2	4	$\frac{1}{2}WL$

If the column feet are pinned then, since $M_A = 0$ we may omit the first row and column for moment A in the matrix table in the usual way. For example, to solve the problem in Fig. 22.6,

$$K_1 = \frac{4}{I}\frac{2I}{8} = 1 \text{ and } \frac{3}{8}WL = 30$$

so we may use the data statement

  1000 DATA 4,1,2,4,30,60

Incorporating this in MATIN provides the results

$$M_B = 4.29 \text{ and } M_C = 12.86 \text{ kNm}$$

This construction is clearly most unstable since a slight amount of eccentricity or a lateral load might be sufficient to cause the frame to collapse.

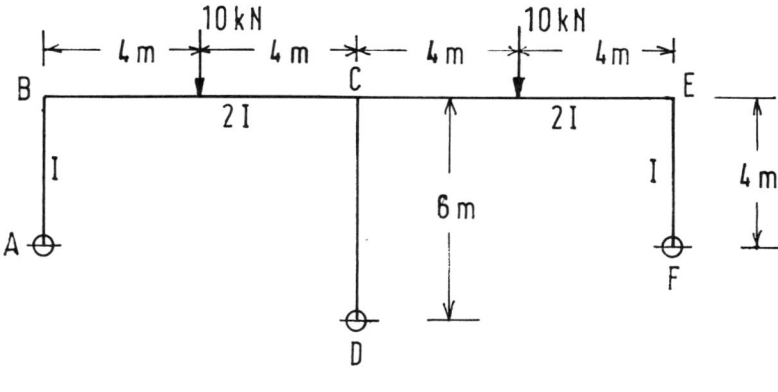

Fig. 22.6.

The next section includes an examination of lateral movement in portals, referred to as 'sway', incurred by asymmetrical conditions.

Portals subject to sway

So far we have dealt with symmetrical frames in which the joint moments are due entirely to bending under the applied loading. Where conditions are not symmetrical because of frame geometry, loading or support characteristics, another effect has to be considered. This is the sway phenomenon illustrated in Fig. 22.7. Because the columns are of unequal height, unequal moments occur at nodes B and C. These moments at the column heads produce an unbalanced thrust that tends to sway the frame to the right, i.e. away from the node where the greater bending moment occurs.

For lateral stability, the total reaction at the column heads, including that from any externally applied load, must balance. Where the frame supports vertical loads only, by taking moments about the column heads we have

$$H_A H_1 = M_B - M_A \quad \text{(see Fig. 22.7)}$$

and $\quad H_D H_2 = M_C - M_D$

But $\quad H_D - H_A = 0$

$$\therefore H_D - H_A = \frac{M_C - M_D}{H_2} - \frac{M_B - M_A}{H_1} = 0$$

If, in addition to the vertical loading, the frame has an external horizontal thrust, T, acting at the head of the columns as shown, then $T = H_D - H_A$, so we can write:

E.1:
$$\frac{M_A - M_B}{H_1} + \frac{M_C - M_D}{H_2} = T \quad (22.1)$$

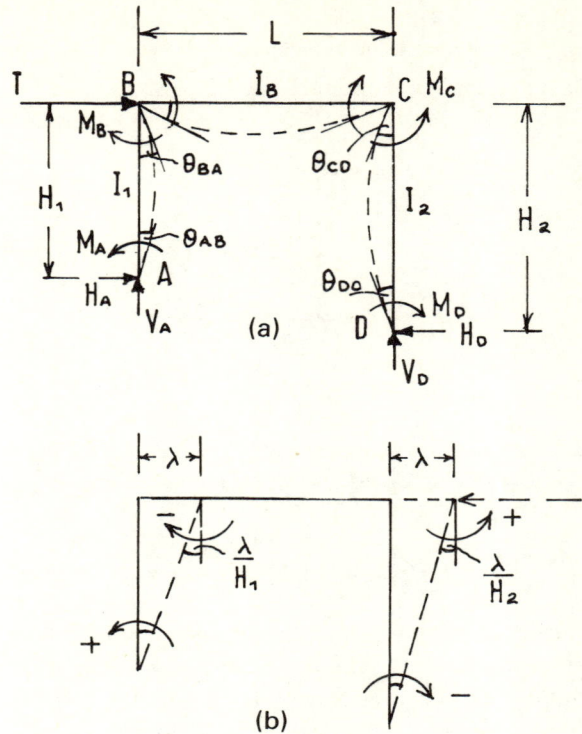

Fig. 22.7.

This is the equilibrium equation for unbalanced shear in the columns.

The deformation diagram showing the angular changes due to bending at the joints on release will be the same as before. This is reproduced in Fig. 22.7(a) for the applied loads, with arrows indicating the direction of the release moments assuming distributed loading on both columns and beam BC. The deformation diagram in (b) shows the angular changes due to sway only in the columns. Release moments are applied to the columns to reduce the angular displacements. Where these act in the same direction as the 'loads' moments in Fig. 22.7(a), they are given a positive sign; where they act in the opposite direction they are marked negative. For stability, when releases are introduced at the nodes, a temporary horizontal prop must be inserted at C to hold the frame in its displaced position. This is indicated by the dashed arrow in Fig. 22.7(b).

If the sway at the column heads is denoted by λ, then, since the lateral displacement, λ, is the same at B and C, the angular changes due to sway equal $\dfrac{\lambda}{H_1}$ in column AB and $\dfrac{\lambda}{H_2}$ in column CD. From the deformation diagrams it will be seen that the total angular changes at the nodes are:

At A: $\theta_{AB} + \dfrac{\lambda}{H_1}$

At B: $\theta_{BA} - \dfrac{\lambda}{H_1}$

At C: $\theta_{CD} + \dfrac{\lambda}{H_2}$

At D: $\theta_{DC} - \dfrac{\lambda}{H_2}$

This information allows the angular compatibility equations for the portal frame to be assembled using the same notation as before.

A.A: $M_A \dfrac{H_1}{3EI_1} + M_B \dfrac{H_1}{6EI_1} = \theta_{AB} + \dfrac{\lambda}{H_1}$

Multiplying across by $P = \dfrac{6EI_B}{L_B}$

$2K_1 M_A + K_1 M_B = \dfrac{1}{4} K_1 W_1 L_1 + P \dfrac{\lambda}{H_1}$

A.B: $M_A \dfrac{H_1}{6EI_1} + M_B \dfrac{H_1}{3EI_1} + M_B \dfrac{L_B}{3EI_B} + M_C \dfrac{L_B}{6EI_B} = \theta_{BA} + \theta_{BC} - \dfrac{\lambda}{H_1}$

$K_1 M_A + 2K_1 M_B + 2M_B + M_C = \dfrac{1}{4} K_1 W_1 L_1 + \dfrac{1}{4} W_B L_B - P\dfrac{\lambda}{H_1}$

$K_1 M_A + 2(1+K_1) M_B + M_C = \dfrac{1}{4} (K_1 W_1 L_1 + W_B L_B) - P\dfrac{\lambda}{H_1}$

Table 22.4 Portal frame subject to sway

Equation	Moment				$\overline{W}$	$\overline{\lambda}$
	A	B	C	D		
A.A	$2K_1$	K_1			$\dfrac{1}{4} K_1 W_1 L_1$	$\dfrac{1}{H_1}$
A.B	K_1	$2(1+K_1)$	1		$\dfrac{1}{4}(K_1 W_1 L_1 + W_B L_B)$	$-\dfrac{1}{H_1}$
A.C		1	$2(1+K_2)$	K_2	$\dfrac{1}{4}(W_B L_B + K_2 W_2 L_2)$	$\dfrac{1}{H_2}$
A.D			K_2	$2K_2$	$\dfrac{1}{4} K_2 W_2 L_2$	$-\dfrac{1}{H_2}$
E.1	$\dfrac{1}{H_1}$	$-\dfrac{1}{H_1}$	$\dfrac{1}{H_2}$	$-\dfrac{1}{H_2}$	T	

A.C: $M_B \dfrac{L_B}{6EI_B} + M_C \dfrac{L_B}{3EI_B} + M_C \dfrac{H_2}{3EI_2} + M_D \dfrac{H_2}{6EI_2} = \theta_{CB} + \theta_{CD} + \dfrac{\lambda}{H_2}$

$M_B + 2(1+K_2)M_C + K_2 M_D = \dfrac{1}{4}(W_B L_B + K_2 W_2 L_2) + P\dfrac{\lambda}{H_2}$

A.D: $M_D \dfrac{H_2}{3EI_2} + M_C \dfrac{H_2}{6EI_2} = \theta_{DC} - \dfrac{\lambda}{H_2}$

$K_2 M_C + 2K_2 M_D = \dfrac{1}{4} K_2 W_2 L_2 - P\dfrac{\lambda}{H_2}$

These equations together with the equilibrium equation, E.1, are incorporated in Table 22.4. The $\overline{W}$ vector terms must be altered as required, of course, to suit the actual loading on the frame.

The set of equations is very similar in form to those derived for the solution of beam frameworks in Chapter 16, with the sway vector, $\overline{\lambda}$, corresponding to the previous deflection vector, $\overline{\Delta}$. It is to be expected, therefore, that the same method of analysis and computer program could be used to find the moments in this case also. The XBEAM program is quite suitable for this purpose, in fact.

The value of T to be used in preparing the data is the thrust or reaction due to the external loads only. The true value, including that due to bending in the frame, is calculated and printed out when the program is run. The program determines the 'loads' moments and the unit sway moments and the factor by which the latter must be multiplied to obtain the actual sway moments; these are then added algebraically to the loads moments to obtain the final moment, as described earlier in Chapter 16. Use of the XBEAM program can be demonstrated by the example in

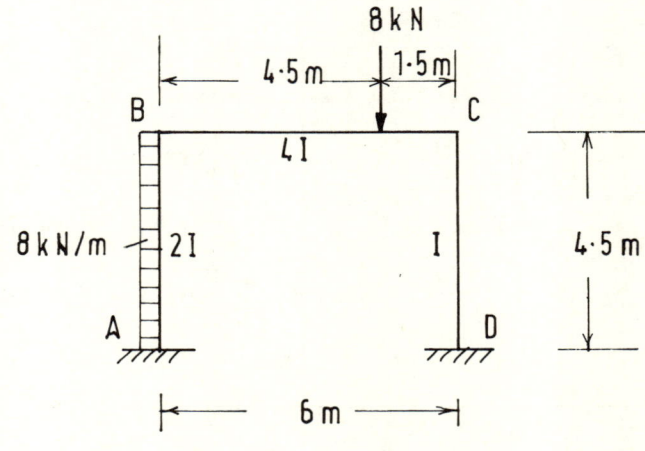

Fig. 22.8.

Fig. 22.8 which includes both vertical and horizontal loading on the portal. Values required for the data table for this problem are:

Factors: $K_1 = \dfrac{4.5}{2I} \dfrac{4I}{6} = 1.5$; $K_2 = \dfrac{4.5}{I} \dfrac{4I}{6} = 3$; $T = 18$

$\overline{W}$ terms: angular change at A for load on AB $= \dfrac{1}{4} \times \dfrac{3}{2} \times 36 \times \dfrac{9}{2}$

$$= 60.75$$

$$\text{at B for load on BA} = 60.75$$

$$\text{at B for load on BC} = \dfrac{8 \times 4.5 \times 1.5}{36} \times 7.5$$

$$= 11.25$$

$$\text{at C for load on CB} = 11.25 \times \dfrac{10.5}{7.5} = 15.75$$

The data table for use with XBEAM is shown in Table 22.5; results obtained when the program is RUN with N = 4 and U = 1 are shown in the bottom line. Other problems with answers will be found in Appendix A.

Portal with cantilever arm

If the portal beam is extended beyond a column to form a cantilever arm, as in Fig. 22.9, a moment M_E is applied at E so that M_{CB} is no longer equal to M_{CD}. In this case, in fact, we have

$$M_{CD} = M_{CB} + M_{CE}$$
or $$M_{CB} = M_{CD} - M_{CE}$$

Using the same equations as before, but substituting for M_{CB}, gives:

Table 22.5 Data table for Fig. 22.8

Ref.	Moment				
	A	B	C	D	
A.A	3	1.5			
A.B	1.5	5	1		
A.C		1	8	3	
A.D			3	6	
$\overline{W}$	60.75	72	15.75	0	
$\overline{\lambda}$	0.2222	−0.2222	0.2222	−0.2222	
E.1	0.2222	−0.2222	0.2222	−0.2222	18
Results	40.90	−10.17	14.77	−15.17	kNm

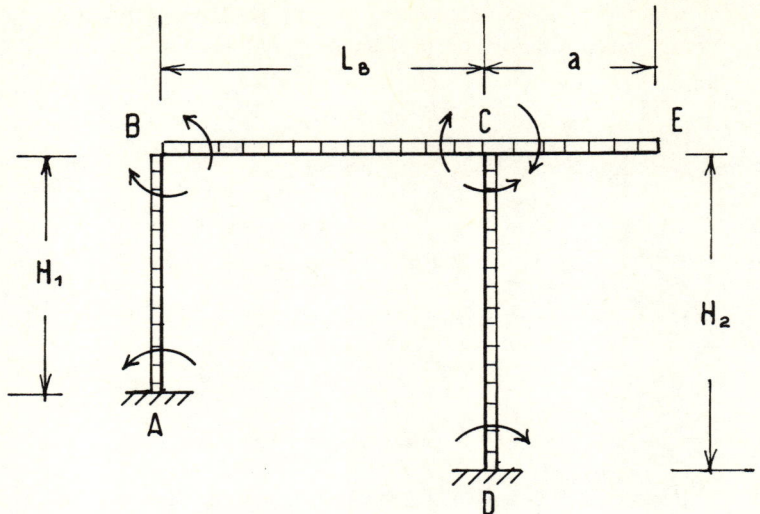

Fig. 22.9.

A.B: $\quad K_1 M_A + 2(1 + K_1) M_B + M_{CB} = \frac{1}{4}(K_1 W_1 L_1 + W_B L_B) - P\frac{\lambda}{H_1}$

$\quad\quad\quad K_1 M_A + 2(1 + K_1) M_B + M_{CD} = \frac{1}{4}(K_1 W_1 L_1 + W_B L_B)$

$$- P\frac{\lambda}{H_1} + M_{CE}$$

A.C: $\quad M_B + 2M_{CB} + 2K_2 M_{CD} + K_2 M_D = \frac{1}{4}(W_B L_B + K_2 W_2 L_2) + P\frac{\lambda}{H_2}$

$\quad\quad\quad M_B + 2(1 + K_2) M_{CD} + K_2 M_D = \frac{1}{4}(W_B L_B + K_2 W_2 L_2)$

$$+ P\frac{\lambda}{H_2} + 2M_{CE}$$

The only change to Table 22.4 resulting from these equations is that the $\overline{W}$ vector contains two new terms, namely M_{CE} opposite A.B and $2M_{CE}$ opposite A.C. If the cantilever is located at B then $2M_{BE}$ and M_{BE} should be substituted for M_{CE} and $2M_{CE}$ respectively. These values can be determined and inserted with the loads data before the XBEAM program is run.

Portals with inclined columns

The similarity between the analysis of portal frames with vertical legs and continuous beams was mentioned at the start of the chapter. In point of

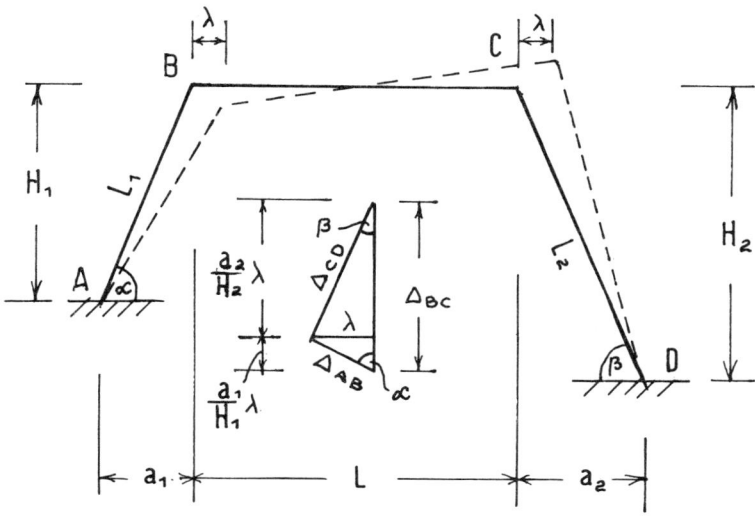

Fig. 22.10.

fact, disregarding the $\bar{\lambda}$ vector column, Table 22.4 is very similar to Table 5.1 for solving continuous beams, except that in the latter the first span was adopted as the reference element instead of beam BC selected for the portal frame. If the portal columns are splayed at some intermediate angle, as shown in Fig. 22.10, one would expect the coefficient matrix and $\overline{W}$ terms to remain unchanged. The angular changes represented by the $\bar{\lambda}$ vector are affected, however, as may be seen from Fig. 22.10.

The angular changes at A and D remain the same, but the lateral displacement, λ, of the column heads causing them to rotate about their bases will produce a lateral and downwards displacement at B and a lateral and upwards displacement at C. The relative displacements are embodied in the classical 'triangle of displacements', which is reproduced in Fig. 22.10. Displacements are drawn at right angles to the members, sufficiently accurate for small displacements. From similar triangles we can deduce

$$\frac{\Delta_{AB}}{\lambda} = \frac{L_1}{H_1}; \frac{\Delta_{CD}}{\lambda} = \frac{L_2}{H_2}; \frac{\Delta_{BC}}{\lambda} = \frac{a_1}{H_1} + \frac{a_2}{H_2}$$

where L_1 and L_2 are the lengths of the inclined columns AB and CD and Δ_{AB}, etc. are the displacements perpendicular to the members.

The angular changes at the nodes are, therefore:

$$\frac{\Delta_{AB}}{L_1} = \frac{\lambda}{H_1}, \frac{\Delta_{CD}}{L_2} = \frac{\lambda}{H_2} \text{ and } \frac{\Delta_{BC}}{L} = \left(\frac{a_1}{H_1} + \frac{a_2}{H_2}\right)\frac{\lambda}{L}$$

From this it will be seen that the *total* angular changes when releases are introduced at the joints are:

At A: $\dfrac{\lambda}{H_1}$

At B: $-\dfrac{\lambda}{H_1} - \left(\dfrac{a_1}{H_1} + \dfrac{a_2}{H_2}\right)\dfrac{\lambda}{L}$

At C: $\dfrac{\lambda}{H_2} + \left(\dfrac{a_1}{H_1} + \dfrac{a_2}{H_2}\right)\dfrac{\lambda}{L}$

At D: $-\dfrac{\lambda}{H_2}$

For unit displacements, as assumed initially in the XBEAM program, $\lambda = 1$. This substitution gives the $\bar{\lambda}$ vector terms.

The equilibrium equation needs adjustment also, since H_A is now not necessarily equal to H_D. To determine the new equilibrium conditions first take moments for the left-hand side of the frame about joints B and C, Fig. 22.11. The external loading is represented by a vertical resultant, W, acting at a distance x from C and a horizontal thrust, T, acting along BC as shown.

Moments about C: $\quad H_A H_1 = V_A (L+a_1) - Wx - M_A + M_C$

Moments about B: $\quad H_A H_1 = V_A a_1 - M_A + M_B$

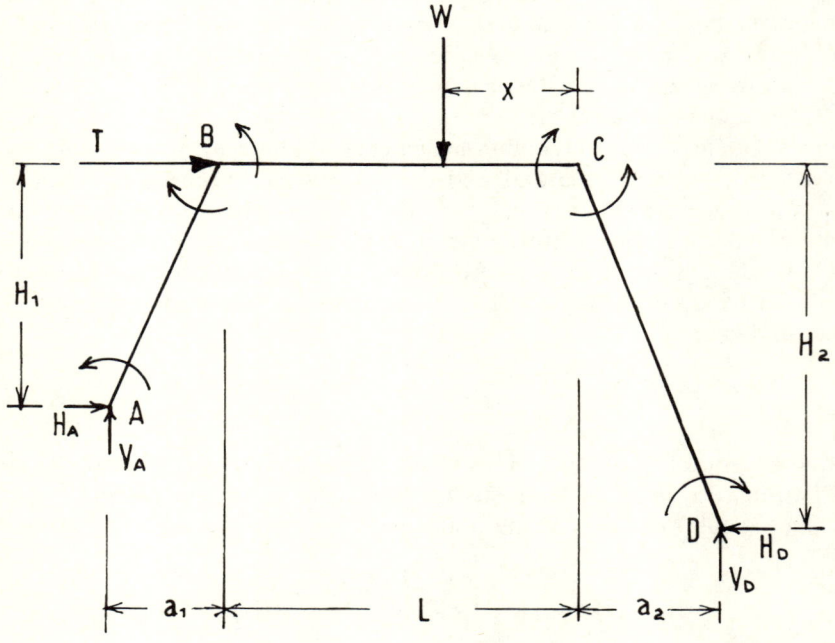

Fig. 22.11.

Subtracting,
$$0 = V_A L - Wx + M_C - M_B$$
$$V_A = \frac{M_B - M_C + Wx}{L}$$

Moments about C for the right-hand side of the frame:
$$H_D H_2 = V_D a_2 + M_C - M_D$$
$$H_D = \frac{a_2}{H_2}(W - V_A) + \frac{M_C - M_D}{H_2}$$

also
$$H_A = \frac{a_1}{H_1} V_A + \frac{M_B - M_A}{H_1}$$

$$H_D - H_A = T = \frac{M_A - M_B}{H_1} + \frac{M_C - M_D}{H_2} -$$
$$V_A \left(\frac{a_1}{H_1} + \frac{a_2}{H_2}\right) + \frac{a_2}{H_2} W$$

Substituting for V_A gives

E.1: $M_A \dfrac{1}{H_1} - M_B \left(\dfrac{1}{H_1} + \dfrac{a_1}{H_1 L} + \dfrac{a_2}{H_2 L}\right) + M_C \left(\dfrac{1}{H_2} + \dfrac{a_1}{H_1 L} + \dfrac{a_2}{H_2 L}\right) -$
$M_D \dfrac{1}{H_2} = T + W \left(\dfrac{a_1 x}{H_1 L} + \dfrac{a_2 x}{H_2 L} - \dfrac{a_2}{H_2}\right)$

These findings are incorporated in Table 22.6. (The $\overline{W}$ vector applies only where BC carries a uniformly distributed load.) The equations may

Table 22.6 Portal frame with inclined columns

Equation	Moment A	B	C	D	$\overline{W}$	$\overline{\lambda}$
A.A	$2K_1$	K_1				$\dfrac{1}{H_1}$
A.B	K_1	$2(1 + K_1)$	1		$\dfrac{1}{4}WL$	$-\left(\dfrac{1}{H_1} + \dfrac{a_1}{H_1 L} + \dfrac{a_2}{H_2 L}\right)$
A.C		1	$2(1 + K_2)$	K_2	$\dfrac{1}{4}WL$	$\dfrac{1}{H_2} + \dfrac{a_1}{H_1 L} + \dfrac{a_2}{H_2 L}$
A.D			K_2	$2K_2$		$-\dfrac{1}{H_2}$
E.1	$\dfrac{1}{H_1}$	$-\left(\dfrac{1}{H_1} + \dfrac{a_1}{H_1 L} + \dfrac{a_2}{H_2 L}\right)$	$\dfrac{1}{H_2} + \dfrac{a_1}{H_1 L} + \dfrac{a_2}{H_2 L}$	$-\dfrac{1}{H_2}$		$T + W\left(\dfrac{a_1 x}{H_1 L} + \dfrac{a_2 x}{H_2 L} - \dfrac{a_2}{H_2}\right)$

be simplified for special cases such as portals in which $a_1 = a_2$, $a_1 = 0$, etc. For the case where $a_1 = a_2 = 0$ the table becomes coincident with Table 22.3 for rectilinear portal frames. An example of one of these special cases is shown in Fig. 22.12, where $a_1 = a_2 = 10$ m. The frame supports both vertical and horizontal loading.

Principle values required for the program are:

$$K_1 = \frac{22.36}{2I} \frac{3I}{20} = 1.667; \quad K_2 = \frac{18.03}{I} \frac{3I}{20} = 2.705$$

$$\frac{a}{L}\left(\frac{1}{H_1} + \frac{1}{H_2}\right) = \frac{10}{20}(0.05 + 0.667) = 0.0584$$

$$\frac{Wax}{L}\left(\frac{1}{H_1} + \frac{1}{H_2}\right) = \frac{20 \times 10 \times 15}{20} \times 0.1167 = \quad 17.500$$

$$-\frac{Wa_2}{H_2} = -\frac{20 \times 10}{15} = -13.333$$

$$T \qquad\qquad\qquad\qquad\qquad\qquad\qquad\qquad = -10.000$$

$$\frac{Wax}{L}\left(\frac{1}{H_1} + \frac{1}{H_2}\right) - \frac{Wa_2}{H_2} + T \qquad\qquad = \quad -5.833$$

$$\frac{Wab}{L^2}(L + a) = \frac{20 \times 15 \times 5}{400} \times 35 = \quad 131.25$$

$$\frac{Wab}{L^2}(L + b) = \frac{20 \times 15 \times 5}{400} \times 25 = \quad 93.75$$

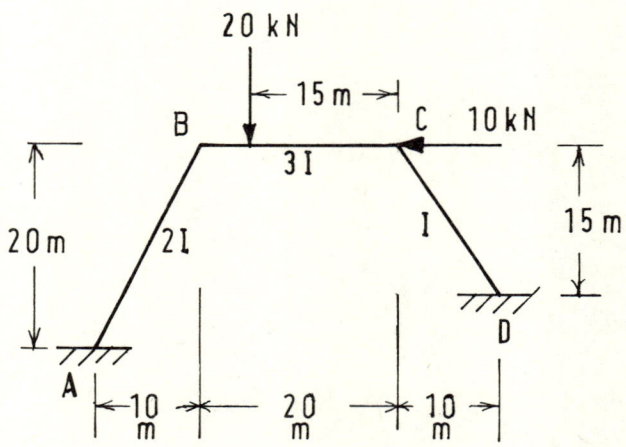

Fig. 22.12.

Table 22.7 Data table for Fig. 22.12

Ref.	A	B	C	D	
A.A	3.355	1.677			
A.B	1.677	5.355	1		
A.C		1	7.41	2.705	
A.D			2.705	5.41	
$\overline{W}$		131.25	93.75		
$\overline{\lambda}$	0.05	−0.1084	0.1251	−0.0667	
E.1	0.05	−0.1084	0.1251	−0.0667	−5.833
Loads	−13.30	26.61	11.08	−5.54	
Sway	−11.61	12.81	−11.25	9.93	
Total	−24.91	39.42	−0.17	4.39	

These values are included in the data table showing statements to be entered in XBEAM before it is run. Results are shown at the bottom of Table 22.7.

Pitched portals

The angular changes in a regular pitched portal frame subject to vertical loads are shown in Fig. 22.13. It will be seen that joint B is displaced outwards and downwards and joint C vertically downwards. The relative displacements are shown in the triangle of displacements, constructed as before. From similar triangles,

$$\frac{\Delta_{BC}}{\lambda} = \frac{L_{BC}}{R} = \frac{\Delta_{CD}}{\lambda}$$

$$\therefore \frac{\Delta_{BC}}{L_{BC}} = \frac{\lambda}{R}, \text{ for a half frame}$$

$$= \frac{2\lambda}{R}, \text{ for a full frame}$$

The angular changes are marked on Fig. 22.13, as well as arrows showing the direction of the release moments assuming all members are loaded as before. The total angular change at each node is given below. The signs indicate whether the angular displacements are opposed or augmented by the release moments; the former carry a positive sign.

At A: $\quad -\dfrac{\lambda}{H_1}$

At B: $\quad \dfrac{\lambda}{H_1} + \dfrac{\lambda}{R}$

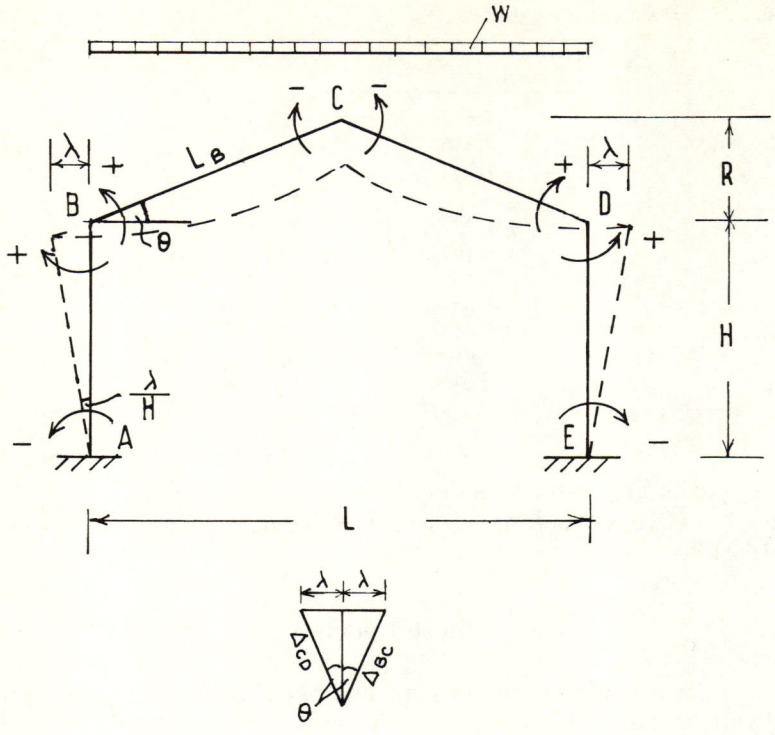

Fig. 22.13.

At C: $\quad -\dfrac{2\lambda}{R}$

At D: $\quad \dfrac{\lambda}{H_1} + \dfrac{\lambda}{R}$

At E: $\quad -\dfrac{\lambda}{H_1}$

For symmetrical conditions we need consider only one-half of the pitched portal, so the sway vector terms for unit lateral displacement reduce to

$$-\frac{1}{H}, \quad \frac{1}{H} + \frac{1}{R} \quad \text{and} \quad -\frac{1}{R} \quad \text{for the half-frame with } \lambda = 1$$

The equilibrium equation may be determined by first considering member BC as an isolated 'free body' (see Fig. 22.14), so that

$$V_B L_B = M_B - M_C + \frac{WL}{8}$$

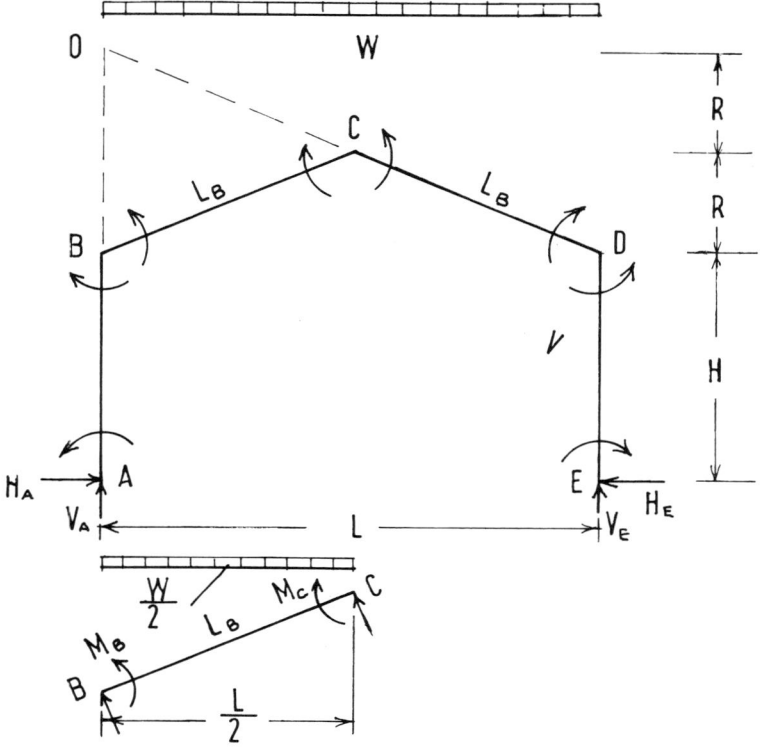

Fig. 22.14.

also $H_A H = M_B - M_A$

Taking moments about the intersection point, 0, for the portal section ABCD,

$$H_A (H + 2R) = \frac{WL}{2} - M_A + M_D - V_D 2L_B$$

For symmetrical conditions, $V_D = V_B$, so substituting for V_D and H_A,

E.1: $\qquad -M_A \dfrac{1}{H} + M_B \left(\dfrac{1}{R} + \dfrac{1}{H} \right) - M_C \dfrac{1}{R} = \dfrac{WL}{8R}$

This is incorporated in the half-frame matrix in Table 22.8. Note that, since the rafter BC has been selected as the 'reference member', the angular changes at B and C are measured from the rafter slope in each case, so the applied load is to be taken as acting perpendicular to the rafter. Since W is the total vertical load on the portal, then the rafter load

$$W_B = \frac{W}{2}\frac{L}{2L_B} = \frac{WL}{4L_B}$$

The $\overline{W}$ vector term $\frac{1}{4} W_B L_B = \frac{1}{4}\frac{WL}{4L_B} L_B = \frac{1}{16} WL$

Table 22.8 Half-frame matrix for a pitched portal with uniformly distributed load

Equation	Moment			$\overline{W}$	$\overline{\lambda}$
	A	B	C		
A.A	$2K_1$	K_1			$-\dfrac{1}{H}$
A.B	K_1	$2(1+K_1)$	1	$\dfrac{1}{16}WL$	$\dfrac{1}{H}+\dfrac{1}{R}$
A.C		1	2	$\dfrac{1}{16}WL$	$-\dfrac{1}{R}$
E.1	$-\dfrac{1}{H}$	$\dfrac{1}{H}+\dfrac{1}{R}$	$-\dfrac{1}{R}$	$\dfrac{WL}{8R}$	

Its practical use can be demonstrated by considering Fig. 22.15, in which the columns have a moment of inertia three times that of the rafter member. Thus,

$$K_1 = \frac{12}{3I}\frac{I}{13} = 0.3077;\quad \frac{1}{16}WL = \frac{24\times 24}{16} = 36;\quad \frac{WL}{8R} = 14.4$$

Suitable data statements for XBEAM are:

    1000 DATA 0.6154, 0.3077, 0, 0.3077, 2.6154, 1, 0, 1, 2
    1000 DATA 0, 36, 36, −0.0833, 0.2833, −0.2
    1005 DATA −0.0833, 0.2833, −0.2, 14.4

By running the program with input N = 3 and U = 1, the bending moments are obtained as follows:

	M_A	M_B	M_C
Loads:	−4.59	9.18	13.41
Sway:	−27.86	24.77	−23.81
Total:	−32.45	33.95	−10.40 kNm

Table 22.8 may also be used to solve the problem where a concentrated load is applied at the apex of the pitched portal. In this case, omit the $\overline{W}$ vector and alter WL/8R in the equilibrium equation to WL/4R to conform

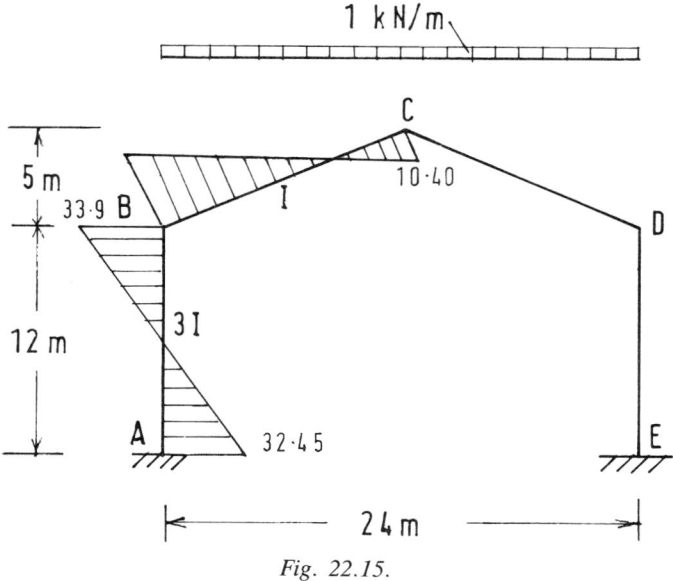

Fig. 22.15.

with the previous equations; it will be more convenient to use the GRID program in this instance.

Lateral loads on pitched portals

The portal frame in Fig. 22.16 is subject to a lateral load, T, applied at B,

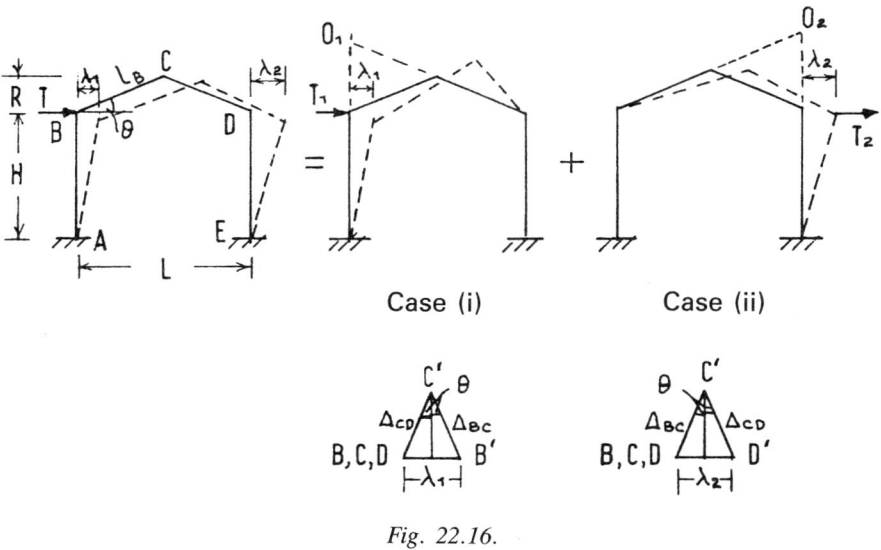

Fig. 22.16.

resulting in horizontal displacements λ_1 and λ_2 at the eaves, as shown. The frame displacement may be divided into two component cases, designated (i) and (ii) in the figure. The displacement vector for case (i) can be ascertained by reference to the triangle of displacements reproduced below the frame diagram. BB' and CC' represent the displacements of joints B and C respectively; therefore, $B'C' = \Delta_{BC}$ and $DC' = \Delta_{DC}$.

From similar triangles,

$$\Delta_{BC} = \Delta_{CD} = \frac{\lambda_1}{2 \sin \theta} = \frac{\lambda_1 L_B}{2R}$$

Change in slope of BC = change in slope of CD = $\dfrac{\lambda_1 L_B}{2RL_B} = \dfrac{\lambda_1}{2R}$

Total angular change at joint B = $\dfrac{\lambda_1}{H} + \dfrac{\lambda_1}{2R}$

at joint C = $2\dfrac{\lambda_1}{2R} = \dfrac{\lambda_1}{R}$

at joint D = $\dfrac{\lambda_1}{2R}$

To obtain the displacement vector, put $\lambda_1 = 1$ and add signs to show the relationship of the angular changes to the release moments marked on the main diagram. For joints A, B, C, D and E this gives:

$$\frac{1}{H}, \ -\left(\frac{1}{H} + \frac{1}{2R}\right), \ \frac{1}{R}, \ -\frac{1}{2R}, \ 0$$

For case (ii), where relative displacements act in the opposite direction as shown in the triangle of displacements, the vector is reversed, so that for joints A, B, C, D and E we obtain:

$$0, \ \frac{1}{2R}, \ -\frac{1}{R}, \ \frac{1}{H} + \frac{1}{2R}, \ -\frac{1}{H}$$

The coefficient matrix for the frame is unchanged, so the tabular matrix for joint moments under load T appears as shown in Table 22.9.

The equilibrium equations for each case can be derived by considering moments about joints B and D and intersection points 0_1 and 0_2 and treating BC and CD as 'free body' elements in the same way as before. Results for the two cases are:

Case (i): $M_A \dfrac{1}{H} - M_B \left(\dfrac{1}{H} + \dfrac{1}{2R}\right) + M_C \dfrac{1}{R} - M_D \dfrac{1}{2R} = T_1$

Case (ii): $M_B \dfrac{1}{2R} - M_C \dfrac{1}{R} + M_D \left(\dfrac{1}{H} + \dfrac{1}{2R}\right) - M_E \dfrac{1}{H} = T_2$

These are also entered in the table to form the equilibrium equations E.1 and E.2. Where a horizontal load T is applied directly at B, as shown in

Table 22.9 Pitched portal with lateral loading

Equation	Moment					$\bar{\lambda}$	
	A	B	C	D	E	1	2
A.A	$2K$	K				$\dfrac{1}{H}$	
A.B	K	$2(1+K)$	1			$-\left(\dfrac{1}{H}+\dfrac{1}{2R}\right)$	$\dfrac{1}{2R}$
A.C		1	4	1		$-\dfrac{1}{R}$	$-\dfrac{1}{R}$
A.D			1	$2(1+K)$	K	$-\dfrac{1}{2R}$	$\dfrac{1}{H}+\dfrac{1}{2R}$
A.E				K	$2K$		$-\dfrac{1}{H}$
E.1	$\dfrac{1}{H}$	$-\left(\dfrac{1}{H}+\dfrac{1}{2R}\right)$	$\dfrac{1}{R}$	$-\dfrac{1}{2R}$		T_1	
E.2		$\dfrac{1}{2R}$	$-\dfrac{1}{R}$	$\dfrac{1}{H}+\dfrac{1}{2R}$	$-\dfrac{1}{H}$	T_2	

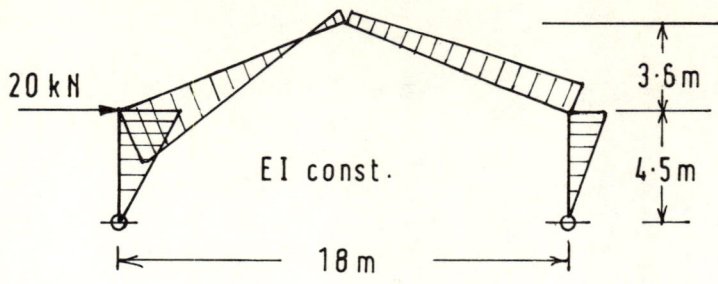

Fig. 22.17.

Fig. 22.16, then $T_1 = T$ and $T_2 = 0$.

Table 22.9 may be used with the GRID program to determine the joint moments, as illustrated by the problem in Fig. 22.17. The data table for this problem is reproduced in Table 22.10; the answers obtained with GRID are shown in the bottom row. Reactions, axial forces and shears for any member may be determined quite easily from first principles once the joint moments are known. (The printout heading in line 300 of GRID should be altered to "NODAL LOADS:−", as mentioned in Chapter 18.) Eaves displacements can be determined if we know the properties of the column, AB. If this is a 305 × 127 × 42 UB for which $I_x = 8140$ cm^4 and where $E = 210$ kN/mm^2, then for a pin-ended column

$$P = \frac{6 \times 210 \times 8140 \times 10^4}{4500} = 2279.2 \times 10^4 \text{ kN/mm}$$

$$\lambda = \frac{MH^2}{3EIP}$$

Table 22.10 Data table for the problem in Fig. 22.17

Ref.	Moment		
	B	C	D
A.B	2.928	1	
A.C	1	4	1
A.D	0	1	2.928
λ1	−0.3611	0.2778	−0.1389
λ2	0.1389	−0.2778	0.3611
E.1	−0.3611	0.2778	−0.1389
E.2	0.1389	−0.2778	0.3661
T	20	0	
Results	−57.37	13.37	31.92 kNm

where I_C = moment of inertia of the column.

For column AB,
$$\lambda_1 = \frac{57.37 \times 10^3 \times 4500 \times 4500}{3 \times 210 \times 8140 \times 10^4} = 22.6 \text{ mm}$$

For column DE,
$$\lambda_2 = \frac{31.92}{57.37} \times 22.6 = 12.6 \text{ mm}$$

Plastic design

Plastic methods of design have been developed for structures such as portal frames, where the main design criterion is that of strength rather than stiffness. In this method, loads on the structure are increased above their working value until the yield strength of the material is exceeded. Initially this occurs at the outer fibres of the section, but with increase in load yielding extends over the whole cross-section to form a 'plastic hinge' where rotation can take place without restraint. Additional hinges develop under further increased loading until sufficient number have been created to form a mechanism, whereupon collapse takes place. Once the failure mode has been established, it is possible to predict the collapse load quite accurately.

For safe design, a load factor is selected to give the required reserve of strength. Rules for plastic design are included in BS 5950, for instance, which specifies the partial factors for use under different load conditions. Computer packages for plastic analysis (sometimes referred to as 'ultimate load', 'load factor' or 'collapse load' analysis) are now available. Plastic design can give some slight economy over elastic methods. Because higher strains are permissible than in the elastic method, it is important to check for overall and local instability and to ensure that deflections are not excessive. Deflection checks usually require that an elastic analysis be carried out; the methods described in the previous sections can be used for this purpose.

The location of the plastic hinges and the collapse mode of failure may be ascertained graphically by combining the 'free' moment and 'reactant' moment diagrams. An example is shown in Fig. 22.18 for a fixed-ended beam carrying a uniformly distributed load. Plastic hinges develop initially at the supports A and B. Increasing the load causes another hinge to develop near mid-span, producing collapse. It is clear from the BM diagram that, at that stage,

$$M_A = M_B = M_C = \frac{1}{2}\frac{WL}{8} = \frac{WL}{16}$$

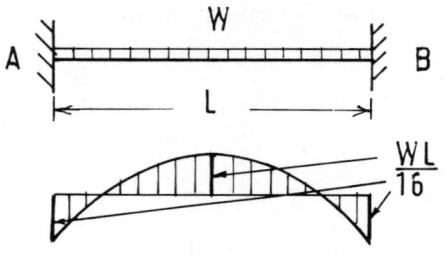

Fig. 22.18.

The same approach can be used for the analysis of portal frames. In Fig. 22.19 the moment diagrams for a frame with pin-ended columns are combined as shown. The locations of the plastic hinges are indicated by heavy vertical lines in the figure. For the example where L = 30 m, R = 2.7 m, H = 6 m and W = 700 kN, the following comparative results were obtained for a steel portal fabricated from Grade 50 high-yield steel and with fixed-column feet:

Plastic design
Plastic moment = 950 kNm
Suitable section: 533 × 210 × 109 UB

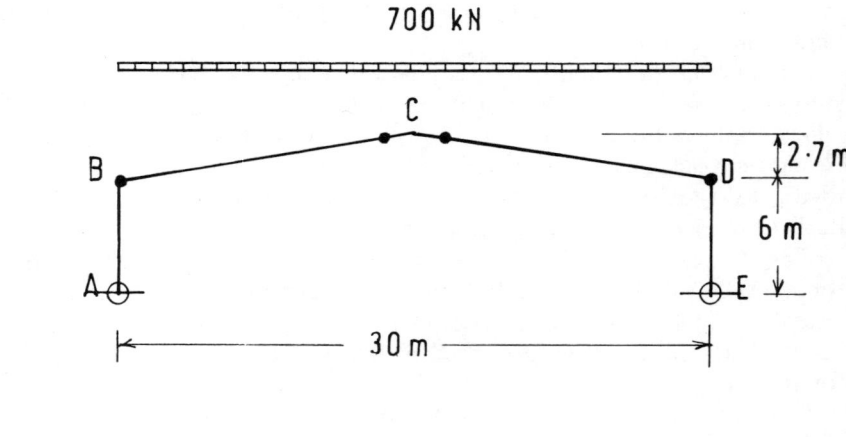

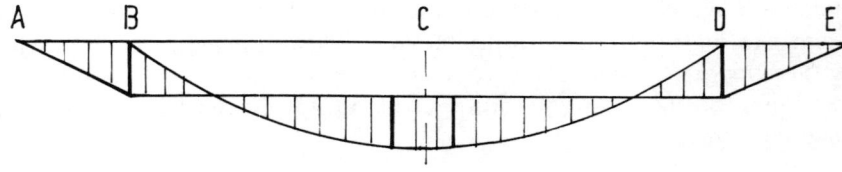

Fig. 22.19.

Elastic design
Maximum moment = 1209 kNm (from release-deformation analysis)
Suitable section: 610 × 229 × 140 UB
For a fixed column, the eaves deflection is given by

$$\Delta = \frac{MH^2}{6EI}$$

For the elastic design section,

$$\Delta = \frac{1209 \times 6 \times 6 \times 10^3 \times 10^6}{6 \times 210 \times 112\,000 \times 10^4} = 31 \text{ mm}$$

(assuming no haunch).

Therefore, although a lighter section seems permissible under the plastic design method, the eaves deflection will be higher; and also stability calculations to BS 5950 show the need for additional restraints near the hinge positions, which will affect the potential saving in steel.

23 Multi-storey and multi-bay frames

Symmetrical frames

A single-bay symmetrical multi-storey frame, such as that in Fig. 23.1, can be regarded as a series of rectilinear portals erected one on top of each other. Release moments may be applied at the joints when these are released, as indicated by arrows in the figure. For symmetrical conditions, analysis of one-half of the frame will suffice for our purposes.

Equations for joints A and C will obviously be the same as those derived previously for the single-storey portal. At joint B, equations may be constructed separately for ABC and for configuration AB−BE. From joint equilibrium conditions at B, $M_{BA} = M_{BC} + M_{BE}$. Also, from symmetry, $M_{EB} = M_{BE}$. These substitutions are made in the following equations.

A.ABC: $\quad M_A \dfrac{H_1}{6EI_1} + M_{BA} \dfrac{H_1}{3EI_1} + M_{BC} \dfrac{H_2}{3EI_2} + M_C \dfrac{H_2}{6EI_2} = 0$

Multiplying across by $\dfrac{6EI_B}{L_B}$,

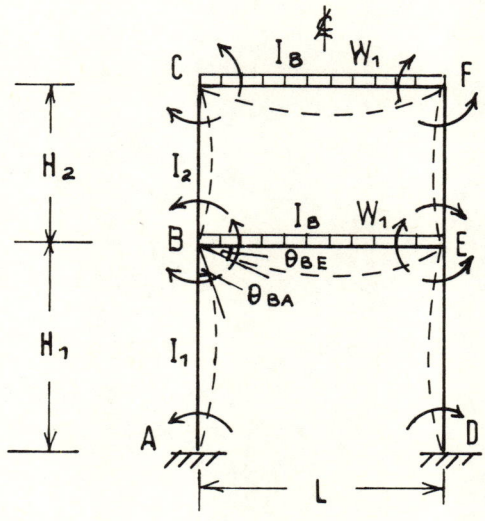

Fig. 23.1.

A.ABE: $K_1M_A + 2K_1M_{BA} + 2K_2M_{BC} + K_2M_C = 0$

$M_A \dfrac{H_1}{6EI_1} + M_{BA} \dfrac{H_1}{3EI_1} + M_{BE} \dfrac{L_B}{3EI_B} + M_{EB} \dfrac{L_B}{6EI_B} = \theta_{BA} + \theta_{BE}$

$K_1M_A + 2K_1M_{BA} + 2M_{BE} + M_{EB} = \dfrac{1}{4}W_1L$

$K_1M_A + 2K_1M_{BA} + 3M_{BE} = \dfrac{1}{4}W_1L$

$K_1M_A + (3 + 2K_1)M_{BA} - 3M_{BC} = \dfrac{1}{4}W_1L$

The equations for the half-frame are summarised in Table 23.1.

Clearly the method can be extended to solve rigid frames with any number of storeys. Table 23.2 may be useful in this regard. This shows the tabular matrix for a four-storey symmetrical frame. To cover the situation where the beams at different storey heights may have different stiffnesses, a term K_B has been introduced which can be adjusted to suit the relative flexibilities of the beams.

As an example, the frame in Fig. 23.2 has columns and beams of varying section, as shown, so that

$K_1 = \dfrac{I}{10} \dfrac{4 \times 2}{I} = 0.8 = K_2; \quad K_3 = \dfrac{I}{10} \dfrac{4 \times 4}{I} = 1.6 = K_4;$

$K_B = \dfrac{I}{10} \dfrac{10 \times 2}{I} = 2 \text{ (roof only)}$

The moments shown below were obtained using the MATIN program and the following data statements:

```
1000 DATA 1.6,0.8,0,0,0,0,0,0
1005 DATA 0.8,3.8,-3,0,0,0,0,0
1010 DATA 0.8,1.6,1.6,0.8,0,0,0,0
1015 DATA 0,0,0.8,4.6,-3,0,0,0
```

Table 23.1 Half-frame matrix for a two-storey symmetrical frame

Equation	Moment				$\overline{W}$
	A	BA	BC	C	
A.A	$2K_1$	K_1			
A.ABE	K_1	$3 + 2K_1$	-3		$\dfrac{1}{4}W_1L$
A.ABC	K_1	$2K_1$	$2K_2$	K_2	
A.C			K_2	$2K_2$	$\dfrac{1}{4}W_1L$

Table 23.2 Half-frame matrix for a four-storey symmetrical frame

Equation A	Moment							$\overline{W}$
	A	BA	BC	CB	CD	DC	DE	E	
A	$2K_1$	K_1							
ABG	K_1	$3+K_1$	-3						$\frac{1}{4}W_1L$
ABC	K_1	$2K_1$	$2K_2$	K_2					
BCH			K_2	$3K_B + 2K_2$	$-3K_B$				$\frac{1}{4}K_BW_2L$
BCD			K_2	$2K_2$	$2K_3$	K_3			
CDJ					K_3	$3K_B + 2K_3$	$-3K_B$		$\frac{1}{4}K_BW_3L$
CDE					K_3	$2K_3$	$2K_4$	K_4	
E							K_4	$3K_B + 2K_4$	$\frac{1}{4}K_BW_4L$

```
1020 DATA 0,0,0.8,1.6,3.2,1.6,0,0
1025 DATA 0,0,0,0,1.6,6.2,-3,0
1030 DATA 0,0,0,0,1.6,3.2,3.2,1.6
1035 DATA 0,0,0,0,0,0,1.6,9.2
1040 DATA 0,100,0,100,0,100,0,100
```

Results: $-6.44, 12.88, -18.73, 18.14, -10.51, 12.25, -13.62, 13.24$ kNm

These results were used to construct the BM diagram for half the frame.

If the column bases are pinned, then the first row and column of the matrix are omitted in the usual fashion.

Lateral loads

Lateral loading on a multi-storey frame can induce sways of differing amounts in each storey. In Fig. 23.3, the sway in the first storey is denoted by λ_1, in the second by λ_2 and in the top storey by λ_3. Under lateral loading from the forces T_1, T_2 and T_3 assumed applied at each floor level, the unloaded beams in the frame are bent in double curvature, indicated by dashed lines. In a symmetrical frame a point of contraflexure will occur at their mid-span, so it will be sufficient to consider one-half of the frame only to determine the bending moments.

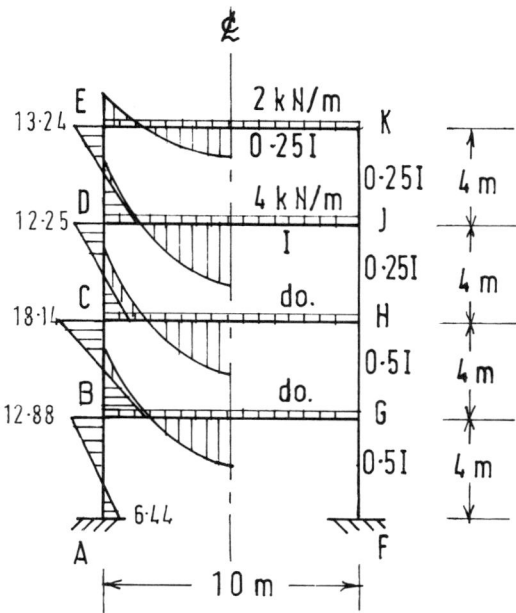

Fig. 23.2.

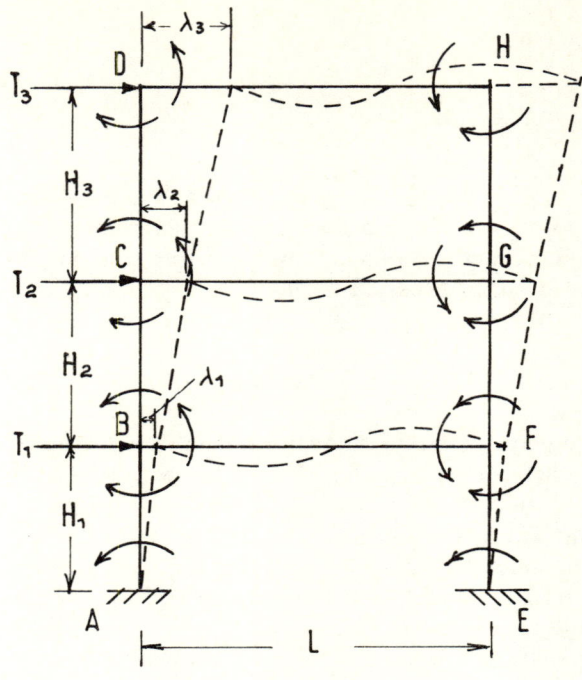

Fig. 23.3.

The arrows in the diagram show the direction of the release moments in line with previous convention. It will be seen that the moments in the two half-frames are of opposite sign so that, for instance, $M_{FB} = -M_{BF}$, $M_{GC} = -M_{CG}$ and $M_H = -M_D$. Also, from joint equilibrium conditions,

$$M_{BA} = M_{BC} + M_{BF}$$
$$M_{CB} = M_{CD} + M_{CG}$$

The criterion for lateral stability in the columns is the same as that described previously for rectilinear portal frames and expressed in equation (22.1) for unbalanced shear in the columns. This equation can be applied to the columns in the top storey and, since we are dealing with one-half of the frame only, it reduces to:

E.3: $\quad \dfrac{M_{CD} - M_D}{H_3} = \dfrac{T_3}{2}$

The unbalanced shear in the storey below this is $T_2 + T_3$, hence

E.2: $\quad \dfrac{M_{BC} - M_{CB}}{H_2} = \dfrac{T_2 + T_3}{2}$

Similarly, for the bottom storey,

E.1: $\dfrac{M_A - M_{BA}}{H_1} = \dfrac{T_1 + T_2 + T_3}{2}$

The other set of equations is obtained, as usual, by considering the angular displacements when the joints in the frame are released. Figure 23.4 indicates the relevant angles and the direction of the release moments. The angular changes at the joints are summarised in Table 23.3. These can be identified directly from the diagram; as usual, where the release moments tend to increase the angular discontinuities, they are given a negative sign. The first column in the table is used to compile the compatibility equations set out below. (The other columns are referred to later.)

A.A: $2K_1M_A + K_1M_{BA} = P\dfrac{\lambda_1}{H_1}$

A.ABF: $K_1M_A + 2K_1M_{BA} + 2M_{BF} + M_{FB} = -P\dfrac{\lambda_1}{H_1}$

$K_1M_A + 2K_1M_{BA} + M_{BF} = -P\dfrac{\lambda_1}{H_1}$

$K_1M_A + (1 + 2K_1)M_{BA} - M_{BC} = -P\dfrac{\lambda_1}{H_1}$

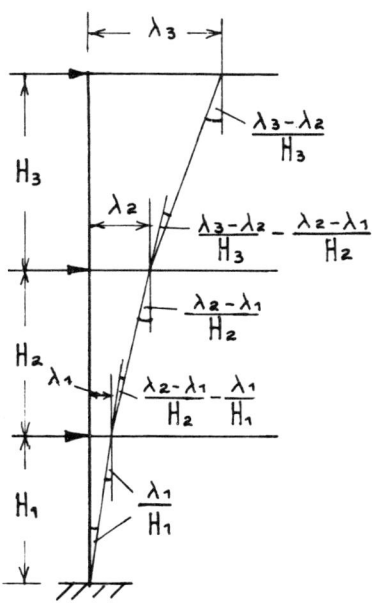

Fig. 23.4.

Table 23.3 Angular changes and sway vectors

Joint	Angular change	Case (i) $\lambda_1 = \lambda_2 = \lambda_3$	Case (ii) $\lambda_1 = 0, \lambda_2 = \lambda_3$	Case (iii) $\lambda_1 = \lambda_2 = 0$
A	$\dfrac{\lambda_1}{H_1}$	$\dfrac{\lambda}{H_1}$	0	0
ABF	$-\dfrac{\lambda_1}{H_1}$	$-\dfrac{\lambda}{H_1}$	0	0
ABC	$\dfrac{\lambda_2 - \lambda_1}{H_2} - \dfrac{\lambda_1}{H_1}$	$-\dfrac{\lambda}{H_1}$	$\dfrac{\lambda}{H_2}$	0
BCG	$-\left(\dfrac{\lambda_2 - \lambda_1}{H_2}\right)$	0	$-\dfrac{\lambda}{H_2}$	0
BCD	$\dfrac{\lambda_3 - \lambda_2}{H_3} - \dfrac{\lambda_2 - \lambda_1}{H_2}$	0	$-\dfrac{\lambda}{H_2}$	$\dfrac{\lambda}{H_3}$
D	$-\left(\dfrac{\lambda_3 - \lambda_2}{H_3}\right)$	0	0	$\dfrac{\lambda}{H_3}$

A.ABC: $K_1 M_A + 2K_1 M_{BA} + 2K_2 M_{BC} + K_2 M_{CB} = P\left(\dfrac{\lambda_2 - \lambda_1}{H_2} - \dfrac{\lambda_1}{H_1}\right)$

A.BCG: $K_2 M_{BC} + 2K_2 M_{CB} + 2K_B M_{CG} + K_B M_{GC} = -P\left(\dfrac{\lambda_2 - \lambda_1}{H_2}\right)$

$K_2 M_{BC} + 2K_2 M_{CB} + K_B M_{CG} = -P\left(\dfrac{\lambda_2}{H_2} - \dfrac{\lambda_1}{H_2}\right)$

$K_2 M_{BC} + (K_B + 2K_2) M_{CB} - K_B M_{CD} = -P\left(\dfrac{\lambda_2}{H_2} - \dfrac{\lambda_1}{H_2}\right)$

A.BCD: $K_2 M_{BC} + 2K_2 M_{CB} + 2K_3 M_{CD} + K_3 M_D =$
$\qquad P\left(\dfrac{\lambda_3 - \lambda_2}{H_3} - \dfrac{\lambda_2 - \lambda_1}{H_2}\right)$

A.D: $\quad K_3 M_{CD} + 2K_3 M_D + 2K_B M_D + K_B M_{HD} = -P\left(\dfrac{\lambda_3 - \lambda_2}{H_3}\right)$

$K_3 M_{CD} + (K_B + 2K_3) M_D = -P\left(\dfrac{\lambda_3 - \lambda_2}{H_3}\right)$

The coefficient matrix and equilibrium equations are displayed in Table 23.4, while the right-hand side vector appears in Table 23.3 under the heading 'Angular change'.

Computer program for lateral loads

To solve these equations, we can adopt the same procedure described in Chapter 18 for solving grillages. Included in this is the process of initially

Table 23.4 Three-storey frame with lateral loads

Equation	Moment						$\bar{\lambda}$		
	A	BA	BC	CB	CD	D	1	2	3
A.A	$2K_1$	K_1					$\dfrac{1}{H_1}$		
A.ABF	K_1	$1 + 2K_1$	-1				$-\dfrac{1}{H_1}$		
A.ABC	K_1	$2K_1$	$2K_2$	K_2			$-\dfrac{1}{H_1}$		
A.BCG			K_2	$K_B + 2K_2$	$-K_B$			$\dfrac{1}{H_2}$	
A.BCD			K_2	$2K_2$	$2K_3$	K_3		$-\dfrac{1}{H_2}$	
A.D				K_3	K_3	$K_B + 2K_3$		$-\dfrac{1}{H_2}$	$\dfrac{1}{H_3}$ $-\dfrac{1}{H_3}$
E.1	$\dfrac{1}{H_1}$	$-\dfrac{1}{H_1}$							$\dfrac{T_1 + T_2 + T_3}{2}$
E.2			$\dfrac{1}{H_2}$	$-\dfrac{1}{H_2}$					$\dfrac{T_2 + T_3}{2}$
E.3					$\dfrac{1}{H_3}$	$-\dfrac{1}{H_3}$			$\dfrac{T_3}{2}$

inducing unit displacements at each node in turn capable of movement; then, from consideration of the equilibrium conditions, determining the factors by which the unit moments should be multiplied to give the final values.

Unit displacement in the case of multi-storey frames is equivalent to unit sway at the head of each column taken storey by storey. The physical interpretation of sway in this context is shown in Fig. 23.5. The bottom storey alone is subject to sway in diagram (i); the second storey in diagram (ii) and the top storey only in diagram (iii). If the sways are λ_1, λ_2 and λ_3, as shown, then it will be seen that the relationships are:

Case (i) : $\lambda_1 = \lambda_2 = \lambda_3$
Case (ii) : $\lambda_1 = 0, \lambda_2 = \lambda_3$
Case (iii): $\lambda_1 = \lambda_2 = 0$

Table 23.3 embodies these three cases, showing the effect of substituting the λ values in turn in the total angular displacement terms. The vectors for unit sway are obtained, therefore, simply by putting $\lambda = 1$ in each case. Table 23.4 reproduces these and also displays the coefficient matrix derived from the compatibility equations above. It will be seen that the table is similar in form to those shown in Chapter 18 for grillages (compare with Table 18.2, for example). It is to be expected, therefore, that the GRID computer program used previously should prove equally suitable for determining the moments in the multi-storey frame. To check this out, a simple example is shown in Fig. 23.6, in which the three-storey frame has lateral loads applied at floor and roof level. The columns have a value of

$$K = \frac{3}{2I} \frac{3I}{6} = 0.75$$

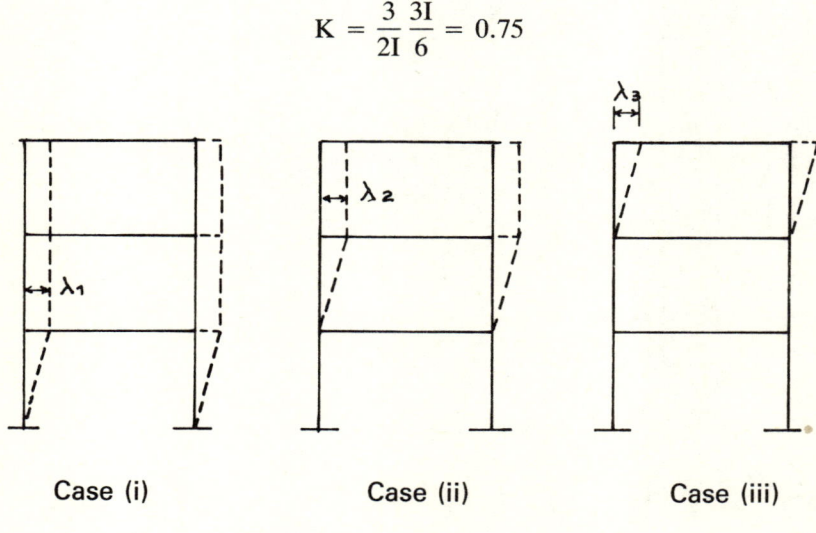

Fig. 23.5.

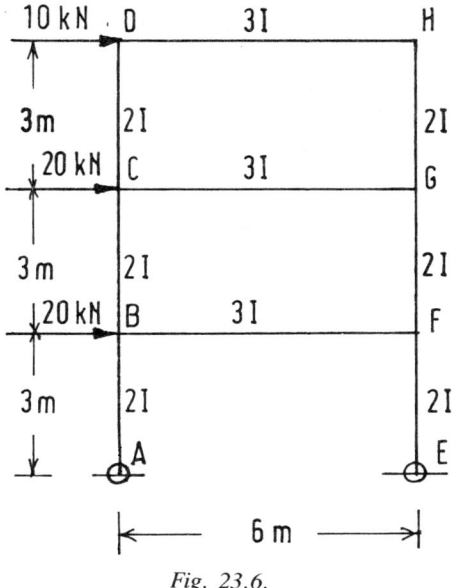

Fig. 23.6.

Unbalanced shears for the half-frame are 25, 15 and 5 respectively.

Before running the program, it is advisable to make one change in GRID so as to correct the printout by altering line 300 to read:

300 PRINT "STOREY SHEARS, HALVED:-"

The printout for the problem is reproduced below, with some typical headings typed in before the program was RUN.

HIGH ST. PROJECT

FRAMES A1−2, B1−2, C1−2, D1−2.
LOADING − WIND LOAD

MATRIX SIZE? 5

DATA:−

2.5	−1	0	0	0
1.5	1.5	0.75	0	0
0	0.75	2.5	−1	0
0	0.75	1.5	1.5	0.75
0	0	0	0.75	2.5
−0.333	−0.333	0	0	0

? ANOTHER DEFLECTED NODE. FOR YES PRESS 1
? 1

0	0.333	−0.333	−0.333	0

? ANOTHER DEFLECTED NODE. FOR YES PRESS 1
? 1

0	0	0	0.333	−0.333

? ANOTHER DEFLECTED NODE. FOR YES PRESS 1
? 0

EQUILIBRIUM COEFFS:−

−0.333	0	0	0	0
0	0.333	−0.333	0	0
0	0	0	0.333	−0.333

STOREY SHEARS, HALVED:−
1 25
2 15
3   5

RESULTS:−
1   −75.08   FINAL MOMENT
2   11.33   FINAL MOMENT
3   −33.72   FINAL MOMENT
4   2.32   FINAL MOMENT
5   −12.69   FINAL MOMENT

The moments are output in the same sequence as the column headings — in this case, M_{BA}, M_{BC}, M_{CB}, M_{CD} and M_D.

Situations can arise where the columns in the bottom storey are of different height. The tabular matrix may be extended quite easily to cope with this eventuality using the same approach. Since the frame is no longer symmetrical, however, the 'full frame' analysis has to be carried out using a suitably derived tabular matrix (see Problem A.29). In a similar fashion, the tabular matrix can be extended by inspection to deal with any number of additional storeys.

Uniform portals and rigid frames

With portals of uniform height or rigid multi-storey frames in which the storey height, H, is constant, data preparation can be simplified to some extent by putting $H = 1$ in the sway vectors and equilibrium equations. For example, for a three-storey single bay frame with lateral loads, the tabular matrix would appear as shown in Table 23.5. A similar substitution may be made for any *uniform* rigid frame structure.

Table 23.5 Three-storey uniform frame with lateral loads

Equation	Moment						$\bar{\lambda}$		
	A	BA	BC	CB	CD	D	1	2	3
A.A	$2K_1$	K_1	-1				1		
A.ABF	K_1	$1 + 2K_1$	$2K_2$	K_2			-1	1	
A.ABC	K_1	$2K_1$	K_2	$K_B + 2K_2$	-1		-1	-1	
A.BCG			K_2	$2K_2$	$2K_3$	K_3		-1	1
A.BCD					K_3	$K_B + 2K_3$			-1
A.D									
E.1	1	-1					$\dfrac{(T_1 + T_2 + T_3) H}{2}$		
E.2			1	-1			$\dfrac{(T_2 + T_3) H}{2}$		
E.3					1	-1	$\dfrac{T_3 H}{2}$		

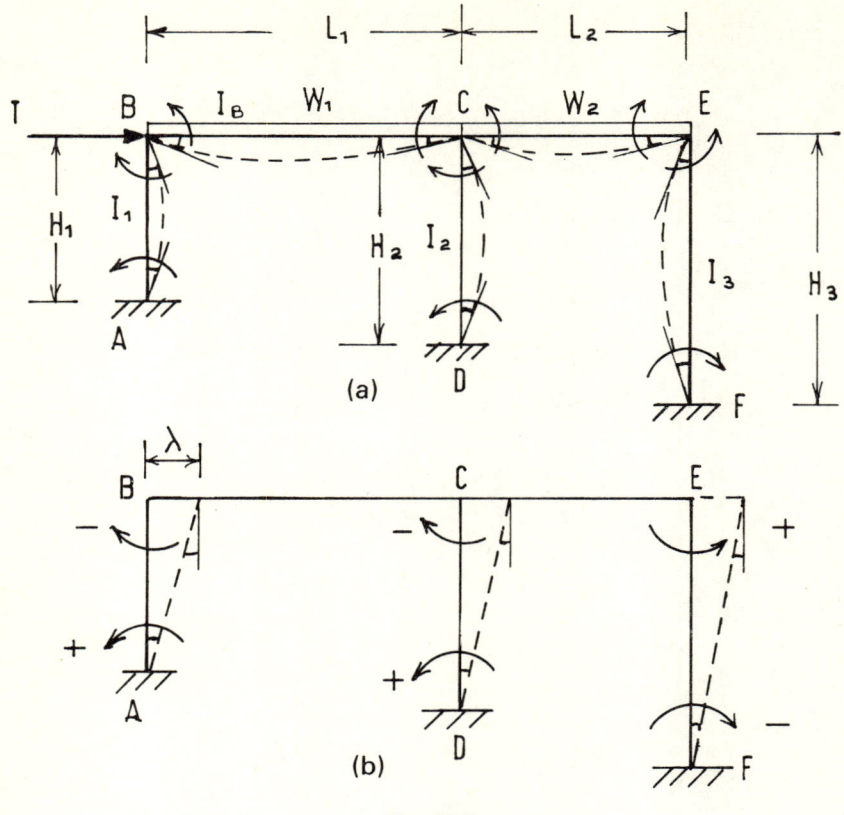

Fig. 23.7.

Multi-bay frames

Figure 23.7 illustrates a two-bay rigid frame carrying uniformly distributed loading on each span and a lateral load, T, applied at the head of the column AB. The deformation diagram is shown dashed in the figure, on which are marked the angular rotations from bending and sway on release of the joints. The direction of the release moments is indicated by arrows, as usual. Sway at the column heads is equal to λ in each case. The moments of inertia of the members are marked on the diagram. Beam BC is selected as the reference member, as previously. Beam CE is given the suffix, B, as it can be of different section. From consideration of unbalanced shear in the columns,

E.1: $\quad \dfrac{M_A - M_B}{H_1} + \dfrac{M_D - M_{CD}}{H_2} + \dfrac{M_E - M_F}{H_3} = T$

From joint equilibrium conditions at C,

$$M_{CE} = M_{CB} + M_{CD}$$

This substitution is made in the following angular compatibility equations:

A.A: $\quad 2K_1M_A + K_1M_B = P\dfrac{\lambda}{H_1}$

A.B: $\quad K_1M_A + 2(1 + K_1)M_B + M_{CB} = \dfrac{1}{4}W_1L_1 - P\dfrac{\lambda}{H_1}$

A.BCE: $\quad M_B + 2M_{CB} + 2K_BM_{CE} + K_BM_E = \dfrac{1}{4}W_1L_1 + \dfrac{1}{4}K_BW_2L_2$

$\quad\quad\quad M_B + 2(1 + K_B)M_{CB} + 2K_BM_{CD} + K_BM_E = \dfrac{1}{4}W_1L_1 + \dfrac{1}{4}K_BW_2L_2$

A.BCD: $\quad M_B + 2M_{CB} - 2K_2M_{CD} - K_2M_D = \dfrac{1}{4}W_1L_1 + P\dfrac{\lambda}{H_2}$

A.D: $\quad 2K_2M_2 + K_2M_{CD} = P\dfrac{\lambda}{H_2}$

A.E: $\quad K_BM_{CE} + 2K_BM_E + 2K_3M_E + K_3M_F = \dfrac{1}{4}K_BW_2L_2 + P\dfrac{\lambda}{H_3}$

$\quad\quad K_BM_{CB} + K_BM_{CD} + 2(K_B + K_3)M_E + K_3M_F = \dfrac{1}{4}K_BW_2L_2 + P\dfrac{\lambda}{H_3}$

A.F: $\quad K_3M_E + 2K_3M_F = -P\dfrac{\lambda}{H_3}$

The equations are summarised in Table 23.6.

As an example, consider the two-bay frame in Fig. 23.8, subject to both vertical and lateral loading. From the information in the figure, it will be seen that:

$K_1 = \dfrac{8}{2I}\dfrac{I}{12} = 0.333; \quad K_2 = \dfrac{8}{I}\dfrac{I}{12} = 0.667 = K_3; \quad K_B = \dfrac{12}{2I}\dfrac{I}{12} = 0.5$

$T = 12 \times 4 = 48$

$\overline{W}$ terms may be determined from Table 23.6. The values are incorporated in the data table shown in Table 23.7. Output from the XBEAM program is shown in the bottom lines of the table.

Vierendeel girders

Vierendeel girders differ from lattice girders in that they feature vertical members with rigid joints capable of resisting bending moments in place

Table 23.6 Two-bay portal frame

Equation	A	B	CB	CD	D	E	F	$\overline{W}$	$\overline{\lambda}$
A.A	$2K_1$	K_1							$\frac{1}{H_1}$
A.B	K_1	$2(1+K_1)$	1						$-\frac{1}{H_1}$
A.BCE		1	$2(1+K_1)$	$2K_B$		K_B		$\frac{1}{4}W_1L_1 + \frac{1}{4}K_BW_2L_2$	0
A.BCD		1	2	$-2K_2$	$-K_2$			$\frac{1}{4}W_1L_1$	$\frac{1}{H_2}$
A.D				K_2	$2K_2$				$-\frac{1}{H_2}$
A.E			K_B	K_B		$2(K_B + K_3)$	K_3	$\frac{1}{4}K_BW_2L_2$	$\frac{1}{H_3}$
A.F						K_3	$2K_3$		$-\frac{1}{H_3}$
E.1	$\frac{1}{H_1}$	$-\frac{1}{H_1}$	0	$-\frac{1}{H_2}$	$\frac{1}{H_2}$	$\frac{1}{H_3}$	$-\frac{1}{H_3}$	T	

(Moment columns: A, B, CB, CD, D, E, F)

MULTI-STOREY AND MULTI-BAY FRAMES 315

Table 23.7 Data table for the problem in Fig. 23.8

Ref.	Moment							
	A	B	CB	CD	D	E	F	
A	0.667	0.333						
B	0.333	2.667	1	1				
BCE		1	3	−1.333	−0.667	0.5		
BCD		1	2	0.667	1.333			
D				0.5				
E			0.5			2.333	0.667	
F						0.667	1.333	
$\overline{W}$	64	136	504	72	0	432	0	48
λ	0.125	−0.125	0	0.125	0.125	0.125	−0.125	
E	0.125	−0.125	0	−0.125	0.125	0.125	−0.125	
Beam moments	94.80	2.30	98.28	126.97	−63.53	159.75	−79.93	XBEAM N = 7 U = 1
Secondary moments	61.65	−26.29	17.23	−40.43	44.51	30.09	−39.34	
Results	156.45	−23.99	115.51	86.54	−19.02	189.84	−119.27	kNm

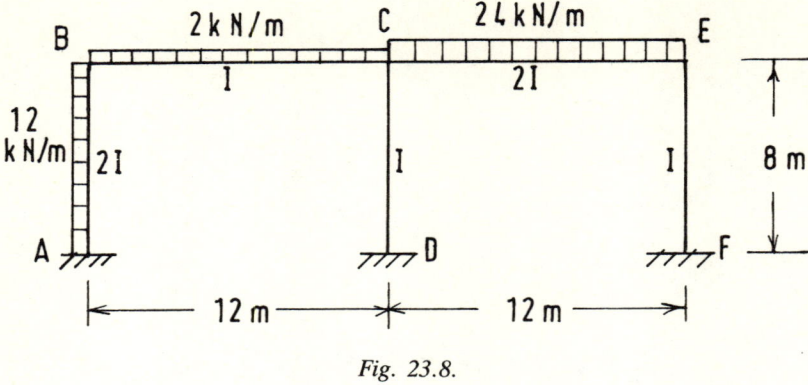

Fig. 23.8.

of web members with pinned joints (Fig. 23.9). Loads are resisted by moments developed at the joints; axial forces in the members are neglected. In almost all cases the girder is symmetrical about the 0—0 axis, so that bending in the verticals produces points of contraflexure at mid-height. By inserting notional hinges at these points, we need only consider the behaviour of the lower section of the girder. This is shown by the dashed deformation diagram indicating the new shape of the girder following release of the joints in accordance with release-deformation convention. It will be seen that vertical deflections, Δ, occur at all joints except the supports.

In asymmetrical or asymmetrically loaded girders a lateral sway, λ, will occur also at the top flange joints due to unbalanced shears in the verticals. The bottom chord, which is restrained at the supports, suffers vertical deflections only, however, so there is little loss of accuracy in ignoring the effect of the sway movement.

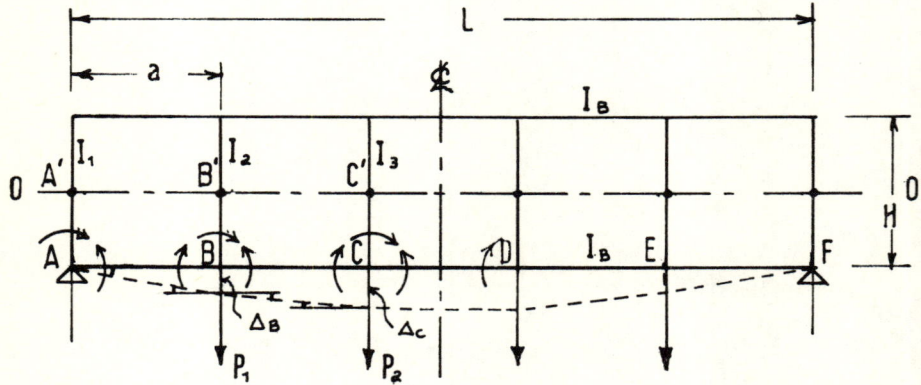

Fig. 23.9.

The moment of inertia of the top and bottom chord members, which are assumed to be uniform, is denoted by I_B. The chord, of panel length a, is taken as the reference member. The moment of inertia of the verticals is denoted by I_1, I_2, etc. If the span of the girder is L and depth of the girder H, then,

$$P = \frac{6EI_B}{a} \text{ and } K_1 = \frac{I_B}{a}\frac{H}{I_1}, \text{ etc.}$$

The diagram in Fig. 23.9 shows a symmetrical five-bay Vierendeel girder with loads applied at the panel points B, C, D and E. From the diagram it will be seen that,

$$\text{At panel B, } M_{BA} + M_{BB'} = M_{BC}$$
$$\text{At panel C, } M_{CB} + M_{CC'} = M_{CD}$$
$$\text{From symmetry, } M_{CD} = M_{DC}$$
$$\Delta_C = \Delta_D$$

Only one-half of the girder need be examined. The equations for angular compatibility and equilibrium are obtained in the usual way.

A.A: $\quad M_{AB}\left(\dfrac{a}{3EI_B} + \dfrac{H}{2 \times 3EI_1}\right) + M_{BA}\dfrac{a}{6EI_B} = \dfrac{\Delta_B}{a}$

Multiplying by P,

$$M_{AB} + (2 + K_1) + M_{BA} = \frac{P}{a}\Delta_B$$

A.ABB′: $\quad M_{AB}\dfrac{a}{6EI_B} + M_{BA}\dfrac{a}{3EI_B} - M_{BB'}\dfrac{H}{2 \times 3EI_2} = -\dfrac{\Delta_B}{a}$

$$M_{AB} + 2M_{BA} - K_2(M_{BC} - M_{BA}) = -\frac{P}{a}\Delta_B$$

$$M_{AB} + (2 + K_2)M_{BA} - K_2 M_{BC} = -\frac{P}{a}\Delta_B$$

A.ABC: $\quad M_{AB} + 2M_{BA} + 2M_{BC} + M_{CB} = \dfrac{P}{a}(-2\Delta_B + \Delta_C)$

A.BCC′: $\quad M_{BC} + (2 + K_3)M_{CB} - K_3 M_{CD} = -\dfrac{P}{a}(\Delta_C - \Delta_B)$

$$= \frac{P}{a}(\Delta_B - \Delta_C)$$

A.BCD: $\quad M_{BC} + 2M_{CB} + 2M_{CD} + M_{DC} = -\dfrac{P}{a}(\Delta_C - \Delta_B) +$

$$\frac{P}{a}(\Delta_D - \Delta_C)$$

But $\Delta_C = \Delta_D$ and $M_{DC} = M_{CD}$

$$M_{BC} + 2M_{CB} + 3M_{CD} = -\frac{P}{a}(\Delta_C - \Delta_B) = \frac{P}{a}(\Delta_B - \Delta_C)$$

E.1: $\quad \dfrac{M_{AB} - M_{BA}}{a} + \dfrac{M_{CB} - M_{BC}}{a} = \dfrac{P_1}{2}$

$$M_{AB} - M_{BA} - M_{BC} + M_{CB} = P_1\frac{a}{2}$$

E.2: $\quad M_{BC} - M_{CB} - M_{CD} + M_{DC} = P_2\dfrac{a}{2}$

$$M_{BC} - M_{CB} = P_2\frac{a}{2}$$

The equations are summarised in Table 23.8. As the bays are of uniform length, data preparation is simplified by putting $a = 1$ in the displacement vectors and equilibrium coefficients, as done previously in the preparation of Table 23.5. This set of equations can be solved, as might be expected, by means of the listed GRID program. Since only half the girder is considered, the statement in line 300 should be altered in much the same way as described previously when dealing with horizontal loads on rigid frames, so that it reads:

300 PRINT "NODAL LOAD X HALF BAY LENGTH:-"

Use of the program and table may be illustrated by the problem in which a load of 200 kN is applied at each of the bottom nodes in the span and where $a = 3$ m, $H = 3$ m, $L = 15$ m and EI is constant for the girder. The data table and results obtained by running GRID with $N = 5$ and $U = 1, 0$ may be seen in Table 23.9.

Table 23.8 Five-bay Vierendeel girder

Equation	Moment					$\overline{\Delta}$	
	AB	BA	BC	CB	CD	B	C
A.A	$K_1 + 2$	1				1	
A.ABB'	1	K_2+2	$-K_2$			-1	
A.ABC	1	2	2	1		-2	1
A.BCC'			1	K_3+2	$-K_3$	1	-1
A.BCD			1	2	3	1	-1
E.1	1	-1	-1	1		$P_1a/2$	
E.2			1	-1		$P_2a/2$	

Table 23.9 Data table for the Vierendeel girder problem

Equation	Moment				
	AB	BA	BC	CB	CD
A.A	3	1			
A.ABB'	1	3	−1		
A.ABC	1	2	2	1	
A.BCC'			1	3	−1
A.BCD			1	2	3
Δ_B	1	−1	−2	1	1
Δ_C			1	−1	−1
E.1	1	−1	−1	1	
E.2			1	−1	
$\overline{P}$	300	300			
Results	313.5	−286.5	108.2	−191.8	−48 kNm

Moments in the uprights are equal to the algebraic sum of the moments in the horizontal members meeting at the joint. Thus,

$$M_{AA'} = 313.5 \text{ kNm}$$
$$M_{BB'} = -286.5 + 108.2 = -178.3 \text{ kNm}$$
$$M_{CC'} = -191.8 - 48 = -239.8 \text{ kNm}$$

Table 23.8 can be extended by inspection to suit Vierendeel girders with any number of uneven bays. For a three-bay frame one can see by writing down the relevant equations that the tabular matrix will take the form shown in Table 23.10.

Vierendeel girder with even number of bays

Where the Vierendeel girder has an even number of bays, the tabular matrix requires adjustment because the axis of symmetry will pass through the central vertical member. For instance, assuming a four-bay girder

Table 23.10 Three-bay Vierendeel girder

Equation	Moment			$\overline{\Delta}$
	AB	BA	BC	B
A.A	$K_1 + 2$	1		1
A.ABB'	1	$K_2 + 2$	$-K_2$	−1
A.ABC	1	2	3	−1
E.1	1	−1		$P_1 a/2$

(Fig. 23.10), the equations for nodes A and B will be exactly as before, but for node C we have:

A.BCC′: $\quad M_{BC} + 2M_{CB} - K_3 M_{CC'} = -\dfrac{P}{a}(\Delta_C - \Delta_B)$

But $M_{CC'} = 0$, from symmetry

∴ $\quad\quad\quad M_{BC} + 2M_{CB} = \dfrac{P}{a}(\Delta_B - \Delta_C)$

E.2: $\quad\quad\quad M_{BC} - M_{CB} - M_{CD} + M_{DC} = P_2 \dfrac{a}{2}$

But $M_{CD} = M_{CB}$ and $M_{DC} = M_{BC}$

$\quad\quad\quad 2M_{BC} - 2M_{CB} = P_2 \dfrac{a}{2}$

These equations are included in Table 23.11. This table can also be extended by inspection for girders with any number of even bays. To take an example, the printout for a two-bay Vierendeel girder of height H = 3 m, bay length a = 3 m and with moment of inertia of the top and bottom chords twice that of the verticals and carrying a load of 200 kN on the centre node might appear as set out below. Some headings have been added to identify the problem.

2–BAY VIERENDEEL, A = 3, H = 3, W = 200 ON CENTRE NODE.
MOMENT OF INERTIA OF MAIN CHORDS = I
MOMENT OF INERTIA OF VERTICALS = I/2

DATA:–
4   1
1   2

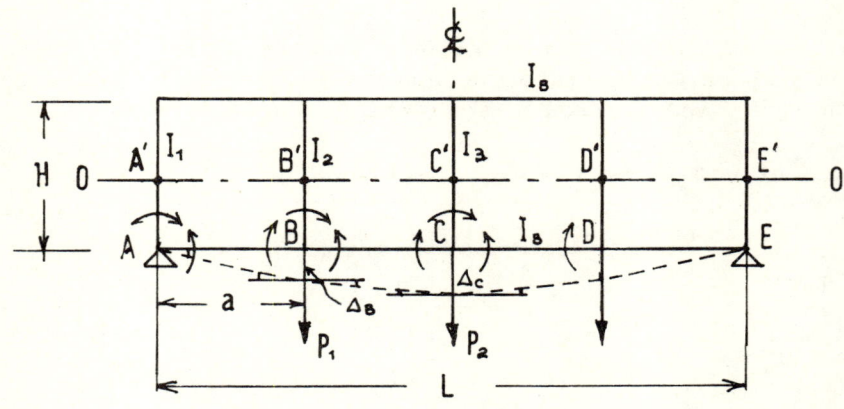

Fig. 23.10.

```
1
-1
```

? ANOTHER DEFLECTED NODE. FOR YES PRESS 1.
? 0

EQUILIBRIUM COEFFS:–
```
1    -1
```

NODAL LOAD X HALF BAY LENGTH:-
150

RESULTS:-
```
1      56.25       FINAL MOMENT
2     -93.75       FINAL MOMENT
```

The printout gives moments M_{AB} and M_{BA} – sufficient to enable the BM diagram to be constructed and for shears, etc. to be estimated. Note that the nodal load by half the bay length in fact equals $P_1 a/4$ in this case, since the YY axis passes through the vertical; therefore only half the load is considered and multiplied by half the bay length. The BM diagram for this problem is shown in Fig. 23.11.

Asymmetrically loaded Vierendeel girders

If the loading applied to the Vierendeel girder is not symmetrical, the sway effect, λ, indicated in Fig. 23.12, has to be taken into consideration. This affects the angular changes on release of the joints between the verticals and bottom chord members, i.e. ABB′ and BCC′, and the joints at supports A and D. The angular displacements due to sway are indicated in the figure. For small displacements, the angle in each case equals λ/H. The directions of the release moments applied at A, B, C and D are indicated by arrows. As the loading is asymmetrical, the matrix for the whole girder has to be considered, as reproduced in Table 23.12.

Table 23.11 Four-bay Vierendeel girder

Equation	Moment				$\overline{\Delta}$	
	AB	BA	BC	CB	B	C
A.A	$K_1 + 2$	1			1	
A.ABB′	1	$K_2 + 2$	$-K_2$		-1	
A.ABC	1	2	2	1	-2	1
A.BCC′			1	2	1	-1
E.1	1	-1	-1	1	$P_1 a/2$	
E.2			2	-2	$P_2 a/2$	

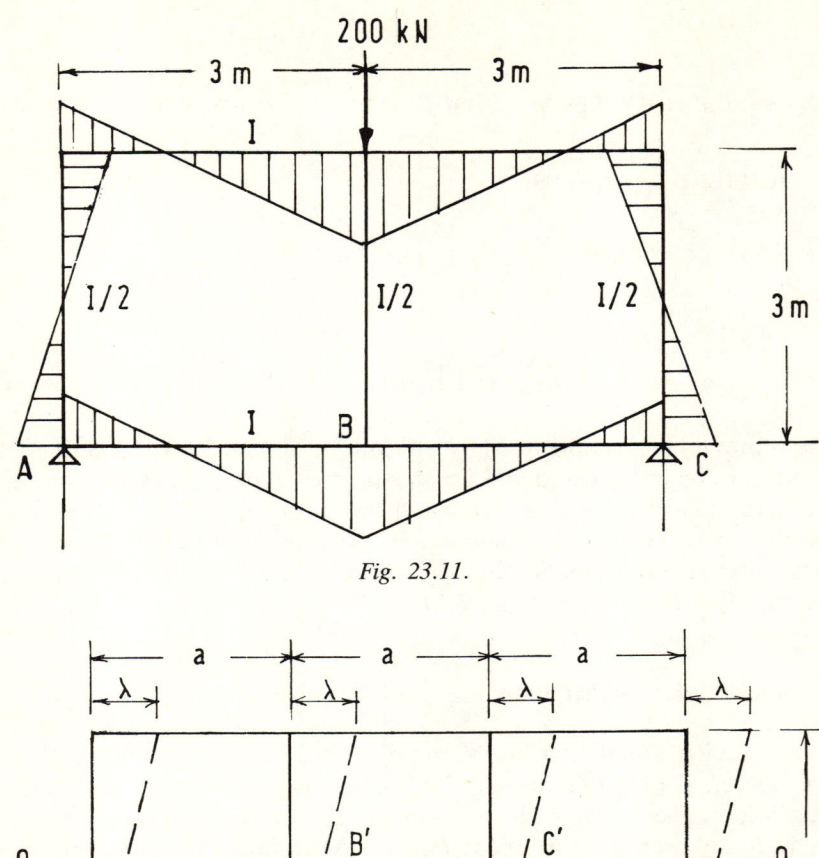

Fig. 23.11.

Fig. 23.12.

It will be seen from Fig. 23.12 that the release moments increase the angular change due to sway at B, C and D and tend to counteract it at support A. This determines the sign of the sway terms. Taking as a typical example the equation for node A, we now have:

A.A: $\quad M_A \left(\dfrac{a}{3EI_B}\right) + \dfrac{H}{2 \times 3EI_B} + M_{BA} \left(\dfrac{a}{6EI_B}\right) = \dfrac{\Delta_B}{a} + \dfrac{\lambda}{H}$

Table 23.12 Asymmetrical three-bay Vierendeel girder

Equation	\multicolumn{6}{c}{Moment}	$\overline{\Delta}$		$\overline{\lambda}$
	AB	BA	BC	CB	CD	D	B	C	
A.A	$K_1 + 2$	1					1		$\dfrac{a}{H}$
A.ABB'	1	$K_2 + 2$	$-K_2$				-1		$-\dfrac{a}{H}$
A.ABC	1	2	2	1			-2	1	
A.BCC'			1	$K_3 + 2$	$-K_3$		1	-1	$-\dfrac{a}{H}$
A.BCD			1	2	2	1	1	-2	
A.D					1	$K_4 + 2$		1	$-\dfrac{a}{H}$
E.1	1	-1	-1	1			$P_1 a/2$		
E.2			1	-1	-1	1	$P_2 a/2$		
E.0	1	-1	1	-1	1	-1	0		

$$M_A (2 + K_1) + M_{BA} = \frac{P}{a} \Delta_B + \frac{P}{H} \lambda$$

$$= \frac{P}{a}\left(\Delta_B + \frac{a}{H} \lambda\right)$$

The sway terms are inserted in an additional vector column in the matrix table. By considering each equation in turn, these can be established as:

$$\frac{a}{H}, \ -\frac{a}{H}, \ 0, \ -\frac{a}{H}, \ 0, \ -\frac{a}{H}$$

(see Table 23.12). If the girder has square panels so that $H = a$, the sway vector reduces to:

$$1, \ -1, \ 0, \ -1, \ 0, \ -1.$$

The new vector requires an additional equilibrium equation for a solution. This can be obtained by considering the total horizontal shear at the bottom of the verticals, which must equal zero. This gives rise to the equation

$$\frac{M_{AB} - M_{BA}}{a} + \frac{M_{BC} - M_{CB}}{a} + \frac{M_{CD} - M_D}{a} = 0$$

These findings are incorporated in Table 23.12. This provides the data for solving Vierendeel girders loaded asymmetrically using the GRID program (see Problem A.32).

Appendix A Problems

Introduction

Problems are the recognised means of checking ability — either by self-testing questions or by formal examination. Solutions to problems in this book involve the following stages.

(1) Draw the deformation diagram to represent the structural problem correctly.
(2) Write down the relevant equations and put into general tabular matrix format.
(3) Insert the actual values and form a data table appropriate to the selected computer program.
(4) Add the data statements to the selected program and RUN to obtain the answers.

Stages 1–3 can be completed without computer assistance, so problems may be brought to that stage without actual recourse to a computer. This may be a suitable approach for many readers.

In the answers to the problems in the Appendix, stage 3 and results from stage 4 only are reproduced, but the reader is strongly advised to go through the preceding stages in each case. Only by sketching out the deformation diagrams and writing down the equations can a firm grasp be acquired of the concepts behind the computerised solution, which is the primary aim of the book.

Note that in the tables, the first column headed 'Ref.' or 'Equation Ref.' contains the relevant angular, linear or equilibrium equation and vector references; deflection, equilibrium equation coefficient and reaction references appear below these, where applicable.

Problems

A.1 Determine the support moments using SBEAM for the beam in Fig. A.1 where the support B suffers a settlement of 25 mm. The beam has a moment of inertia $I_x = 50 \times 10^6$ mm^4 and Young's modulus $E_s = 210$ kN/mm^2.

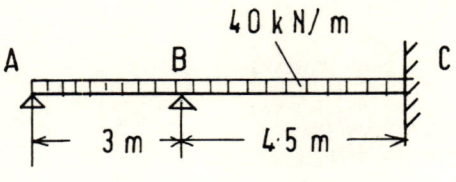

A.1

Ref.	Moment		
	B	C	
A.B	5	1.5	
A.C	1.5	3	
$\overline{W}$	393.75	303.75	
$\overline{\Delta}$	−291.67	116.67	
Results	56.91	72.79	Loads
	−82.35	80.07	Settlement
	−25.44	152.86	Total

A.2 Determine the support moments in the continuous beam in Fig. A.2 using SBEAM where the following settlements occur at the supports: $\Delta_B = 20$ mm, $\Delta_C = 40$ mm and $\Delta_D = 15$ mm. The beam has a moment of inertia $I_X = 90 \times 10^6$ mm^4 and Young's modulus $E_S = 200$ kN/mm^2.

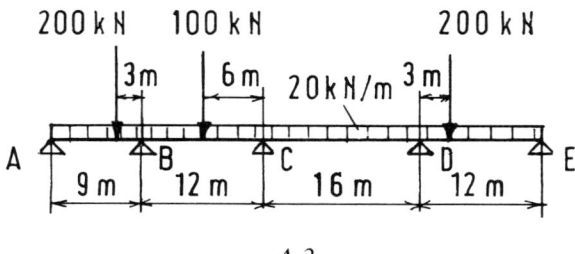

A.2

Ref.	Moment		
	B	C	D
A.B	4.667	1.333	0
A.C	1.333	6.222	1.778
A.D	0	1.778	6.222
$\overline{W}$	2633.5	3835.6	4285.6
$\overline{\Delta}$	−6.67	−38.75	3.75
Results	464.73	348.53	589.19	Loads
	0.60	−7.11	2.63	Settlement
	465.33	341.42	591.82	Total

A.3 A column load of 100 kN is applied at the centre of a foundation 1 m wide × 8 m long × 300 mm deep, as shown in Fig. A.3. The clay soil has a modulus of subgrade reaction of 20000 kN/m³. By dividing the foundation into eight sections determine the values needed to draw the BM and pressure diagrams using the MATIN program. (Assume $E_C = 30 \times 10^6$ kN/m².)

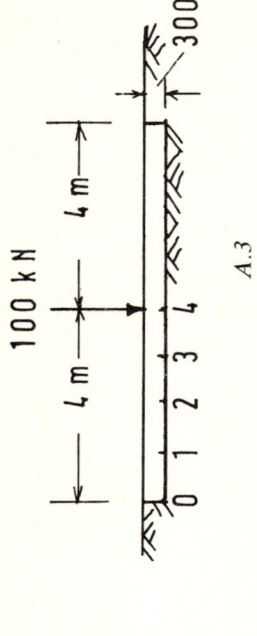

A.3

Ref.	0		1		2		3		4		p, kN/m² M, kNm
	p		p	M	p	M	p	M	p	M	
E.0	1.5		0.5								
A.1	98			−4	98						
E.1	1			4	1	1					
A.2			2	8	−196	−4	98				
E.2			98		2	4	1	−4			
A.3			1	−4	98	8	−196	4	98	1	
E.3					1	1	2	8	1	−4	
A.4						−4	98	1	−98	2	
E.4							1	−4	1	4	
W										200	
Results	2.84		8.25	2.10	13.44	12.38	17.74	35.89	19.67	76.54	

A.4 The concrete foundation in Fig. A.4 is 1 m wide and rests on a soil for which the coefficient of subgrade reaction $q = 12\,000$ kN/m³. Assuming $E_C = 28 \times 10^6$ kN/m², use MATIN or another suitable program to determine the moments and ground pressures at 1 m intervals along the foundation.

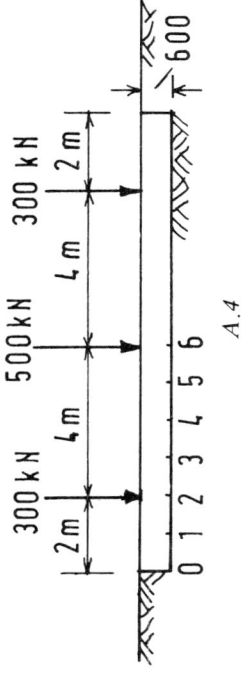

Fig. A.4

Ref.	0		1		2		3		4		5		6		p, kN/m² M, kNm
	p	M	p	M	p	M	p	M	p	M	p	M	p	M	
E.0	1.5		0.5	−4											
A.1	252		−504	4	252	1									
E.1	1		2	8	1	−4									
A.2			252	1	−504	4	252	1							
E.2			1	−4	2	8	1	−4							
A.3					252	1	−504	4	252	1					
E.3					1	−4	2	8	1	−4					
A.4							252	1	−504	4	252	1			
E.4							1	−4	2	8	1	−4			
A.5									252	1	−504	4	252	1	
E.5									1	−4	2	8	1	−4	
A.6											252	1	−504	2	
E.6											1	−4	2	4	
W							1200						1000		
Results	63.63	33.29	75.43	86.15	141.74	94.34	35.71	101.31	23.72	107.32	112.80	110.21	308.41		

A.5 The concrete foundation in Fig. A.5 is 1 m wide and 500 mm deep and is supported on a soil with a coefficient of subgrade reaction $q = 8000$ kN/m^3. Determine the moments and ground pressures at 1 m intervals under the given load system using MATIN ($E_C = 28 \times 10^6$ kN/m^2).

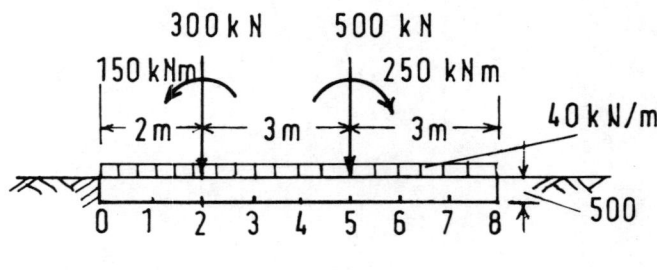

A.5

PROBLEMS

	0		1		2		3		4		5		6		7		8	
Ref.	p	M	p	M	p	M	p	M	p	M	p	M	p	M	p	M	p	M
E.0	1.5		0.5															
A.1	234	−4	−468	4	234	1												
E.1	1	4	2	8	1	−4												
A.2		8	234	1	−468	4	234	1										
E.2		1	1	−4	2	8	1	−4										
A.3		−4			234	1	−468	4	234	1								
E.3					1	−4	2	8	1	−4								
A.4							234	1	−468	4	234	1						
E.4							1	−4	2	8	1	−4						
A.5									234	1	−468	4	234	1				
E.5									1	−4	2	8	1	−4				
A.6											234	1	−468	4	234	1		
E.6											1	−4	2	8	1	−4		
A.7													234	1	−468	4	234	1
E.7													1	−4	2	8	1	−4
E.8															0.5		1.5	
W̄	80	160	300	160	1960	−500	150	0	160	−500	1160	−250	1160	0	160	80		
Results	126.9	44.2	133.2	138.0	181.3	141.0	−34.0	144.3	16.7	146.8	171.5	143.3	181.5	134.7	43.6	124.7		

p, kN/m²
M, kNm

A.6 A concrete foundation 1 m wide and varying in depth from 200 mm to 600 mm, as shown in Fig. A.6, carries a central column load of 200 kN. If the coefficient of subgrade reaction q = 12 000 kN/m³, then, using MATIN, determine the moments and ground pressures at 0.5 m intervals ($E_C = 30 \times 10^6$ kN/m²).

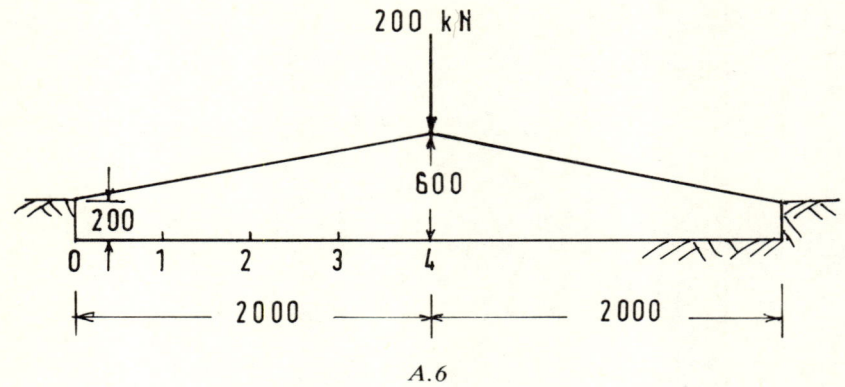

A.6

Ref.	0	1		2		3		4		
	p	p	M	p	M	p	M	p	M	
E.0	0.75									
A.1	78	0.125	−4							
E.1	0.25	−156 0.5	2.73 8							
A.2		78 0.25	0.36 −4	78 0.25	0.36 −4					
E.2				−156 0.5	1.07 8					
A.3				78 0.25	0.17 −4	78 0.25	0.17 −4			
E.3						−156 0.5	0.53 8			
A.4						78 0.25	0.09 −4	78 0.25	0.09 −4	
E.4								−78 0.25	0.19 4	
$\overline{W}$									200	
Results	42.93	44.93	9.45	46.47	30.11	47.41	62.35	47.74	106.40	p, kN/m² M, kNm

A.7 A column load of 2000 kN is supported centrally on a 6 m × 4 m reinforced concrete foundation, 300 mm thick, resting on a soil with a coefficient of subgrade reaction q = 5000 kN/m³. Assuming E_C = 30 × 10⁶ kN/m², determine the moments and ground pressures at points 1, 2, 3, 4, 5 and 6 in Fig. A.7 using the MATIN program.

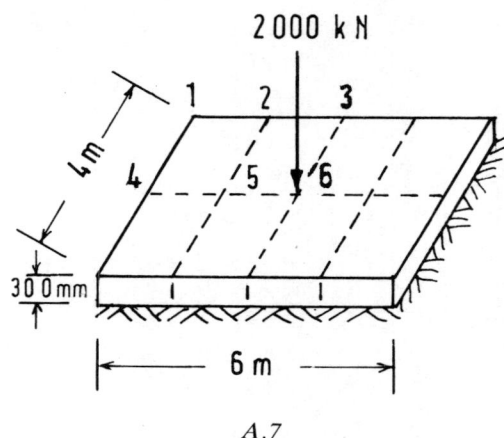

A.7

	Equation	1		2	3		4		5			6		
	Ref.	M_X	p	p	M_X	p	M_Y	p	M_X	M_Y	p	M_X	M_Y	p
	E.1	−1	0.563	0.281			−0.75	0.281						
	A.2	4	144	−288	1	144								
	E.2	2	0.375	1.125	−1	0.375				−0.75	0.375		−0.75	0.375
	A.3	2		288	4	−288								
	E.3	−2		0.75	2	1.125								
	A.4		60.75				4	−60.75						
	E.4		0.75				1.5	1.125			0.375			72
	A.5X			60.75				72	−1		−144	1		
	A.5Y			1.125					4	4	−60.75			0.563
	E.5							0.563	2	1.5	2.25	−1		−144
	A.6X					60.75			2		144	4	4	−60.75
	A.6Y					1.125						2	1.5	2.25
	E.6								−2		1.125			3000
	$\overline{W}$													
Results		149.4	66.8	77.5	265.3	82.2	−97.4	60.3	7.9	283.2	96.1	845.5	569.9	119.7
		74.7			132.7		−64.9		4.0	188.8		422.8	379.9	

$$\frac{p}{b} \quad \begin{array}{c} M_X \\ M_Y \end{array} \text{ or } \frac{M_X \; M_Y}{a}$$

A.8 A 250 mm diameter tubular steel pile ($I = 4000$ cm^4) is driven 8 m below ground. A lateral load of 150 kN and a moment of 100 kNm are applied at 2 m above ground level, as shown in Fig. A.8. If the soil has a coefficient of lateral restraint $q = 6000$ kN/m^3 and $E = 200 \times 10^6$ kN/m^2, determine the moments and deflections in the pile at 1 m intervals using MATIN or a similar program.

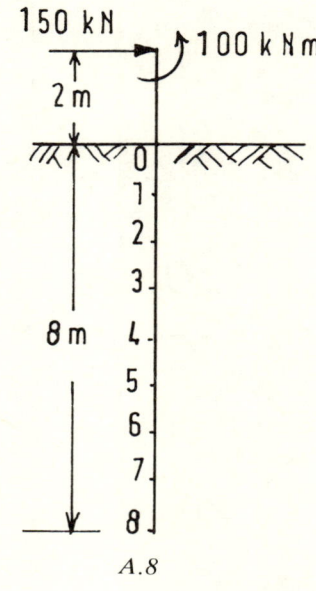

A.8

		0		1		2		3		4		5		6		7		8		
Ref.	Δ	p	M	p	M	p	M	p	M	p	M	p	M	p	M	p	M	p	M	
A.0	−24000	12		−8	1															
E.0		0.375		0.125	4															
A.1		−8		16	4	−8	1													
E.1		0.25		0.5	−8	0.25	4													
A.2				−8	1	16	4	−8	1											
E.2				0.25	4	0.5	−8	0.25	4											
A.3						−8	1	16	4	−8	1									
E.3						0.25	4	0.5	−8	0.25	4									
A.4								−8	1	16	4	−8	1							
E.4								0.25	4	0.5	−8	0.25	4							
A.5										−8	1	16	4	−8	1					
E.5										0.25	4	0.5	−8	0.25	4					
A.6												−8	1	16	4	−8	1			
E.6												0.25	4	0.5	−8	0.25	4			
A.7														−8	1	16	4	−8	1	
E.7														0.25	4	0.5	−8	0.25	4	
E.8																0.125	4	0.375		
W	−600	800		−100	−400															
Results	0.21	516.9		238.2	144.1	59.4	122.4	−30.2	80.3	−59.2	41.9	−55.1	16.3	−37.1	3.5	−15.4	−0.1	6.77		M, kNm
	210	86		40		10		−5		−10		−9		−6		−3		1		p, kN/m² Δ, mm

A.9 The steel sheet piling in Fig. A.9, for which $I_X = 2000$ cm^4 per metre of wall, retains a granular soil with properties $w_e = 17$ kN/m^3, $\emptyset = 30°$ and $q_H = 10\,000$ kN/m^3. If $E_S = 210 \times 10^6$ kN/m^2, use MATIN to determine the moments and deflections in the pile at 2 m intervals.

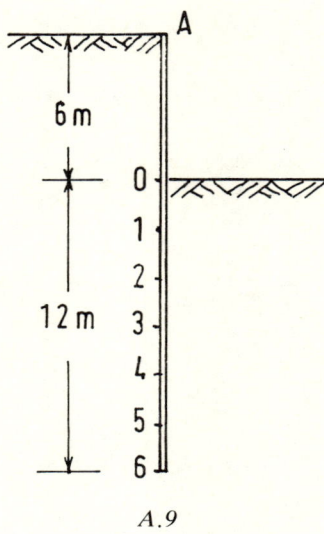

A.9

Ref.	A Δ	0 p	0 M	1 p	1 M	2 p	2 M	3 p	3 M	4 p	4 M	5 p	5 M	6 p	6 M		M, kNm p, kN/m² Δ, mm
A.0	−210000	84		−63	1												
E.0		6		2	4												
A.1		−63		126	4	−63	1										
E.1		4		8	−8	4	4										
A.2		−63		1	126	4	−63	1									
E.2		4		4	8	−8	4	4									
A.3						−63	4	126	−63	1							
E.3						4	126	4	4	4							
A.4							−63	−8	8	−8	−63	1					
E.4							4	1	−63	1	4	4					
A.5								4	4	4	126	4	−63	1			
E.5											8	−8	4	4			
E.6												2		4	6		
$\overline{W}$	−1142.4	1088		−204	−816												
Results	0.021 21	45.9 4.6		12.6 1.3	197.1	−4.1 −0.4	124.3	−8.4 −0.8	56.2	−6.9 −0.7	15.9	−3.4 −0.3		0.3 0.03	1.2		

A.10 Determine the member forces and joint displacements in the K-bracing in Fig. A.10 used as flying shoring between two rigid buildings, where W = 200 kN. The timber bracing members are 200 × 100 size with $E_T = 20$ kN/mm².

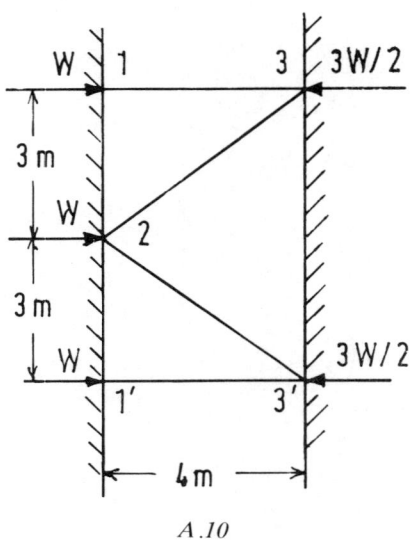

A.10

	F		Δ		
Ref.	23	13	1X	2X	
E.2X	0.8				
E.3X	0.8	1			
L.13		−1	100		
L.23	−1.25			80	
W̄	100	300			
Results (MATIN)	125	200	2	1.95	kN mm

A.11 The trench sheeting in Fig. A.11 retains a sandy soil with properties $w_e = 21$ kN/m^3, $\phi = 30°$ and $q = 8000$ kN/m^3. The steel trench sheeting has a moment of inertia per metre length $I_x = 12 \times 10^{-6}$ m^4 and is supported by stiff walings and struts at 3 m intervals. Assuming $E_S = 200 \times 10^6$ kN/m^2, determine values for the soil pressure and bending moments in the sheeting using the SBEAM program.

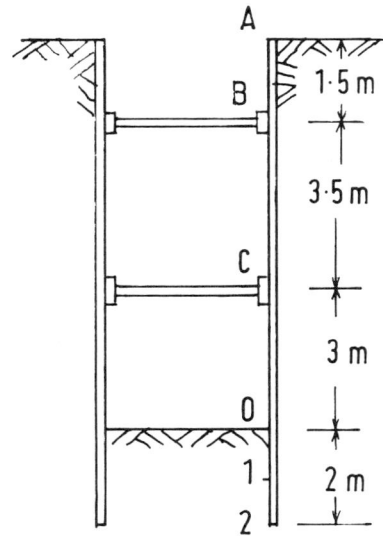

A.11

Ref.	M			p			
	C	0	1	0	1	2	
A.C	13	3					
A.0	3	8	1	2.4	−1.8		
A.1		1	4	1.8	−3.6	1.8	
E.0	−0.333	1.333	−1	−0.375	−0.125		
E.1		−1	2	−0.25	−0.5	−0.25	
E.2			−1		−0.125	−0.375	
W	480.67	245.75					
Δ	−71	71.04					
Results	33.17	16.48	7.67	6.86	5.41	−22.25	Loads
	−7.25	7.74	5.65	16.32	7.76	−17.64	Displacement
	25.92	24.22	13.31	23.18	13.17	−39.89	Total

A.12 Formwork for a bridge deck is supported on a simply supported beam AB propped at mid-span by two raking struts (Fig. A.12) founded on rock. If the beam section is a 250 × 150 × 8 rectangular hollow section and each strut is a 150 × 150 × 10 square hollow steel section and $\alpha = 45°$, find the member forces, nodal deflection and moments where the decking imposes a distributed load of 25 kN/m on the beam ($E_S = 200$ kN/mm², $A_B = 61.1$ cm², $I_X = 5170$ cm⁴, $A_S = 55.5$ cm²).

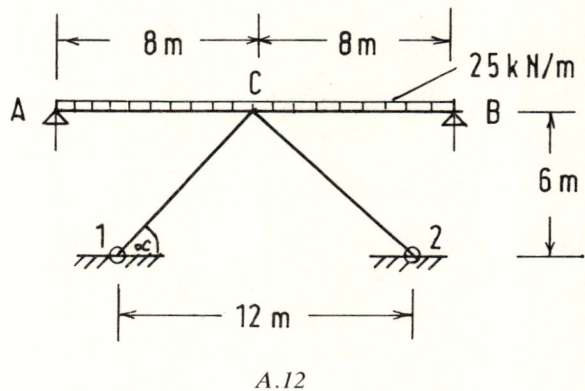

A.12

Ref.	F_{1C}	Δ_C	M_C	
E.C	0.7071		−0.000125	
L.C1	−0.584	54		
A.C		484.75	1	
$\overline{W}$	100		200 000	
	494.5			kN
Results		5.3		mm
(MATIN)			1997	kNm

A.13 The strutted beam in Fig. A.13 is supported on slender columns; it is composed of a 305 × 127 × 42 UB and two 32 mm diameter tie bars acting against the rigid strut 24. The beam supports a distributed load of 5 kN/m. Given $E_S = 200$ kN/mm^2, $A_S = 53.2$ cm^2 and $I_X = 8140$ cm^4, find the member forces, bending moment at joint 2 and nodal deflections using MATIN.

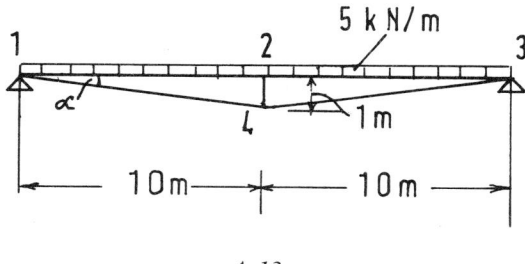

A.13

Ref.	F			Δ		M
	12	14	24	1X	2Y	2
E.1X	−1	0.9951				
E.1Y		0.0996				−0.0001
E.2Y	0.0996		−0.5			
L.12	−0.5			53.2		
L.14		−1.663		−52.94	5.30	
A.2					488.4	1
W		200				500 000
Results	260.3	261.5	51.8	2.4	106.5	10.49	kN mm kNm

A.14 The indeterminate pin-jointed truss in Fig. A.14 is freely supported at joints 1 and 2; the cross-braces are not connected at their intersection. Prepare a table incorporating the equilibrium and linear displacement equations for the frame and use it to solve the case where $W = 1000$ kN, $L = 20$ m, $H = 6$ m, $l = 8$ m, and the cross-sectional area of the main members is 50 cm² and that of the cross braces is 25 cm² ($E_S = 200$ kN/mm²).

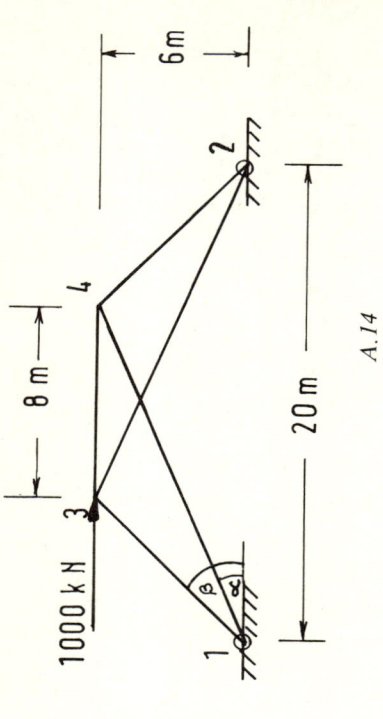

A.14

	F						Δ			
Ref.	23	13	14	24	34	3X	3Y	4X	4Y
E.3X	$\cos \alpha$	$\cos \beta$							
E.3Y	$\sin \alpha$	$-\sin \beta$							
E.4X			$-\cos \alpha$	$-\cos \beta$					
E.4Y			$-\sin \alpha$	$\sin \beta$					
L.34					1				
L.23	$-S$				1	r	$r \sin \alpha$	$-r$	$r \sin \alpha$
L.13		$-S$				$r \cos \alpha$	$-r \sin \beta$		
L.14			$-S$			$r \cos \beta$		$r \cos \alpha$	$r \sin \alpha$
L.24				$-S$				$r \sin \beta$	$-r \cos \beta$
$\overline{W}$	W								

Ref.	F					Δ				
	23	13	14	24	34	3X	3Y	4X	4Y	
E.3X	0.9191	0.7071			1					
E.3Y	0.3939	−0.7071								
E.4X			−0.9191	−0.7071	1					
E.4Y			−0.3939	0.7071						
L.34					−1	125		−125		
L.23	−3.808					114.89	49.24			
L.13		−1.061				88.39	−88.39			
L.14			−3.808					114.89	49.24	
L.24				−1.061				88.39	−88.39	
$\overline{W}$	1000									
Results (MATIN)	446.5	248.7	315.1	175.6	413.8					kN
						11.3	8.3	7.9	5.8	mm

A.15 The composite roof frame in Fig. A.15 on slender columns consists of a steel 254 × 102 × 25 UC tie member AB, 200 × 75 timber rafters A3 and B3 and 150 × 50 timber members elsewhere. It carries a distributed load of 6 kN/m on member 15 and panel point loads indicated in kN in the figure. If $\alpha = 40°$, $E_T = 10$ kN/mm² and $E_S = 200$ kN/mm², write out the general matrix for determining the axial forces, nodal deflections and moments using MATIN and the results. (From the 'blue book', $I_X = 3410$ cm⁴ and $A = 32.2$ cm².)

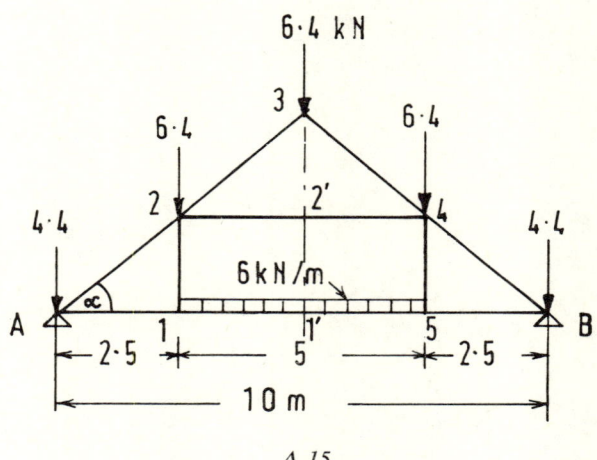

A.15

Equation	F					Δ					M	$\overline{W}$
Ref.	A1'	A2	24	12	23	AX	2Y	2X	1Y	3Y		
E.AX	1	$-\cos\alpha$										$R_A - W_A$
E.AY		$\sin\alpha$									$\dfrac{1}{a}$	
E.2X		$\cos\alpha$	-1									
E.1Y				1							$\dfrac{1}{a}$	W_1
E.3Y					$\sin\alpha$							$W_{3/2}$
L.A1'	$-S$					r						
L.A2		$-S$				$-r\cos\alpha$	$r\sin\alpha$	$-r\cos\alpha$				
L.22'			$-S/2$				$-r$	r				
L.12				$-S$			$-r\sin\alpha$	$r\cos\alpha$				
L.23					$-S$				$\dfrac{P}{a}$	$r\sin\alpha$		
A.1											$\dfrac{2a+b}{L}$	$\dfrac{wb^2}{8}$
Results	36.37	47.47	36.37	19.92	4.98	0.28	3.39	1.21	3.95	2.11	12 292	

A.16 Use the force matrix for the Warren lattice girder in Fig. A.16 to determine the member forces. Write down the displacement matrix and use it to determine the nodal displacements for half the truss using the MATIN program. Given $W = 100$ kN, $a = 4$ m, $\alpha = 45°$ and $E_s = 200$ kN/mm², The cross-sectional area of the internal diagonal members is half that of all the other members for which $A = 2000$ mm².

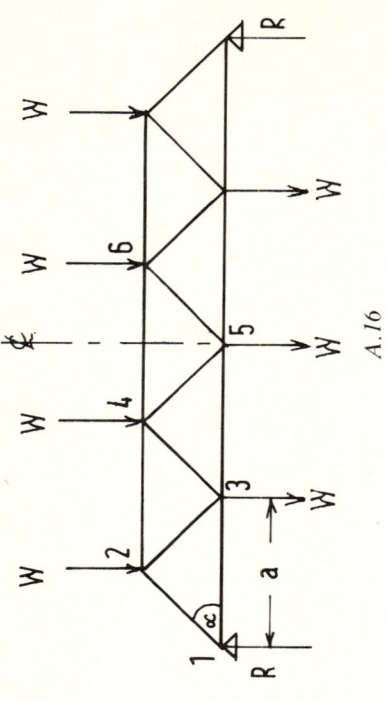

A.16

Equation				rΔ					F	$\dfrac{L_1}{L_n}$	$\dfrac{A_n}{A_1}$	$\overline{FS}$
Ref.	1X	2X	2Y	3X	3Y	4X	4Y	5Y				
L.13	1								350	1	1	350
L.24		1		−1					600	1	1	600
L.12	−0.7071	−0.7071	0.7071						495	0.707	1	350
L.35				1		−1			750	1	1	750
L.23		−0.7071	−0.7071	−0.7071	0.7071				354	0.707	2	501
L.46						1			800	1	1	800
L.34						−0.7071	0.7071		212	0.707	2	300
L.45				−0.7071	−0.7071	−0.7071	−0.7071	0.7071	70.7	0.707	2	100
Results	1100	1400	2995	750	5854	800	7828	8769	rΔ			
	11	14	30	7.5	58.5	8	78.28	87.7	Δ mm			

A.17 The simply supported torsionless beam framework in Fig. A.17 carries concentrated loads of 90 kN on member 12 and 180 kN on member 23. All members have the same E and I values. Construct a matrix table and using XBEAM solve for the nodal moments in the frame.

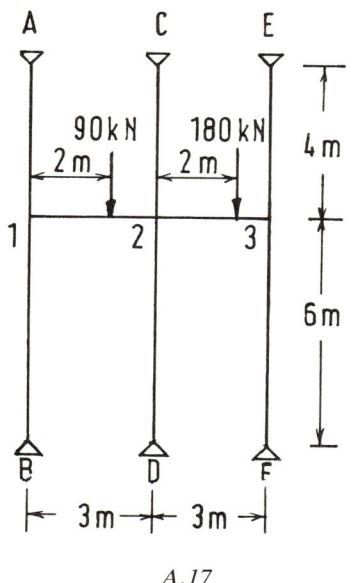

A.17

Ref.	Moment			
	1Y	2Y	3Y	2
A.A1B	5			
A.C2D		5		
A.E3F			5	
A.123				6
$\overline{W}$				260
Node		Deflection and equilibrium coefficients		
1	−0.417			0.333
2		−0.417		−0.667
3			−0.417	0.333
$\overline{R}$	30	120	120	
Results	−91.1	−249.3	−307	−24.0 kNm

A.18 The simply supported frame in Fig. A.18 has members with constant E and I values. Assuming no torsion at the joints, prepare a data table for use with XBEAM to determine the moments at the joints.

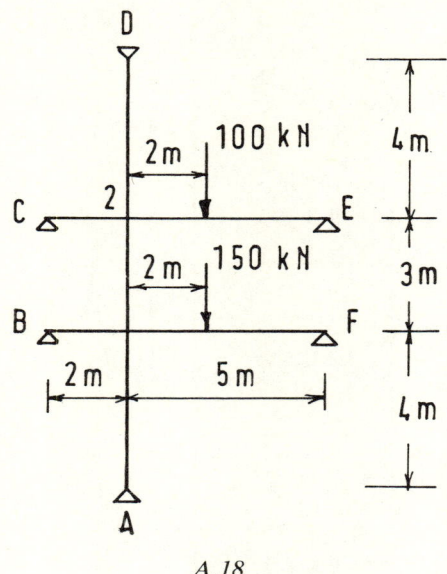

A.18

Equation Ref.	Moment			
	1	1Y	2	2Y
A.B1F	7			
A.A12		7		
A.C2E			7	
A.12D				7
$\overline{W}$	288		192	
Node	Deflection and equilibrium coefficients			
1	−0.7	−0.583		0.333
2		0.333	−0.7	−0.583
$\overline{R}$	90	60		
Results	−94.7	−58.4	−87.6	−31.2 kNm

A.19 The beam framework in Fig. A.19 has torsion-free joints. The members have a constant E value but the moment of inertia, I, in one direction is one-third of that in the other. Prepare a data table to be used with XBEAM to solve for the nodal moments in the frame.

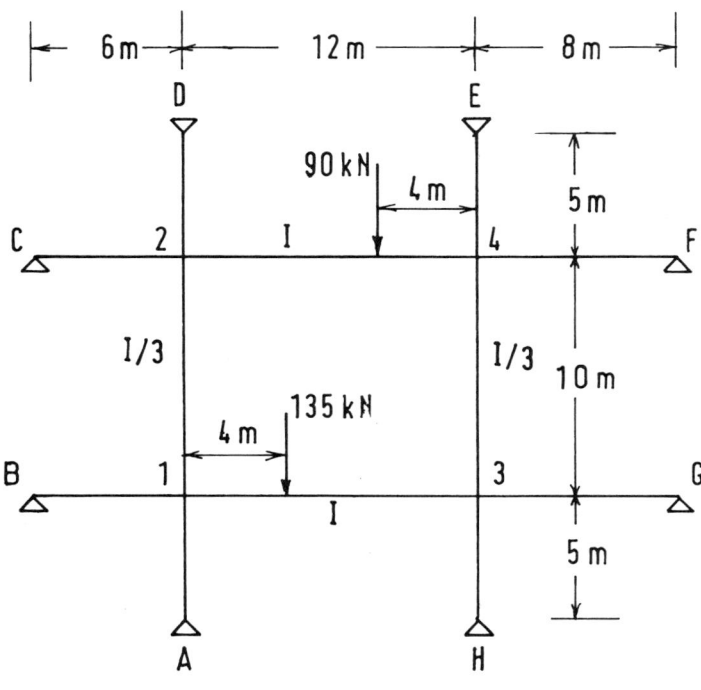

A.19

Equation	Moment							
Ref.	1	1Y	2	2Y	3	3Y	4	4Y
A.B12	6	15	2					
A.A13	2		6.667					
A.126				15		5		5
A.H24					6		2	
A.C34		5			2	15	6.667	15
A.13D				5				
A.34F								
A.24F								
$\overline{W}$	600		480		320		400	
Node	Deflection and equilibrium coefficients							
1	−0.25	−0.3	0.083					
2	0.083		−0.208	−0.3	−0.25	0.1	0.083	0.1
3		0.1		0.1	0.083	−0.3	−0.208	−0.3
4				60				
$\overline{R}$	90	45	30					
Results	−231.4	−178.3	−156.6	−154.9	−162.3	−83.3	−213.7	−148.3 kNm

A.20 The rectangular grillage in Fig. A.20 has a load of 100 kN applied at nodes 1, 1', 3 and 3'. The moment of inertia of members in the XX direction is four times that of members in the YY direction. Prepare a data table for one-quarter of the frame to enable the nodal moments to be determined using the GRID program.

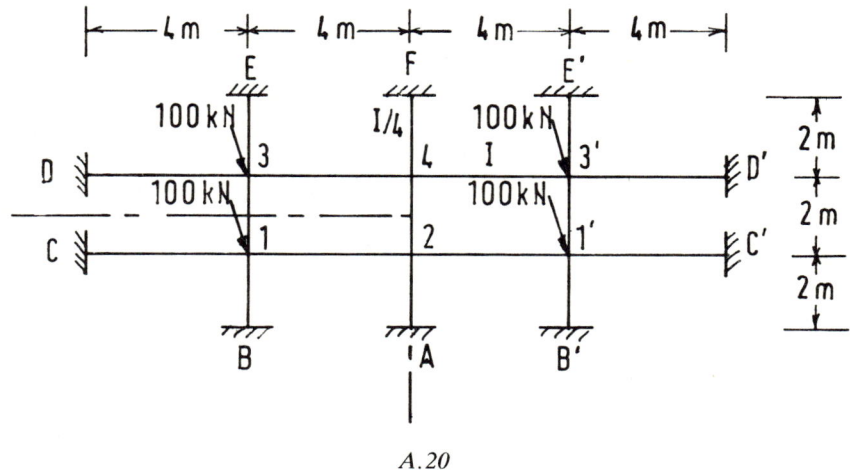

A.20

Equation	Moment						
Ref.	A	B	C	1	1Y	2	2Y
A.A	4						
A.B		4			2		2
A.C			2	1			
A.C12			1	4		1	
A.B13		2			10		
A.121'				1		2	
A.A24	2						10
Node			Deflection and equilibrium coefficients				
1		2	1	−2	−2	1	
2	2			1		−1	−2
$\overline{R}$	400	0					
Results	34.41	55.91	70.97	−58.06	−27.96	45.16	−17.2 kNm

A.21 The transverse member in the rectangular grillage in Fig. A.21 has a moment of inertia equal to one-third of those in the opposite direction. Find the nodal moments by writing down the data table for use with the GRID program for the given load system.
(If necessary, use a sub-routine based on PIVOT, rearranging the data to suit.)

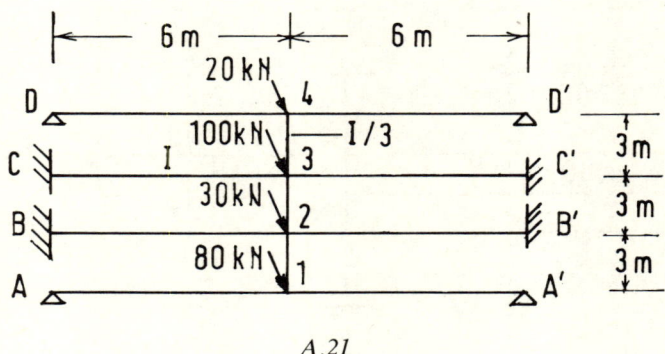

A.21

Equation						Moment				
Ref.	B	C	B'	C'	1	2X	2Y	3X	3Y	4
A.B	2									
A.C		2				1		1		
A.B'			2			1		1		
A.C'				2						
A.A1A'					4					
A.B2B'	1		1			4				
A.123							6		1.5	
A.C3C'		1		1			1.5	4	6	
A.234										
A.D4D'										4
Node					Deflection and equilibrium coefficients					
1					−2					
2	1		1			−2	2		2	
3		1		1			−4		−4	
4							2	−2	2	−2
$\overline{R}$	480	180	600	120						
Results	119.06	108.34	119.06	108.34	−169.03	−119.06	70.97	−108.34	−6.17	−66.17

A.22 Determine the support moments in the continuous beam in Fig. A.22 with a cantilever moment at support C using PIVOT.

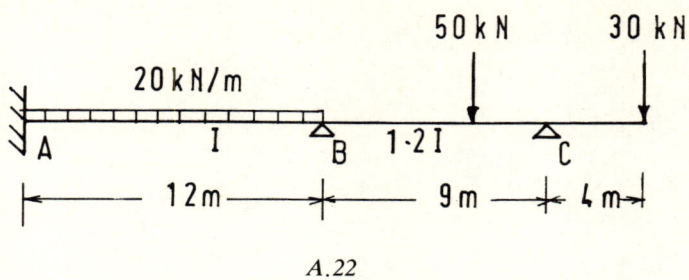

A.22

Equation Ref.	Moment		$\overline{W}$
	A	B	
A.A	2	1	720
A.B	1	3.25	720 + 83.33 − 75 = 728.33
Results	293.03	133.94	$M_c = 120$ kNm

A.23 The cantilever canopy in Fig. A.23 is constructed of two interconnected reinforced concrete beams A1 and B1 and supports a point load of 50 kN at the intersection, 1. If the beams are each 150 mm wide × 300 mm deep with $E_c = 30$ kN/mm² and $\lambda = 0.2$ for the concrete, determine the joint moments and deflection at 1 using the GRID program; include for the effect of torsion at the joints.

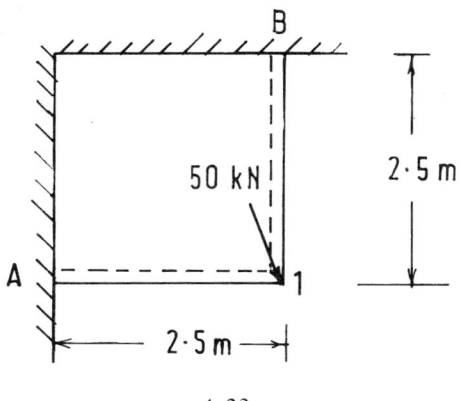

A.23

Equation Ref.	Moment	
	A	B
A.A	2	1
T.A1/1B	1	22.96
$\overline{\Delta}$	0.4	−0.4
E.1	0.4	−0.4
R	25	
Results	55.55	−6.95 kNm
	Δ_1	11 mm

A.24 All joints in the rectangular grillage in Fig. A.24 except the transverse edge connections at 1, 3, 4 and 6 can transmit bending and torsional moments. Construct the tabular matrix for half the grillage and use it to obtain the nodal moments under the given load system using GRID, given that for the beams in the XX direction $GJ = 0.25EI$ and their moment of inertia is four times that of the beams in the YY direction.

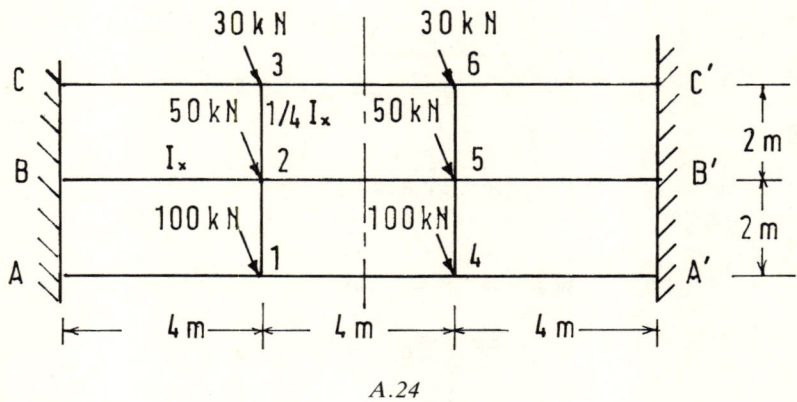

A.24

Equation Ref.	Moment							
	A	B	C	1X	2X	21	23	3X
A.A	2			1				
A.B		2			1			
A.C			2					1
A.A14	1			3				
A.B25		1			3			
A.T.12/2B						28	−24	
A.123						4	4	
A.C36			1					3
Node	Deflection and equilibrium coefficients							
1	1			−1		2	2	
2		1			−1	−2	−4	
3			1				2	−1
R̄	400	200	120					
Results	210.8	144.9	55.7	−158.1	−108.7	−3.5	−11	−41.8 kNm

A.25 Determine the moments at the joints of the rigid portal frame in Fig. A.25 using the XBEAM program.

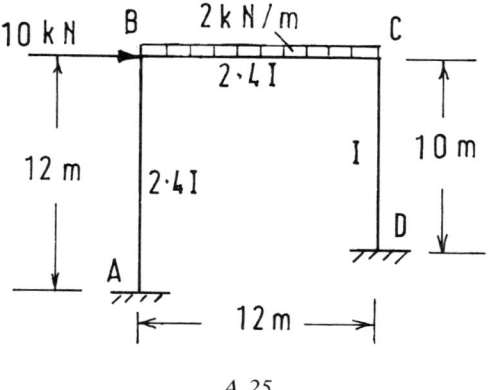

A.25

Equation Ref.	Moment				
	A	B	C	D	
A.A	2	1			
A.B	1	4	1		
A.C		1	6	2	
A.D			2	4	
$\overline{W}$		72	72		
$\overline{\lambda}$	0.0833	−0.0833	0.1	−0.1	
E	0.0833	−0.0833	0.1	−0.1	10
Results	29.97	−10.45	34.32	−32.01	kNm

A.26 The portal frame in Fig. A.26 supports a 5 kN point load at the end of the cantilever canopy EB. If EI is constant for the frame, determine the joint moments using the XBEAM program.

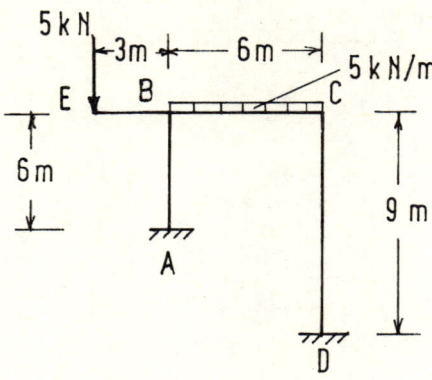

A.26

Equation Ref.	Moment				
	A	B	C	D	
A.A	2	1			
A.B	1	4	1		
A.C		1	5	1.5	
A.D			1.5	3	
$\overline{W} + M'$		60	75		
$\overline{\lambda}$	0.167	−0.167	0.111	−0.111	
E	0.167	−0.167	0.111	−0.111	0
Results	−4.42	11.54	15.57	−8.38	kNm

A.27 Determine the joint moments in the rigid frame in Fig. A.27 using XBEAM.

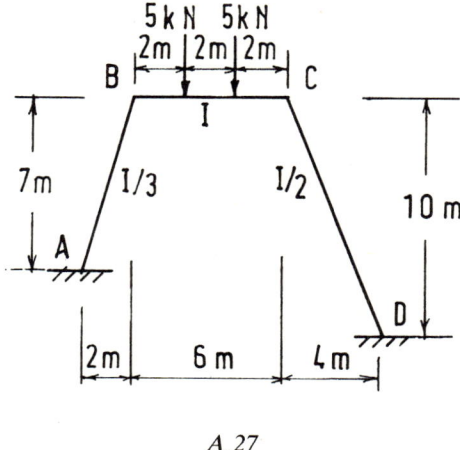

A.27

Equation Ref.	Moment				
	A	B	C	D	
A.A	7.28	3.64			
A.B	3.64	9.28	1		
A.C		1	9.18	3.59	
A.D			3.59	7.18	
$\overline{W}$		20	20		
$\overline{\lambda}$	0.143	−0.257	0.214	−0.1	
E	0.143	−0.257	0.214	−0.1	−0.571
Results	−1.78	3.03	1.82	−0.72	kNm

A.28 The pitched portal in Fig. A.28 has a concentrated load applied at the ridge and a uniformly distributed load over the whole span. Use the XBEAM program to determine the joint moments.

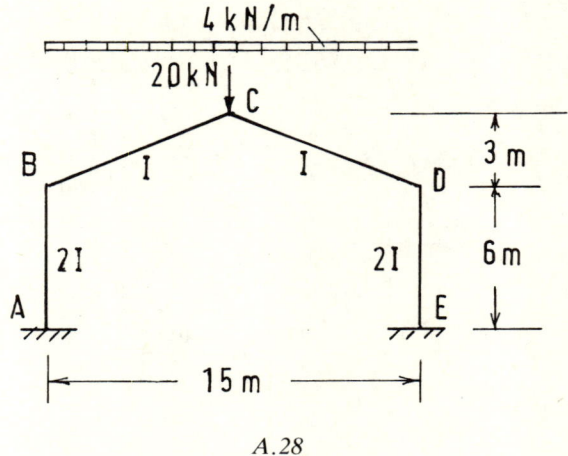

A.28

Equation Ref.	Moment			
	A	B	C	
A.A	0.742	0.371		
A.B	0.371	2.742	1	
A.C		1	2	
$\overline{W}$		56.25	56.25	
$\overline{\lambda}$	−0.167	0.5	−0.333	
E	−0.167	0.5	−0.333	62.5
Results	−75.31	74.97	−37.35	kNm

A.29 The pitched portal frame in Fig. A.29 has a lateral load of 12 kN applied at one of the eaves. Using the GRID program, determine the joint moments.

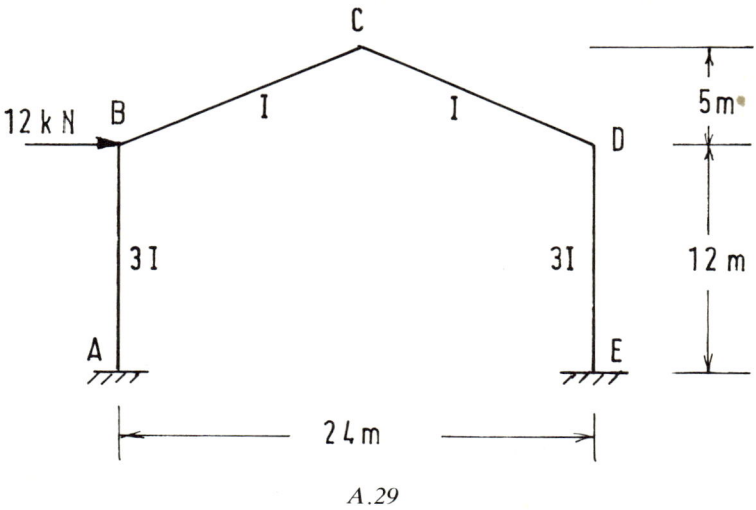

A.29

Equation Ref.	A	B	C	D	E
A.A	0.6154	0.3077			
A.B	0.3077	2.6154	1		
A.C		1	4	1	
A.D			1	2.6154	0.3077
A.E				0.3077	0.6154
Node		Deflection and equilibrium coefficients			
1	0.0833	−0.1833	0.2	−0.1	
2		0.1	−0.2	0.1833	−0.0833
$\overline{T}$	12	0			
Results	66.60	−27.83	10.13	6.75	−42.89

A.30 Write down the tabular matrix for the moments at the joints of the rigid frame in Fig. A.30, which has lateral wind loads applied at each floor level, as shown. Use the XBEAM program to find the moments.

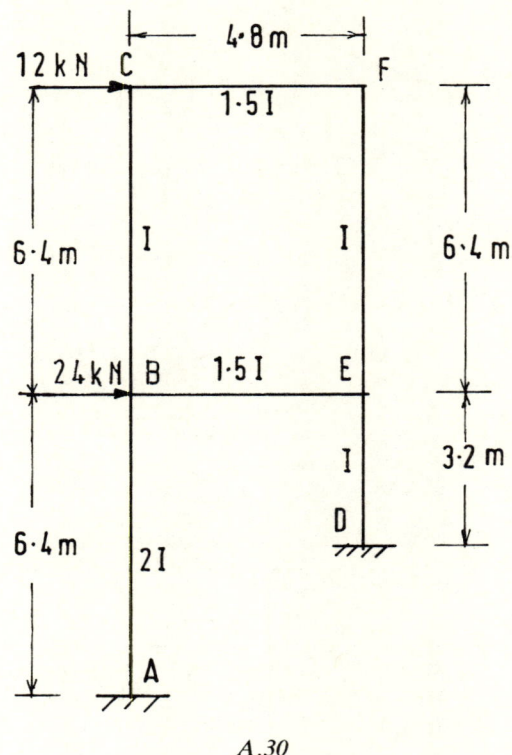

A.30

PROBLEMS

Equation	Moment							
Ref.	A	BA	BC	C	D	ED	EF	F
A.A	2	1						
A.ABE	1	4	−2			−1	1	
A.ABC	1	2	4	2				
A.C			2	6				−1
A.D		−1	1		2	1		
A.DEB					1	4	−2	2
A.DEF					1	2	4	6
A.F				−1			2	0
Δ.1	0.156	−0.156	−0.156	0	0.312	−0.312	−0.312	−0.156
Δ.2	0	0	0.156	−0.156	0	0	0.156	0
E.1	0.156	−0.156	0	0	0.312	−0.312	0	−0.156
E.2	0	0	0.156	−0.156	0	0	0.156	
T	36	12						
Results	27.95	−18.01	21.57	−22.39	56.06	−36.35	13.01	−19.95 kNm

A.31 The two-bay rigid frame in Fig. A.31 is subjected to both lateral wind load and vertical concentrated loads as shown. Determine the moments at the joints using the XBEAM program.

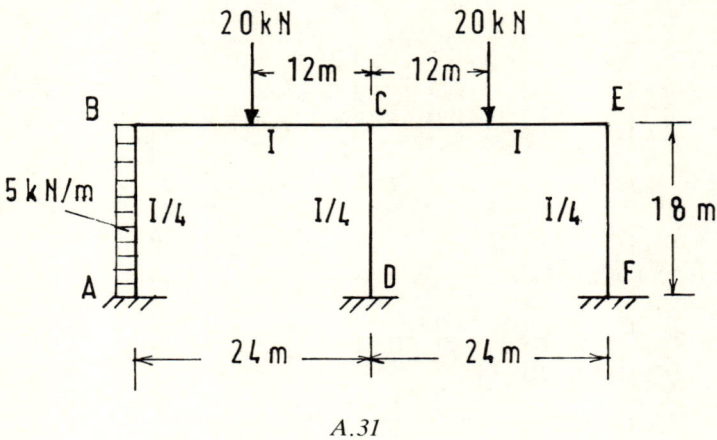

A.31

PROBLEMS

Equation				Moment				
Ref.	A	B	CB	CD	D	E	F	
A.A	6	3	1					
A.B	3	8	6	2				
A.BCE		1	2	-6	-3	1		
A.BCD		1		3	6			
A.D				1				
A.E			1			8	3	
A.F						3	6	
$\overline{W}$	1215	1395	360	180	0	180	0	
$\overline{\lambda}$	0.056	-0.056	0	0.056	0.056	0.056	-0.056	
E.1	0.056	-0.056	0	-0.056	0.056	0.056	-0.056	45
Results	266.3	11.9	82.5	-140.1	139.8	133.1	-136.3	kNm

A.32 The three-bay Vierendeel girder in Fig. A.32 carries panel point loads of 250 kN, as shown. Determine the moments at the joints using GRID.

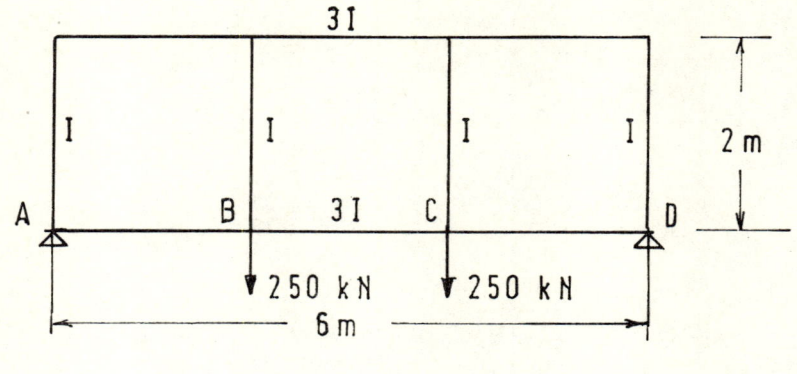

A.32

Equation Ref.	Moment		
	AB	BA	BC
A.A	5	1	
A.ABB'	1	5	−3
A.ABC	1	2	3
Δ_B	1	−1	−1
E.1	1	−1	
$\bar{P}$	250		
Results	107.14	−142.86	−71.43 kNm

A.33 The uniform Vierendeel girder in Fig. A.33 with square panels supports unequal loads at the panel points as shown. List the data statements needed to solve for the moments at the bottom joints using GRID and give the results.

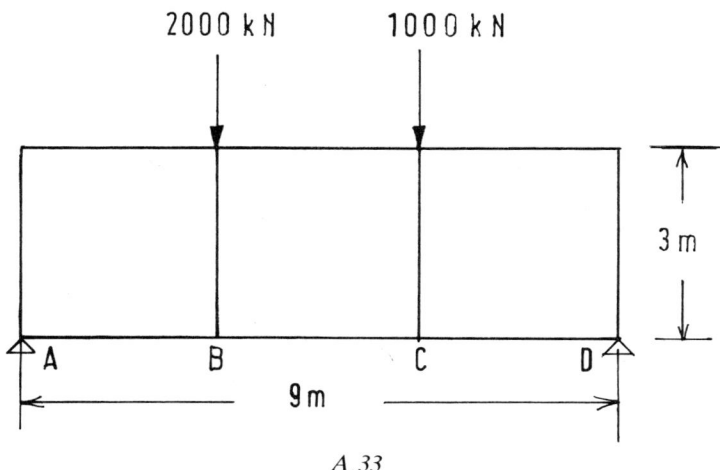

A.33

3,1,0,0,0,0
1,3,−1,0,0,0
1,2,2,1,0,0
0,0,1,3,−1,0
0,0,1,2,2,1
0,0,0,0,1,3
1,−1,−2,1,1,0
0,0,1,−1,−2,1
1,−1,0,−1,0,−1
1,−1,−1,1,0,0
0,0,1,−1,−1,1
1,−1,1,−1,1,−1
3000,1500,0

Results: M_A M_{BA} M_{BC} M_{CB} M_{CD} M_D
        1182   −1318   −540   −40   −1005   995   kNm

A.34 The cable-stayed bridge in Fig. A.34 consists of two main girders 4 m apart, each stayed by single 40 mm diameter cables to twin-legged steel towers fixed at the base. The deck carries a distributed load of 5 kN/m² over the full length of the bridge. The main girders, 11', may be considered as continuous over crosshead supports at 2 and 2' and with a pinned connection so that bending in the deck is not transmitted to the towers. The girders are simply supported on flexible columns at each end.

Determine the member forces, displacements and bending moments at the nodes given Young's moduli, cross-section areas and moments of inertia as follows:

For a single cable — $E = 170$ kN/mm², $A = 10$ cm².
For a single girder — $E = 210$ kN/mm², $A = 230$ cm², $I = 150\,000$ cm⁴.
For a single tower leg — $E = 210$ kN/mm², $A = 270$ cm², $I = 75\,000$ cm⁴.

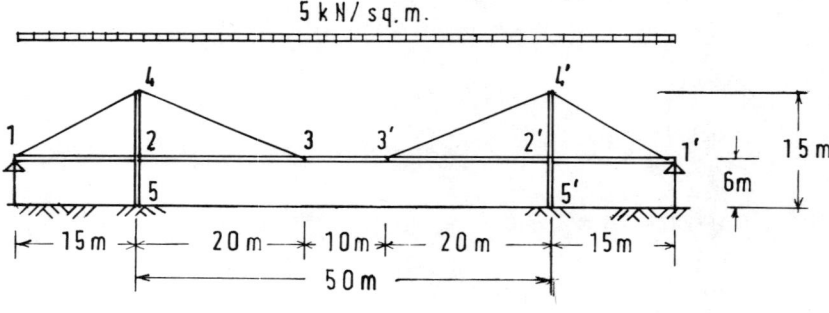

A.34

Equation	Δ				F					M		
Ref.	1X	4X	3Y	4Y	13	14	34	45	5	2	3	
L.13	96 600				−0.7							
L.14	−82 835	82 835				−9.926						
L.34		−88 157	39 645				−12.458					
L.45				−49 700				−0.2556				
E.1X				−39 645	−1							
E.4X				96 600		0.8575	0.9126					
E.3Y						−0.8575	0.4104	1	0.0667			
E.4Y						−0.5145	−0.4104					
A.5									1.2			
A.2		−2520	−94 500							70	20	
A.3			94 500							20	70	
W̄							150			28 438	22 500	
Results	0.002	0.037	0.17	0.0007	254.1	296.3	272.6	264.3	79.3	664.3	−98.3	m
	(2	37	170	1								mm)
												kN
												kNm

Appendix B Programs

MATIN

```
600 REM *********
605 REM * MATIN *
610 REM *********
615 REM PROGRAM TO SOLVE EQUATIONS USING INVERSE OF MATRIX, SIZE (NxN)
620 REM OBTAINED BY PARTITIONING AND THEN BACK-SUBSTITUTING.
625 REM ENTER TABULAR COEFFICIENTS ROW BY ROW
626 REM FOLLOWED BY VECTOR TERMS AS DATA.
630 CLS: REM CLEARS SCREEN
635 REM PRESS KEYS OR INSERT COMMAND TO COPY SCREEN TO PRINTER.
640 DIM   A(25,25),B(25),C(25,25),R(25),D(25),X(25),SUMX(25)
645 INPUT    "MATRIX SIZE "; N
650 PRINT: PRINT "DATA:-"
655 REM READ COEFFICIENTS
660 FOR I=1 TO N: FOR J=1 TO N
665 READ A(J,I): PRINT A(J,I);
670 NEXT J: PRINT: NEXT I:PRINT
675 REM START COMPUTATION OF INVERSE
680 NN=N-1
685 A(1,1)=1/A(1,1)
690 FOR M=1 TO NN
695 K=M+1
700 FOR I=1 TO M
705 B(I)=0
710 FOR J=1 TO M
715 B(I)=B(I)+A(I,J)*A(J,K)
720 NEXT J: NEXT I
725 C=0
730 FOR I=1 TO M
735 C=C+A(K,I)*B(I)
740 NEXT I
745 C=-C+A(K,K)
750 A(K,K)=1/C
755 FOR I=1 TO M
760 A(I,K)=-B(I)*A(K,K)
765 NEXT I
770 FOR J=1 TO M
775 D(J)=0
780 FOR I=1 TO M
785 D(J)=D(J)+A(K,I)*A(I,J)
790 NEXT I: NEXT J
795 FOR J=1 TO M
800 A(K,J)=-D(J)*A(K,K)
805 NEXT J
810 FOR I=1 TO M: FOR J=1 TO M
815 A(I,J)=A(I,J)-B(I)*A(K,J)
820 NEXT J: NEXT I
825 NEXT M
835 REM READ CONSTANTS
840 FOR I=1 TO N
845 READ B(I): R(I)=B(I)
850 PRINT R(I)
855 NEXT I: PRINT
860 REM CARRY OUT BACKSUBSTITUTION
865 FOR I=1 TO N: FOR J=1 TO N
870 C(I,J) =0
875 NEXT J: NEXT I
880 FOR I=1 TO N: FOR J=1 TO N
885 C(I,J)=C(I,J)+A(I,J)*R(I)
890 NEXT J: NEXT I
895 FOR J=1 TO N
900 SUMX(J)=0: NEXT J
905 FOR J=1 TO N: FOR I=1 TO N
910 X(I)=C(I,J)
915 SUMX(J)=X(I)+SUMX(J)
920 NEXT I: NEXT J
925 PRINT "RESULTS:-"
930 FOR I=1 TO N
935 J=I: X(I)=SUMX(J)
940 PRINT I,X(I)
945 NEXT I
950 END
```

PIVOT

Optional program.

```
600 REM *********
605 REM * PIVOT *
610 REM *********
615 REM PROGRAM TO SOLVE N EQUATIONS WITH V VECTORS
616 REM USING GAUSSIAN ELIMINATION.
620 REM PROGRAM PRINTS OUT DATA AND RESULTS
625 REM ALSO FINDS MEAN COEFFICIENT AND DETERMINANT.
630 REM ENTER FULL TABULAR DATA ROW BY ROW IN DATA STATEMENTS.
635 CLS: REM CLEARS SCREEN
640 REM PRESS KEYS OR INSERT COMMAND TO COPY SCREEN TO PRINTER.
645 DIM  A(20,25),B(20,25)
650 PR = .00001: REM DECIDE COEFF. RATIO, PR
655 INPUT   "NO. OF EQUATIONS "; N
660 INPUT   "NO. OF VECTORS "; V
665 M=N+V
670 REM READ TABULATED DATA
675 PRINT: PRINT "TABULAR DATA:-"
680 FOR I= 1 TO N: FOR J=1 TO M
685 READ A(I,J): PRINT A(I,J);
690 NEXT J: PRINT: NEXT I
695 REM CALCULATE MEAN COEFFICIENT MAGNITUDE
700 MC=0: PRINT
705 FOR I=1 TO N: FOR J=1 TO N
710 MC=MC+ABS(A(I,J))
715 MC=MC/(N*N)
720 NEXT J: NEXT I
725 PRINT MC "= MEAN COEFF": PRINT
730 REM BEGIN TRIANGULAR DECOMPOSITION
735 NP1=N+1: NM1=N-1
740 D=1: REM BEGIN DETERMINANT EVALUATION
745 FOR I1=1 TO NM1
750 I2=I1+1
755 REM INTERCHANGE ROWS TO GET MAX PIVOT
760 IP=I1
765 FOR I=I2 TO N
770 IF ABS(A(I,I1))> ABS(A(IP,I1)) THEN IP=I
775 NEXT I
780 IF IP=I1 THEN 815
785 D=-D
790 FOR J=I1 TO M
795 AA=A(I1,J)
800 A(I1,J)=A(IP,J)
805 A(IP,J)=AA
810 NEXT J
815 IF ABS(A(I1,I1)/MC)< PR THEN 845
820 FOR I=I2 TO N
825 C=A(I,I1)/A(I1,I1)
830 FOR J=I2 TO M
835 A(I,J)=A(I,J)-C*A(I1,J)
840 NEXT J: NEXT I: NEXT I1
845 IF ABS(A(N,N)/MC) < PR THEN 960
850 FOR I=1 TO N
855 D=D*A(I,I)
860 NEXT I
865 PRINT D "= DETERMINANT": PRINT
870 REM BEGIN BACK SUBSTITUTION
875 FOR K=NP1 TO M
880 I=N: B(I,K)=A(N,K)/A(N,N)
885 A(N,K)=A(N,K)/A(N,N)
890 FOR L=1 TO NM1
895 I=N-L: I2=I+1
900 C=A(I,K)
905 FOR J=I2 TO N
910 C=C-A(I,J)*A(J,K)
915 NEXT J
920 A(I,K)=C/A(I,I)
925 B(I,K)=A(I,K)
930 NEXT L: NEXT K
935 PRINT "RESULTS:-"
940 FOR K=NP1 TO M: FOR I=1 TO N
945 PRINT I,B(I,K)
950 NEXT I: PRINT: NEXT K
955 END
960 PRINT "ILL CONDITIONED EQUATIONS": END
```

XBEAM

```
5   REM   *********
10  REM   * XBEAM *
15  REM   *********
20  REM   PROGRAM TO SOLVE FRAMES WITH UP TO 4 DEFLECTED NODES
25  REM   AND WITH BEAM LOADS ON MEMBERS USING MATSUB SUB-ROUTINE.
30  REM   DATA INSTRUCTIONS: ENTER COEFF MATRIX, BEAM VECTOR, DEFLECTION
35  REM   VECTORS, EQUILIBRIUM COEFFS AND NODAL LOADS IN SEQUENCE.
40  F=0: REM DEFLECTED NODES COUNTER
45  DIM   W(20),U1(20),U2(20),U3(20),U4(20),L(20,20),M(20,20)
50  DIM   F1(20),F2(20),F3(20),F4(20)
55  GOSUB 630: REM BEAM MOMENTS
60  PRINT "BEAM MOMENTS:-": Z = N
65  FOR I=1 TO N
70  W(I)=X(I): PRINT I,W(I)
75  NEXT I: PRINT: GOTO 120
80  FOR I=1 TO N
85  U1(I)=X(I): NEXT I: GOTO 125
90  FOR I=1 TO N
95  U2(I)=X(I): NEXT I: GOTO 125
100 FOR I=1 TO N
105 U3(I)=X(I): NEXT I: GOTO 125
110 FOR I=1 TO N
115 U4(I)=X(I): NEXT I: GOTO 155
120 PRINT "?DEFLECTED NODE. FOR YES, PRESS 1": GOTO 130
125 PRINT "?ANOTHER DEFLECTED NODE. FOR YES, PRESS 1"
130 INPUT    U: PRINT
135 F=F+U
140 IF U<>1 THEN 155
145 GOSUB 840: REM UNIT MOMENTS
150 ON F GOTO 80,90,100,110
155 FOR I=1 TO N: REM FORM ARRAY OF UNIT MOMENTS
160 J=1: M(I,J)=U1(I)
165 J=2: M(I,J)=U2(I)
170 J=3: M(I,J)=U3(I)
175 J=4: M(I,J)=U4(I)
180 NEXT J
185 PRINT "EQUILIBRIUM COEFFS:-"
190 FOR I=1 TO F: FOR J=1 TO N
195 READ L(I,J): PRINT L(I,J);
200 NEXT J: PRINT
205 NEXT I: PRINT
210 REM MULTIPLY UNIT MOMENTS BY COEFFS
215 FOR L=1 TO N: FOR J=1 TO F
220 A(L,J)=0
225 FOR I=1 TO N
230 A(L,J)=A(L,J)+L(L,I)*M(I,J)
235 NEXT I: NEXT J
240 NEXT L
245 GOSUB 400: REM NODAL REACTIONS
250 N=F: F=-1
255 GOSUB 680: REM UNIT MOMENT FACTORS
260 N=Z
265 PRINT "SECONDARY MOMENTS:-"
270 FOR I=1 TO N
275 F1(I)=X(1)*U1(I)
280 F2(I)=X(2)*U2(I)
285 F3(I)=X(3)*U3(I)
290 F4(I)=X(4)*U4(I): PRINT I,F1(I),F2(I)
295 NEXT I: PRINT
300 PRINT "RESULTS:-"
305 FOR I=1 TO N
310 R(I)=W(I)+F1(I)+F2(I)+F3(I)+F4(I)
315 PRINT I,R(I); " FINAL MOMENT"
320 NEXT I
325 END
400 REM SUB-ROUTINE TO FIND NODAL REACTIONS
405 DIM   N(20),S(20): PRINT
410 PRINT "NODAL LOADS:-"
415 FOR I=1 TO F
420 READ N(I): PRINT I,N(I)
425 NEXT I: PRINT
430 FOR J=1 TO F
435 S(J)=0
440 FOR I=1 TO N
445 S(J)=S(J)+L(J,I)*W(I)
450 NEXT I: NEXT J
455 PRINT "NODAL REACTIONS:-"
460 FOR J=1 TO F
465 R(J)=N(J)-S(J)
470 PRINT J,R(J): NEXT J: PRINT
475 RETURN
```

Note: To be merged with MATSUB before saving.

GRID

```
5 REM ********
10 REM * GRID *
15 REM ********
20 REM PROGRAM TO SOLVE GRID WITH UP TO 9 DEFLECTED NODES
25 REM USING MATSUB SUBROUTINE.
30 REM DATA INSTRUCTIONS: ENTER COEFF MATRIX, DEFLECTION VECTORS
35 REM EQUILIBRIUM COEFFS AND NODAL LOADS IN SEQUENCE.
40 F=0: REM DEFLECTED NODES COUNTER
45 DIM   U1(20),U2(20),U3(20),U4(20),U5(20),U6(20),U7(20),U8(20),U9(20)
50 DIM   L(20,20),M(20,20)
55 DIM   F1(20),F2(20),F3(20),F4(20),F5(20),F6(20),F7(20),F8(20),F9(20)
60 GOSUB 630: REM UNIT MOMENTS
65 FOR I=1 TO N: Z=N
70 U1(I)=X(I): NEXT: GOTO 155
75 FOR I=1 TO N
80 U2(I)=X(I): NEXT: GOTO 155
85 FOR I=1 TO N
90 U3(I)=X(I): NEXT: GOTO 155
95 FOR I=1 TO N
100 U4(I)=X(I): NEXT: GOTO 155
105 FOR I=1 TO N
110 U5(I)=X(I): NEXT: GOTO 155
115 FOR I=1 TO N
120 U6(I)=X(I): NEXT: GOTO 155
125 FOR I=1 TO N
130 U7(I)=X(I): NEXT: GOTO 155
135 FOR I=1 TO N
140 U8(I)=X(I): NEXT: GOTO 155
145 FOR I=1 TO N
150 U9(I)=X(I): NEXT: GOTO 185
155 PRINT "?ANOTHER DEFLECTED NODE. FOR YES, PRESS 1"
160 INPUT   U: PRINT
165 F=F+U
170 IF U<>1 THEN 185
175 GOSUB 840: REM UNIT MOMENTS
180 ON F GOTO 75,85,105,115,125,135,145
185 FOR I=1 TO N: REM FORM ARRAY OF UNIT MOMENTS
190 J=1: M(I,J)=U1(I)
195 J=2: M(I,J)=U2(I)
200 J=3: M(I,J)=U3(I)
205 J=4: M(I,J)=U4(I)
210 J=5: M(I,J)=U5(I)
215 J=6: M(I,J)=U6(I)
220 J=7: M(I,J)=U7(I)
225 J=8: M(I,J)=U8(I)
230 J=9: M(I,J)=U9(I)
235 NEXT I: F=F+1
240 PRINT "EQUILIBRIUM COEFFS:-"
245 FOR I=1 TO F: FOR J=1 TO N
250 READ L(I,J): PRINT L(I,J);
255 NEXT J: PRINT
260 NEXT I: PRINT
265 REM MULTIPLY UNIT MOMENTS BY COEFFS
270 FOR L=1 TO N: FOR J=1 TO F
275 A(L,J)=0
280 FOR I=1 TO N
285 A(L,J) = A(L,J) + L(L,I)*M(I,J)
290 NEXT I: NEXT J
295 NEXT L
300 PRINT "NODAL LOAD X BAY LENGTH:-"
305 FOR I=1 TO F
310 READ B(I): R(I)=B(I)
315 PRINT I,R(I): NEXT I: PRINT
320 N=F: F=-1
325 GOSUB 680: REM DEFLECTION FACTORS
330 N=Z
335 FOR I=1 TO N
340 F1(I)=X(1)*U1(I)
345 F2(I)=X(2)*U2(I)
350 F3(I)=X(3)*U3(I)
355 F4(I)=X(4)*U4(I)
360 F5(I)=X(5)*U5(I)
365 F6(I)=X(6)*U6(I)
370 F7(I)=X(7)*U7(I)
375 F8(I)=X(8)*U8(I)
380 F9(I)=X(9)*U9(I)
385 NEXT I: PRINT
390 PRINT "RESULTS:-"
395 FOR I=1 TO N
400 R(I)=F1(I)+F2(I)+F3(I)+F4(I)+F5(I)+F6(I)+F7(I)+F8(I)+F9(I)
405 R(I)= INT(R(I)*100+.5)/100
410 PRINT I, R(I); " FINAL MOMENT"
415 NEXT I
420 END
```

Note: To be merged with MATSUB before saving.

Bibliography

1. Wang C. K. & Salmon C. G. (1984) *Introductory Structural Analysis*, Prentice-Hall.
2. Terzaghi K. & Peck R. B. (1967) *Soil Mechanics in Engineering Practice*, Wiley.
3. Reynolds C. E. & Steedman J. C. (1981) *Reinforced Concrete Designers Handbook*, Viewpoint Publications.
4. Clear C. A. & Harrison T. A. (1985) *Concrete Pressures on Formwork*, CIRIA Report No. 108.
5. Coates R. C., Coutie M. G. & Kong F. K. (1980) *Structural Analysis*, Van Nostrand Reinhold.
6. Johnson D. (1986) *Advanced Structural Mechanics*, Collins.
7. Bowles J. E. (1982) *Foundation Analysis and Design*, McGraw-Hill.
8. Tomlinson M. J. (1980) *Foundation Design and Construction*, Pitman.
9. Smith J. & Pole A. (1984) *Elements of Foundation Design*, Granada.
10. Holloway R. T. (1985) *Structural Design with the Microcomputer*, McGraw-Hill.
11. Graham L.J. (1985) *Your IBM PC*, Osborne/McGraw-Hill.
12. British Steel General Steels (1989) *Piling Handbook*, British Steel plc, London.
13. Clayton C. R. & Milititsky J. (1986) *Earth Pressure and Earth Retaining Structures*, Surrey University Press.
14. *Formwork – A Guide to Good Practice* (1987) Institution of Structural Engineers, London.
15. BS 5930 (1981) Code of Practice for Site Investigations.
16. BS 5975 (1982) Code of Practice for Falsework.
17. BS 8004 (1986) Code of Practice for Foundations.
18. BS 6031 (1981) Code of Practice for Earthworks.
19. Eurocode A. Actions on Building Structures.
20. Eurocode 2. Design of Concrete Structures.
21. Eurocode 3. Design of Steel Structures.
22. Eurocode 7. Rules for Geotechnical Design.
23. *Qualitive Analysis of Structures* (1989). The Institution of Structural Engineers, London.

Index

abutments, strutted, 140−44
anchored sheet piling, 131−9
angular displacement equations, 9−11

back-substitution, 22
beam framework deflections, 206, 208
beam frameworks, 46, 183−206
   with torsion, 251−61
beams
   continuous, 38−41
   end cantilever to, 239
   interconnected, 183−206
   of varying depth, 42−5, 99, 100
   single span, 37
   strutted, 164−8
bending moment
   convention, 37
   diagrams, 40, 69, 116
boxed culverts, 275−7

cantilever
   balcony, 249
   deflections, 240
   moments, 239
   retaining walls, 126−31
centring, 162−4
cofferdam sheeting, 154−7
column foundations, 73−5
   moments on, 86−90
column on raft foundation, 108−11
concrete
   centring, 162−4
   formwork, 159−62
conjugate beam theory, 11
continuum structures, 42−4
cranked beams, 241−9

data
   preparation, 5
   processing, 85

tables, 18
deformation diagrams, 35
determinant, 22, 25
diaphragm walls, 140
differential settlement
   equations, 11−13
   moments, 49−57
DIM statement, 22
displacements, triangle of, 285, 290, 293
DOS, 29

earth pressure
   active, 126
   at rest, 158
earth theories, fixed and free, 140
equilibrium equations, 9, 15
errors, sources of, 5, 227

falsework, 162−4
file categories, 30
files, merging, 30
finite elements, 42
flexibility
   method, 3
   ratio, relative, 38
footings with distributed loading, 66−9
formatting disks, 29
formwork
   pressure from concrete, 159
   strutted, 159−62
foundation
   alternative loading of, 78
   settlements, 73
   slab elements, 103
   slabs, 101−103
   triangular load on, 71−3
foundations
   analysis of, xiv
   column, 73−5, 80−84

380   INDEX

continuous, 80–84
moment sign in, 68
moments on, 86–90
of varying depth, 96, 98
of varying width, 92–7
rigid analysis of, 64, 90

Gaussiam elimination, 23–5
GRID, use of, 228–34
GRID print-out, 232
GRID program description, 214
grid
deflections, 230, 231
one-way, 216–22, 228–31
shears, 230
two-way, 209–16, 221–6, 231–3
grillages, 46
partitioned, 218–26
symmetrical, 211–14
uniform rectangular, 209–33
with torsion, 262–9

indeterminate structures, 165, 169

keying techniques, 85, 120

laterally loaded piles, signs in, 115
lattice girders, 173–8
limit state design, 4
linear displacement equations, 6–8
listings
GRID, 377
MATIN, 374
MATSUB, 33
PIVOT, 375
PIVOTA, 33
SBEAM, 63
XBEAM, 376
load factors, partial, 62

MATIN flowchart, 21
MATIN program description, 20–22
matrix inversion, 19, 22
matrix notation, 16
matrix partitioning, 19
MATSUB description, 28
mean coefficient, 25
moment distribution theory, xvi
moment-area theorem, 10

moments, redistribution of, 60
multi-bay frames with sway, 312–15
multi-storey frames
laterally loaded, 303–11
uniform, 310
multi-storey symmetrical frames, 300–303

N-girders, 173–8
nodal deflection equations, 14
nodal equations, foundation slab, 101–107
nodal reactions, 190
node types in foundation slabs, 101

output
checks on, xv, 5
in preparing diagrams, xvi

pile caps, 121–4
pile groups, laterally loaded, 121
piles
in elastic medium, 114
laterally loaded, 114–20
with applied moments, 124, 125
pitched portals, 289–93
laterally loaded, 293–7
PIVOT program description, 25
PIVOTA print-out, 32
PIVOTA program description, 29
pivotal condensation, 23
plastic design, 297
portal frames, symmetrical, 273–5
portals
eaves deflections in, 299
multi-bay symmetrical, 277–9
multi-bay, with sway, 312–15
with cantilevers, 283
with inclined columns, 284–9
with sway, 279–83
pressure distribution in soils, 139
printer, output from, 31
problems, 325–71
program, choice of, 26
programs, features of computer, xv

raft foundations, 108–11
alternative loads, 110–13
reaction, coefficient of lateral, 114

reference elements, 3, 35
release-deformation method, 3, 34−6
retaining walls, strutted, 157, 158
rigid frames, 44−6
roof truss, 178−80

SBEAM program description, 59
settlements
   differential, 49−57
   sign convention for, 51
shear forces due to settlement, 58
shear modulus, 235
shears, unbalanced, 279, 304
sheet piles, embedded length, 132
sheet piling, 126−39
shores
   flying, 147−9
   raking, 145−7
sign
   convention, 8, 36
   determination, 51
software
   availability, xvi
   description of, xvi
space frames, 234
spreadsheet format, 17
structural theory, teaching of, xvi
strutted beams, 164−8
subgrade
   coefficient of, 65
   moduli of, 65, 66
sub-routines, 28
support reactions, 58

sway, 279, 303, 321
   equations, 13

tabular matrix format, 17, 18
tie rod anchorages, 131
torsion in
   beam frameworks, 251−61
   circular sections, 235
   cranked beams, 242−9
   rectangular grillages, 262−9
   rectangular sections, 236−8
torsional
   displacements, 16
   properties, 238
   rigidity, 236
trench
   pressure distribution in, 149−51
   sheeting, 149−55
   struts, loads in, 149, 153
trestles, 162−4
triangle of displacements, 285, 290, 293
trusses, 44, 173−81
   indeterminate, 180

vertical beam abutments, 140−44
Vierendeel girders, 313−24
Vierendeel girders with sway, 321−4

XBEAM, use of, 200−206
XBEAM flowchart, 197
XBEAM print-out, 198, 205
XBEAM program description, 195−9